Powertrain

Series Editor
Helmut List
AVL List GmbH, Graz, Austria

More information about this series at http://www.springer.com/series/7569

Kevin Hoag • Brian Dondlinger

Vehicular Engine Design

Second Edition

 Springer

Kevin Hoag
Southwest Research Institute
San Antonio
Texas
USA

Brian Dondlinger
Milwaukee
Wisconsin
USA

ISSN 1613-6349
Powertrain
ISBN 978-3-7091-1858-0   ISBN 978-3-7091-1859-7 (eBook)
DOI 10.1007/978-3-7091-1859-7

Library of Congress Control Number: 2015940330

Springer Wien Heidelberg New York Dordrecht London

Printed on acid-free paper

Springer-Verlag GmbH Wien is part of Springer Science+Business Media (www.springer.com)

I dedicated the first edition of this book to those in the engine development community who have left our company too soon. Sadly, another name has been recently added to that list. This edition is dedicated to Mark Tussing, Director of Engine Design and Development at Southwest Research, a man with soul, lost from our community in July 2014. I consider it my good fortune to have enjoyed working and laughing with Mark for over 20 years.

KLH

San Antonio, Texas

This book is dedicated to my wife Heidi, who served as my chief consultant on the book and key supporter, and to my kids Zoe Elanor and Evan Diesel. It is also dedicated to my parents, Peter and Patricia, for making me the person I am, and to Lynn whose help I couldn't have done it without.

BD

Milwaukee, Wisconsin

Preface

The mechanical engineering curriculum in most universities includes at least one elective course on the subject of reciprocating piston engines. The majority of these courses today emphasize the application of thermodynamics to engine efficiency, performance, combustion and emissions. There is at least one very good text book that supports education in these aspects of engine development.

However, in most companies engaged in engine development there are far more engineers working in the areas of design and mechanical development. University studies should include opportunities that prepare engineers desiring to work in these aspects of engine development as well. My colleagues and I have undertaken the development of a series of graduate courses in engine design and mechanical development. In doing so it becomes quickly apparent that no suitable text book exists in support of such courses.

This book was written in the hopes of beginning to address the need for an engineering-based introductory text in engine design and mechanical development. It is of necessity an overview, and focuses on the initial layout and first principles of engine operation. Its focus is limited to reciprocating piston internal combustion engines—both diesel and spark-ignition engines. Emphasis is specifically on automobile engines although much of the discussion applies to larger and smaller engines as well.

A further intent of this book is to provide a concise reference volume on engine design and mechanical development processes for engineers serving the engine industry. It is intended to provide basic information and most of the chapters include recent references to guide more in-depth study.

A few words should be said concerning the approach taken to the figures presented in this book. To aid understanding, simplified diagrams and plots are presented showing only the features being discussed at the time. Once the concept is illustrated, photos of production components and engines are shown to provide an example of how the theory can be applied in actual practice.

Acknowledgements

We are especially indebted to Dr. Josef Affenzeller who provided the guiding force and many consultations along the way. His guidance was invaluable. Also thank you to Bryce Metcalf and James Lippert who were consulted on the Gaskets and Seals chapter.

Thank you to Springer-Verlag for the professional support and design of this publication. The following companies provided figures as noted throughout the book. Their contributions are greatly appreciated:

AVL List GmbH, BMW GmbH, Daimler AG, Ford Motor Company, Nissan Motor Co., Ltd., Toyota Motor Corporation, Volkswagen AG

A special thank you is reserved for Bruce Dennert at Harley-Davidson. His partnership in many engine design instruction endeavors, and his input and critique throughout the writing process are greatly valued.

Finally, we are indebted to our colleagues at the University of Wisconsin Engine Research Center. Drs. Rolf Reitz, David Foster, Jaal Ghandhi, Christopher Rutland, Scott Sanders, and David Rothammer provide a stimulating environment in which to work, and to develop the ideas contained in this book.

Contents

The Internal Combustion Engine—An Introduction

1.1 Heat Engines and Internal Combustion Engines

It is appropriate to begin with a simple definition of the engine as a device for converting energy into useful work. The goal of any engine is to convert energy from some other form into "mechanical force and motion." The terms "mechanical force" and "motion" are chosen to convey the idea that the interest may be both in work output—how much force can be applied to move something a given distance—and also power output—how quickly the work can be done.

Turning attention to the energy that is being converted to do the desired work, our interest is in the chemical energy contained in the molecular structure of a hydrocarbon fuel. Fundamental to any chemical reaction are the facts that it takes energy to break a chemical bond, and that energy is released when new bonds are formed. If the energy released in forming new bonds is greater than that required in breaking the old bonds the result is an *exothermic* reaction, and net energy available to do work.

Fundamental to any combustion engine is the reaction of a hydrocarbon fuel with oxygen to form carbon dioxide and water. This combustion reaction is highly exothermic—a large amount of energy is released. The goal of the engine will be to utilize that energy *repeatedly*, *efficiently*, and *cost-effectively*. The next question with which the engine designer is faced is that of developing a mechanical device that accomplishes these objectives.

Beginning from this general discussion of combustion engines, one can now make distinctions between various types of engines. These distinctions may be based on thermodynamic process decisions, as well as on the mechanical hardware. The first distinction to be made is that between the *heat engine* and the *internal combustion engine*, as shown in Fig. 1.1—they really are two different things, although they have often been confused or incorrectly identified. By definition a heat engine is an engine in which a working fluid undergoes various state changes through an operating cycle. The working fluid experi-

© Springer Vienna 2016
K. Hoag, B. Dondlinger, *Vehicular Engine Design*, Powertrain,
DOI 10.1007/978-3-7091-1859-7_1

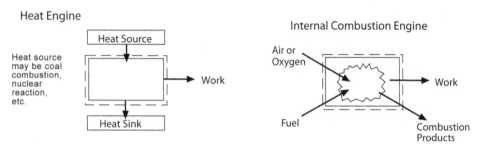

Fig. 1.1 The heat engine versus the internal combustion engine

ences a heat addition, in which its pressure and temperature increase. It then goes through a process converting a portion of its energy to work. Cycle completion requires heat rejection from the fluid to the environment. A Rankine cycle steam turbine, using either coal combustion or a nuclear reaction to provide the heat source (and steam as the working fluid) is a practical example of a heat engine.

The "air standard" Otto cycle and diesel cycle are theoretical representations of processes similar to those of a spark-ignition (SI) or diesel engine, but they assume the working fluid to be air, gaining energy from an external source. In the actual diesel or SI engine, the energy release occurs within the system, and the working fluid undergoes not only a state change but a change in chemical composition. While the mechanical device may undergo a cycle, the working fluid does not. Fuel and air enter the system, go through a series of thermodynamic processes, and are then exhausted from the system. Another example of a practical internal combustion engine is the gas turbine—not to be confused with the air standard Brayton cycle.

Earlier it was emphasized that any practical engine will be expected to produce work repeatedly (or continuously over some period of time), efficiently, and cost-effectively. These terms have been carefully chosen to convey separate expectations—all of which must be met for an engine to be practical. The discussion begins with an emphasis on efficiency and cost-effectiveness; the need for continuous production of work will be taken up shortly. Efficiency is a commonly used engineering term, and the classic definition will suffice here. With this measure one can assess how well the energy available can be converted into useful work. Cost-effectiveness is a more difficult measure to accurately obtain, and is less well understood by engineers. Nevertheless, it is every bit as important, and in many cases far more important, to a successful design. Listed below are the various elements defining the cost-effectiveness of an engine.

- Development, production, and distribution costs
- Maintenance costs
- Fuel costs
- Rebuild costs over useful life

- Disposal costs of parts and fluids
- Minus resale value at end of usage period

The elements listed together on the first line are ultimately reflected in the purchase price of the engine. As such, they provide the most direct measure of whether a given engine will be viable.

The remaining elements are tracked to a greater or lesser extent in particular markets. For example, while many automobile purchasers will consider only the purchase price (from this list) in making their buying decision, the company purchasing several hundred trucks or buses on which their company depends for its economic viability will almost certainly closely track every item on this list. The combination of these measures goes a long way in explaining why the internal combustion engine remains so difficult to replace.

Earlier the internal combustion engine was defined in general terms. Various types of practical engines fit this definition; these types are distinguished by the combination of their combustion process and mechanical configuration. The combustion process may be continuous, as with the gas turbine engine, or intermittent, as with both the diesel and the spark-ignition engine. A mechanical configuration must then be selected that meets the criteria of efficiency and cost-effectiveness, as well as allowing the work to be produced continuously. The objective is to create a mechanical arrangement that contains the combustion process, and utilizes the high pressure and temperature of the combustion products to produce useful work.

1.2   The Reciprocating Piston Engine

While many configurations have been proposed, patented, and demonstrated over the years, few have enjoyed commercial success. Such success results from the ability to address the combination of efficiency and cost-effectiveness discussed previously. In the remainder of this book the discussion will be limited to the reciprocating piston engine. This engine is characterized by a slider-crank mechanism that converts the reciprocating, cyclic motion of a piston in a cylinder into the rotating motion of a crankshaft.

The primary components of the reciprocating engine are shown in Fig. 1.2. The moving piston controls the volume of the combustion chamber between a minimum at *top dead center* (TDC) and a maximum at *bottom dead center* (BDC). The ratio between the volume at BDC and that at TDC is referred to as the *compression ratio*. The change in volume is the *displacement* of the cylinder. The displacement of the cylinder, multiplied by the number of cylinders is the displacement of the engine. The cylinder is sealed opposite from the moving piston by the *cylinder head*. In most engines the *intake* and *exhaust valves* are located in the cylinder head, as shown in the figure.

The piston is linked to the *crankshaft* through a *connecting rod*. As the crankshaft spins about its centerline (the *main bearing bore* in the *cylinder block*) the offset of the

Fig. 1.2 Major operating components of the reciprocating piston internal combustion engine

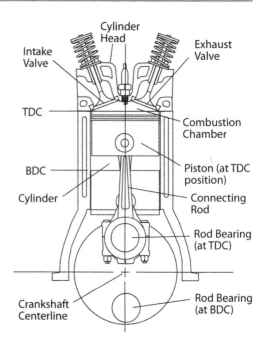

rod bearing from the main bearing determines the travel of the piston. As the crankshaft rotates one half revolution from the position shown in the figure, the piston moves from its TDC to its BDC position. The distance the piston travels is referred to as the **stroke** of the engine. The stroke is equal to twice the offset between the main bearing and rod bearing centerlines of the crankshaft. The diameter of the cylinder is referred to as its **bore**, and the combination of bore and stroke determine the displacement of the cylinder by the equation:

$$Displacement = \frac{\pi(Bore)^2(Stroke)}{4}$$

The crankshaft protrudes through the rear of the engine, where a flywheel and clutch pack or flex plate and torque converter are attached through which the load will be transmitted. Typically at the front of the engine the crankshaft will drive the **camshaft** through a system of gears, a chain, or a cogged belt. The intake and exhaust valves are then actuated by the camshaft(s), either directly or through a **valve train**. Various support systems for cooling and lubricating the engine and for supplying the fuel and igniting the mixture are also required.

Each of the components and sub-systems with the exception of the fuel and ignition systems will be discussed in detail in the chapters that follow. Because of the variety of fuel and ignition systems, and the availability of previously published books devoted to these systems they will not be covered in this book.

1.3 Engine Operating Cycles

Having introduced a particular mechanical mechanism designed to repeatedly extract useful work from the high temperature and pressure associated with the energy release of combustion, we are now ready to look at the specific processes required to complete this task. Shown in Fig. 1.3 is the four-stroke operating cycle, which as the name implies, requires four strokes of the piston (two complete revolutions of the crankshaft) for the completion of one cycle.

In the spark-ignition engine shown, a "charge" of pre-mixed air and fuel are drawn into the cylinder through the intake valve during the intake stroke. The valve is then closed and the mixture compressed during the compression stroke. As the piston approaches TDC, a high energy electrical spark provides the activation energy necessary to initiate the combustion process, forcing the piston down on its power stroke. As the piston nears BDC the exhaust valve opens, and the spent combustion products are forced out of the cylinder during the exhaust stroke. The work output is controlled by a throttle restricting the amount of air-fuel mixture that can pass through the intake valve.

A four-stroke *diesel* engine operating cycle would consist of the same processes. However, air alone would be drawn into the engine and compressed. The spark plug would be replaced with a fuel injector spraying the fuel directly into the cylinder near the end of the compression process. The activation energy would be provided by the high temperature

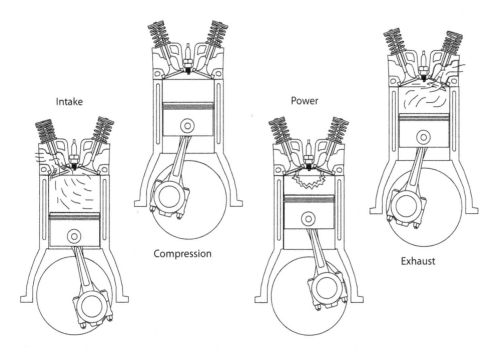

Fig. 1.3 The four-stroke operating cycle shown for a spark-ignition engine

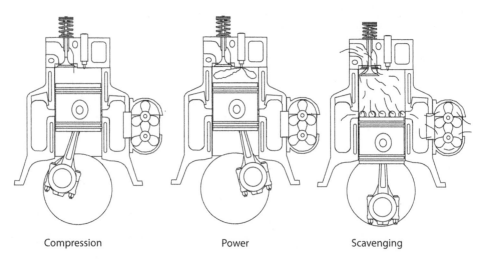

Compression   Power   Scavenging

Fig. 1.4 The two-stroke operating cycle shown for a heavy-duty diesel engine

and pressure of the air into which the fuel is injected. The work output would be controlled by the amount of fuel injected.

An alternative to the four-stroke cycle is the two-stroke cycle shown in Fig. 1.4. As implied, a complete operating cycle is achieved with every two strokes of the piston (one revolution of the crankshaft). The compression and power strokes are similar to those of the four-stroke engine; however the gas exchange occurs as the piston approaches BDC, in what is termed the ***scavenging*** process. During scavenging the intake and exhaust passages are simultaneously open, and the engine relies on an intake supply pressure maintained higher than the exhaust pressure to force the spent products out and fill the cylinder with fresh air or air-fuel mixture.

The engine shown in the figure is a heavy-duty two-stroke diesel. The incoming air is pressurized with a crankshaft-driven compressor, and enters through ***ports*** near the bottom of the cylinder. In this engine exhaust valves similar to those of the four-stroke engine are mounted in the cylinder head. Light-duty two-stroke engines are often crankcase supercharged. In such engines, each time the piston moves upward in the cylinder a fresh charge (mixed with lubricating oil) is drawn into the crankcase. As the piston moves down, the crankcase is sealed and the mixture is compressed—the mixture is then transferred from the crankcase through the intake ports as the piston approaches BDC. This configuration is shown in Fig. 1.5.

In the spark-ignition configuration the engine suffers the disadvantage of sending fresh air-fuel mixture out with the exhaust during each scavenging period. Much recent attention is being given to engines that overcome this problem by injecting the fuel directly into the cylinder after the ports have been sealed.

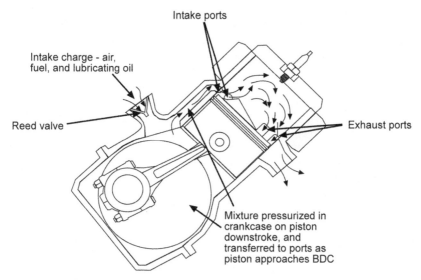

Fig. 1.5 The light-duty two-stroke engine

1.4   Supercharging and Turbocharging

In order to increase the specific power output of an engine (power output per unit of displacement) some form of pre-compression is often considered. This has become standard practice in diesel engines, and is often seen in high-performance spark-ignition engines. Downsized (reduced displacement) engines with turbocharging or supercharging are receiving increased attention as a means of achieving improved part-load fuel economy in automobile engines.

The engine that draws fresh charge into the cylinder at atmospheric pressure, and exhausts directly to the atmosphere is termed *naturally aspirated*. As was shown in Fig. 1.4 a crankshaft-driven compressor may be added to elevate the pressure of the air (or mixture) prior to drawing it into the cylinder. This allows more mixture to be burned in a given cylinder volume. The crankshaft-driven device is generally referred to as a *supercharger* (although this general term is sometimes used to describe a turbocharger as well).

Recognizing that the exhaust gases leaving the engine still contain a significant quantity of energy that was not recovered as work, an alternative is to use a portion of this energy to drive the compressor. This configuration is the *turbocharged* engine.

Whenever the air is compressed its temperature increases. Its density can be further increased (and still more air forced into the cylinder) if it is cooled after compression. The *charge air cooler*, variously termed *intercooler* (cooling between stages of compression) and *aftercooler* (cooling after compression) may be used with either the turbocharger or supercharger.

Most supercharged or turbocharged production applications use a single unit, in many cases followed by a charge air cooler. With some cylinder configurations, two smaller turbochargers may be used instead of one larger unit. The primary reason for such a layout is improved packaging. The smaller turbochargers also reduce acceleration lag, but result in an efficiency penalty due to higher fluid friction losses with the smaller passage sizes.

As the demand for higher specific output grows, both for increased part-load efficiency, and for reduced package volume and weight, more complex configurations are going into production. A turbocharger in series with a supercharger has long been seen on two-stroke diesel engines, formerly in heavy truck applications, and still seen today in some marine engines. The two-stroke requires pressurized intake air under all operating conditions, and this need is met with the crankshaft-driven supercharger. Placing a turbocharger in series with the supercharger allows the specific output to be increased, and improves engine efficiency through utilizing some of the exhaust energy. This arrangement is sometimes seen on four-stroke automotive engines as well.

Two turbochargers may be placed in series, with intake air drawn first through a low-pressure compressor and then a high-pressure compressor. An intercooler may be placed between stages, with an aftercooler following the high-pressure stage. The exhaust gas flows first through the high pressure turbine and then the low pressure turbine. In the series configuration the intake charge and exhaust gas flow through both units under all operating conditions. This configuration is sometimes seen in heavy truck engines, and is common in larger, off-highway diesel engines.

A more recent configuration is referred to as sequential turbocharging. This configuration also uses low- and high-pressure turbochargers, but with intake and exhaust control valves that direct the flow through either the high-pressure or low-pressure turbocharger under most conditions, and both units under only some operating conditions. This configuration allows high specific output to be maintained over a wide range of engine speeds.

At this time production applications of series and sequential turbocharging systems are more prevalent in diesel engines, where the additional boost pressure can be more readily utilized to increase specific output. With careful control and system optimization, sequential turbocharging is now being seen in a few production spark-ignition applications. The combustion system must be optimized within knock limits, but the sequential turbochargers then allow a wider speed range at high specific output.

1.5 Production Engine Examples

In automobile engines cost, weight, and package size are important design parameters in addition to the customer expectations regarding performance and fuel consumption. In most applications the duty cycle is quite light, with the full power output utilized for only a small fraction of the engine's operating time. Engines in these applications will have from three to twelve cylinders, with the majority having between four and eight. The in-line and vee configurations are most common, with horizontally-opposed and 'W' configurations sometimes seen. Spark-ignition examples are depicted in front- and side-view section drawings in Fig. 1.6a, b, and c. Those shown in Figs. 1.6a and b are both double overhead cam (DOHC) in-line four-cylinder engines. The engine shown in Fig. 1.6c is a high-performance, DOHC V-8 incorporating a variable-length intake runner system, and variable valve timing and valve event phasing. A series of photos of a DOHC V-6 engine is presented in Fig. 1.7a, b, c, and d.

Of resurgent interest in automobile applications worldwide is the diesel engine, examples of which are shown in Figs. 1.8a and b. The new diesel engines are universally turbocharged as this has allowed comparable displacements to their spark-ignition counterparts. The primary attraction of the diesel engine in automobile applications is its significantly higher inherent fuel economy. Greater reliability and durability are further attractions. These must be offset against higher initial cost, the challenges of meeting exhaust emission regulations, noise, and cold weather startability.

Heavy-duty engines, as referred to in this book are those used for trucks and buses. In many cases the same engines, or engines of very similar design, are used in agricultural and construction equipment as well as various marine, industrial, and stand-by power applications. These further applications will not be specifically addressed, but much of the engine design discussion will apply equally to these applications. Once the power and torque output requirements have been met the single most important design criterion for many of these engines is that of durability under a highly loaded duty cycle. Cost, weight, and package size are again important, but within the bounds defined by the durability requirement. In other words, these engines are significantly larger, heavier, and more expensive than automobile engines, but these design criteria remain important relative to competitive engines for similar applications. New engines for heavy-duty applications in trucks and buses are almost universally in-line six cylinder engines. They are highly turbocharged diesel or in some cases natural gas engines.

While considerably more time could be spent discussing the engines shown in these figures such discussion is deferred until later in the book. The reader will be asked to refer back to these figures many times in the upcoming chapters as various components and sub-systems are further described.

1.6 Basic Measures

A detailed discussion of engine performance measures is beyond the scope of this book, but a brief review of the most commonly used measures will be necessary to the further discussion.

Torque and Power Fundamental to engine performance is the relationship between work and power. The work, or useful energy output of the engine, is generally referred to as the torque output. Of particular interest in most engine applications is not only how much work can be done but the rate at which it can be done. The power of the engine is the time rate at which work is done. This is simply the product of engine torque and shaft speed,

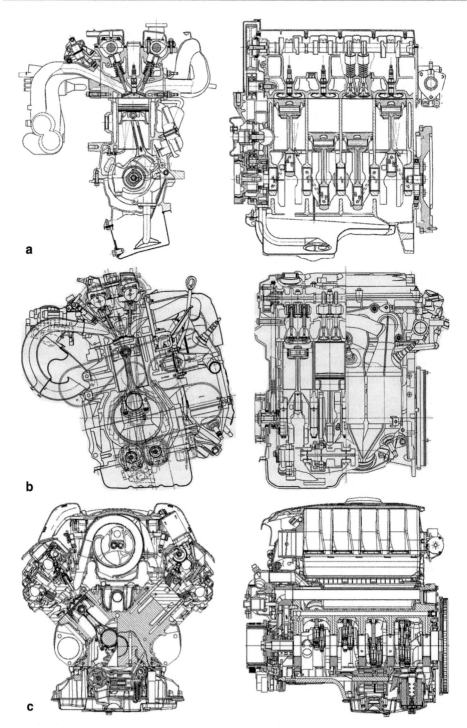

Fig. 1.6 a Four-cylinder, double-overhead-cam spark-ignition automobile engine (Courtesy of Toyota Motor Company). **b** Four-cylinder, double-overhead-cam spark-ignition automobile engine (Courtesy of Ford Motor Company). **c** Double-overhead-cam, high-performance V8 engine (Courtesy of BMW GmbH)

Fig. 1.7 a Double-overhead cam V6 engine, with cut-away showing cam drive chain (Courtesy of Nissan Motor Company). **b** Front view of engine shown in Fig. 1.7a. Serpentine belt accessory drives and cam drive cut-away are shown (Courtesy of Nissan Motor Company). **c** Details of cam drive, showing variable timing device for the intake camshaft. Engine is again that of Fig. 1.7a (Courtesy of Nissan Motor Company). **d** Cylinder head, intake and exhaust manifold cut-away for engine of Fig. 1.7a (Courtesy of Nissan Motor Company)

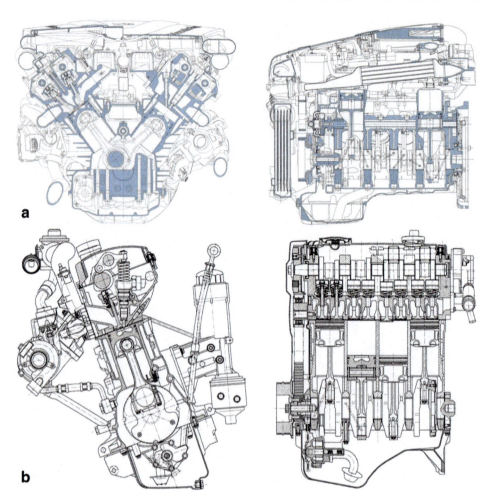

a

b

Fig. 1.8 **a** Four-cylinder, single-overhead-cam, direct-injection turbocharged diesel automobile engine (Courtesy of Volkswagen Audi GmbH). **b** Double-overhead cam V8, direct-injection turbocharged diesel automobile engine (Courtesy of Mercedes Benz GmbH)

$$P = (T)*(N)$$

English:

$$P(bhp) = (T\ ft-lbf)*(N\ rpm)*\left(\frac{1\ min}{60\ sec}\right)*\left(\frac{2\pi}{rev}\right)*\left(\frac{hp-sec}{550\ ft-lbf}\right)$$

$$= \frac{[(T\ ft-lbf)*(N\ rpm)]}{5252}$$

Metric:

$$P(kW) = (T\ N-m)*(N\ rpm)*\left(\frac{1\ min}{60\ sec}\right)*\left(\frac{2\pi}{rev}\right)*10^{-3}$$

$$= (T\ N-m)*(N\ rpm)*\left(0.1047*10^{-3}\right)$$

Mean Effective Pressure There are two physical interpretations that can be used to define the mean effective pressure. First, it is the constant pressure acting over the same volume change (from TDC to BDC) that would produce the same work as the actual engine cycle. Note that since the pressure-volume analysis gives us the net indicated work this interpretation provides the *indicated* mean effective pressure, or *IMEP*. The second interpretation is that mean effective pressure is the work output of the engine divided by its displacement. If indicated work is divided by engine displacement the result is the IMEP. If however the brake work is divided by displacement the result is the *brake* mean effective pressure, or *BMEP*. The calculation follows:

$$BMEP = \left[\frac{(Brake\ Power)*(rev/cycle)}{(D)*(N)}\right]$$

English:

$$BMEP\ (psi) = \left[\frac{(P\ bhp)*(2\ rev/cycle\,(if\ four\text{-}stroke))*396,000}{\left(D\ in^3\right)(N\ rpm)}\right]$$

Metric:

$$BMEP\ (kPa) = \left[\frac{(P\ kW)*(2\ rev/cycle\,(if\ four\text{-}stroke))*10^3}{(D\ liter)(N\ rev/sec)}\right]$$

Specific Fuel Consumption Another important measure of engine performance is its thermal efficiency—how efficiently the fuel energy is being converted into useful work. The measure of efficiency most commonly applied to engines is the specific fuel consumption, or *sfc*. The specific fuel consumption is simply the mass flow rate of fuel divided by the power output. If brake power is used in the calculation the result is the brake specific fuel consumption, or **bsfc**:

$$English:$$

$$BSFC\ (lbm\,/\,HP\text{-}hr) = \left[\dfrac{\dot{m}_{fuel}\,\dfrac{lbm}{hr}}{P\ bhp}\right]$$

$$Metric:$$

$$BSFC\ (gm\,/\,kW\text{-}hr) = \left[\dfrac{\dot{m}_{fuel}\,\dfrac{gm}{hr}}{P\ kW}\right]$$

Similarly, indicated power would be used to calculate indicated specific fuel consumption, or **isfc**. It should be noted that if one knows the energy content per unit mass of the fuel the thermal efficiency can be directly calculated by taking the inverse of the specific fuel consumption and multiplying the mass flow rate of the fuel by the energy per unit mass. The result of these operations is the rate at which work is done (power) divided by the rate at which fuel energy is supplied. The energy of the fuel is generally taken as its Lower Heating Value (LHV) at constant pressure.

Volumetric Efficiency The ability to produce work will always be limited by the ability to supply sufficient oxygen for the combustion process. Therefore an important measure is that of how well the cylinder can be filled during each intake stroke. The measure used is volumetric efficiency. The name is a bit misleading as it is actually a mass ratio. It is the ratio of the actual mass flow rate of air into the engine divided by the ideal mass flow rate—that which would be supplied if the cylinder could be completely filled with air at the supply conditions:

$$\eta_{vol} = \left[\dfrac{\dot{m}_{actual}}{\dot{m}_{ideal}}\right] = \left[\dfrac{\dot{m}_{actual}}{\left(\rho_{ref}\right)*\left(\dfrac{(D)*(N\ rpm)}{2\ rev\,/\,cycle}\right)}\right]$$

Two important points must be noted. First, by convention, the mass flow rates used are that for the air alone. Even though the spark-ignition engine inducts a mixture of air and fuel the mass of the fuel is *not* included. This is an important point as it will have a measurable impact on the result.

Second, one must always be clear on the reference conditions being used for the density calculation. In naturally aspirated engines it is normal to use the density at ambient pressure and temperature. In supercharged or turbocharged engines one may use ambient density (in which case the resulting volumetric efficiency will be well over 100 %), or the density at the intake port pressure and temperature.

1.7   Recommendations for Further Reading

The study of thermodynamics underlying engine performance and efficiency was briefly introduced in this chapter, but the remainder of this book is devoted to the mechanical sciences as applied to engine design. For further reading on engine thermodynamics, performance, combustion and emissions the reader is referred to the following text by John Heywood. This book is an important addition to the library of anyone in the engine development field (see Heywood 1988):

Fuel injection and ignition systems have not been covered in this book. Robert Bosch, Gmbh., publishes comprehensive books on both diesel and spark-ignition fuel injection systems. These books are expected to be regularly revised, and the revisions current at the time of this writing are as follows (see Robert Bosch GmbH 2006a, b).

Engine design continues to evolve as technology and materials advance, and market and regulatory demands change. References on current practices in production engine design soon become dated, but the engine designer can maintain currency by studying the papers published each year on new engine designs. The following papers present summaries of recent engine designs at this writing:

Spark-Ignition Automobile Engines (see Matsunaga et al. 2005; Sandford et al. 2009; Kawamoto et al. 2009; Heiduk et al. 2011; Steinparzer et al. 2011; Okamoto et al. 2011; Kobayashi et al. 2012; Königstedt et al. 2012; Motohashi et al. 2013).

Rotary Spark-Ignition Engine (see Ohkubo et al. 2004).

Diesel Automobile Engines (see Abe et al. 2004; Bauder et al. 2011; Ardey et al. 2011, 2012).

Diesel Heavy-Duty Engines (see Suginuma et al. 2004; Altermatt et al. 1992; Hower et al. 1993; Kasper and Wingart 2011).

References

Abe, T., Nagahiro, K., Aoki, T., Minami, H., Kikuchi, M., Hosogai, S.: Development of New 2.2-liter Turbocharged Diesel Engine for the EURO-IV Standards. SAE 2004-01-1316 (2004)

Altermatt, D., Croucher, S., Shah, V.: The design and development of the IVECO GENESIS diesel engine family. SAE 921699 (1992)

Ardey, N., Hiemesch, C., Honeder, J., Kaufmann, M.: The new BMW 6-Cylinder diesel engine. International Vienna Motor Symposium (2011)

Ardey, N., Wichtl, R., Steinmayr, T., Kaufmann, M., Hiemesch, D., Stütz, W.: The all new BMW top diesel engines. International Vienna Motor Symposium (2012)

Bauder, R., Eiglmeier, C., Eiser, A., Marckwardt, H.: The new high performance diesel engine from Audi, the 3.0 L V6 TDI with dual-stage turbocharging. International Vienna Motor Symposium (2011)

Heiduk, Th., Dornhöfer, R., Eiser, A., Grigo, M., Pelzer, A., Wurms, R.: The new generation of the R4 TFSI engine from Audi. International Vienna Motor Symposium (2011)

Heywood, J.B.: Internal Combustion Engine Fundamentals. McGraw-Hill Book Company, New York (1988)

Hower, M., Mueller, R., Oehlerking, D., Zielke, M.: The New Navistar T 444E Direct-Injection Turbocharged Diesel Engine. SAE 930269 (1993)

Kasper, W., Wingart, H.: MTU Series 2000-06—the next generation of diesel engines for off-highway applications with emission standard EPA Tier 4i. International Vienna Motor Symposium (2011)

Kawamoto, N., Naiki, K., Kawai, T., Shikida, T., Tomatsuri, M.: Development of New 1.8-Liter Engine for Hybrid Vehicles. SAE 2009-01-1061 (2009)

Kobayashi, A., Satou, T., Isaji, H., Takahashi, S., Miyamoto, T.: Development of New I3 1.2 L Supercharged Gasoline Engine. SAE 2012-01-0415 (2012)

Königstedt, J., Assmann, M., Brinkmann, C., Eiser, A., Grob, A., Jablonski, J., Müller, R.: The new 4.0-L V8 TFSI engines from Audi. International Vienna Motor Symposium (2012)

Matsunaga, T., Fujiwara, K., Yamada, T., Doi, I., Yajima, J.: Development of a New 4.0 L V6 Gasoline Engine. SAE 2005-01-1151 (2005)

Motohashi, Y., Kubota, K., Akaishi, N., Ishiki, K., Iwamoto, T., Kinoshita, M.: Development of New L4 2.4 L Gasoline Engine for 2013 Model Year Accord. SAE 2013-01-1734 (2013)

Ohkubo, M., Tashima, S., Shimizu, R., Fuse, S., Ebino, H.: Developed Technologies of the New Rotary Engine (RENESIS). SAE 2004-01-1790 (2004)

Okamoto, T., Kawamura, H., Tsukamoto, K., Nagai, M., Uchida, T.: The new Toyota 4.8? V10 petrol high performance engine for Lexus LFA super car. International Vienna Motor Symposium (2011)

Robert Bosch GmbH: Gasoline Engine Management, 3rd edn. Wiley, Gerlingen, Germany (2006a)

Robert Bosch GmbH: Diesel Engine Management, 4th edn. Wiley, Gerlingen, Germany (2006b)

Sandford, M., Page, G., Crawford, P.: The All New AJV8. SAE 2009-01-1060 (2009)

Steinparzer, F., Unger, H., Brüner, T., Kannenberg, D.: The new BMW 2.0 Litre 4-cylinder S.I. engine with Twin Power Turbo technology. International Vienna Motor Symposium (2011)

Suginuma, K., Muto, H., Nakagawa, H., Yahagi, T., Suzuki, T.: Hino J-Series Diesel Engines Developed for the U.S. 2004 Regulations with Superior Fuel Economy. SAE 2004-01-1314 (2004)

Engine Maps, Customers and Markets

2

2.1 Engine Mapping

In Chap. 1 the relationship between work, power, and engine speed was defined. It is this combination of work, power, and speed that is critical to an engine's performance, the way it responds to changing demands, the way it "feels" to the driver. This subject will now be taken up in further detail. The work measured at the crankshaft of an engine is referred to as the ***brake work.*** The product of the brake work and engine speed (with the appropriate unit conversions) is the brake power. These terms reflect the history of engine testing, since early dynamometers typically consisted of friction brakes clamped around a spinning disk bolted to the engine's crankshaft. While dynamometers have changed a great deal the fundamental principles remain the same. The measurement principles of the dynamometer are shown in Fig. 2.1. The dynamometer allows an engine to be loaded under controlled test conditions, and the twisting force, or torque, produced by the engine to be measured. Regardless of dynamometer type, the engine spins the input shaft, and the outer casing contains elements that resist the spinning of the shaft. The engine must produce work (torque) to overcome the resistance. If the dynamometer casing is placed on low friction bearings, and is then kept from spinning by a force exerted at some radius from the spinning shaft the product of the force and radial distance is equal and opposite to the torque produced by the engine.

The dynamometer just described can be used to determine the work output of the engine over its range of operating speeds. As depicted in Fig. 2.2, a speed can be selected, and the throttle setting can be increased to increment the engine speed upward from idle to the selected speed. The figure depicts the throttle position required to overcome mechanical friction and operate the engine at the selected speed at zero load. If the dynamometer load is now increased the engine again slows down. Incrementally opening the throttle

© Springer Vienna 2016
K. Hoag, B. Dondlinger, *Vehicular Engine Design,* Powertrain,
DOI 10.1007/978-3-7091-1859-7_2

Fig. 2.1 Dynamometer operating principles

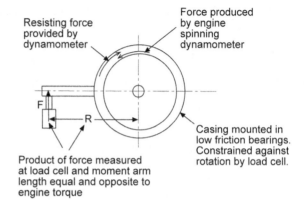

Fig. 2.2 Generating an engine operating map through dynamometer testing

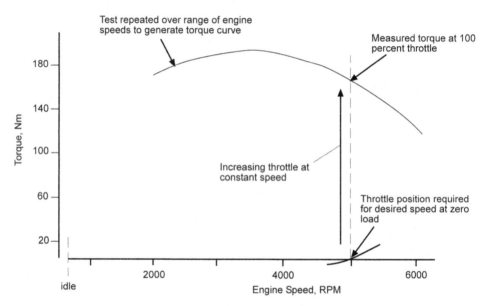

further as the load is increased allows the selected engine speed to be maintained until the maximum, or *full load* work output at that speed is reached.

Repeating the test sequence just described over the range of speeds for which the engine can be operated generates the full load torque curve, thus defining the engine operating map. Example operating maps are shown in Figs. 2.3 and 2.4, for spark-ignition and diesel engines respectively. The operating maps as shown plot brake torque output versus engine speed—the two quantities directly measured during the dynamometer test. Recognizing that power is the product of torque and engine speed, the plots could alternatively be plotted as power output versus engine speed.

Looking further at Figs. 2.3 and 2.4 there is a range of speeds in each figure over which the full-load work output decreases as engine speed increases. Remembering again the

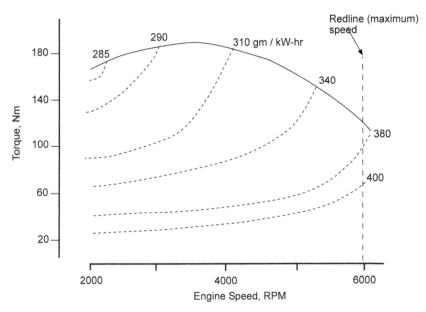

Fig. 2.3 Typical spark-ignition engine operating map, showing lines of constant specific fuel consumption, and maximum engine speed

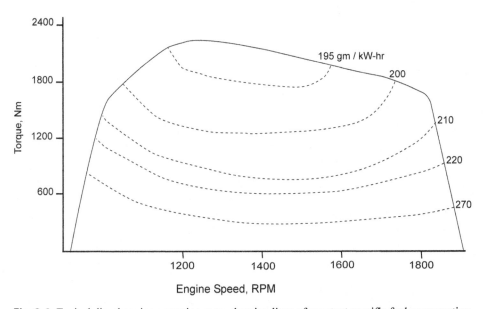

Fig. 2.4 Typical diesel engine operating map, showing lines of constant specific fuel consumption

relationship between power, work, and shaft speed it should be recognized that the slope of the full-load torque curve will determine whether the power is increasing or decreasing as speed is increased. As speed is increased from that of maximum torque the power will typically initially increase. The rate of speed increase is greater than the rate of torque decrease. At some engine speed the torque begins dropping more rapidly, and the power drops with any further speed increase. The speed at which the power reaches its maximum is referred to as *rated speed*, and the maximum power is referred to as *rated power*. The speed range between peak torque and rated power is especially important in applications where the engine is intended to operate for extended periods of time at full load. The negative slope of the torque curve over this range of speed allows the engine to stably respond to changes in load. While operating at full load at any speed between rated power and torque peak if the engine suddenly experiences a further load increase it will slow down, but its work output will increase to address the increased load. If however the engine is operating at a lower speed, where the torque falls with falling speed, a sudden load increase will stall the engine. The engine cannot produce additional work, and therefore cannot address the increased load. The engine can be operated very nicely at part throttle at these lower speeds, but would be quite unstable at full load.

The maximum speed of any engine will be determined by mechanical limitations. Such limitations may be defined by centrifugal or reciprocating forces, or by the rapidly increasing mechanical friction. In some engines a "redline" speed may define the maximum engine speed. The engine could be operated at higher speeds, but the operator risks damaging or destroying the engine. Sound design practice requires that the engine be capable of operating at speeds some margin above that at which maximum power is produced. For example, if an engine produces maximum power at 6600 rpm, a redline speed of 7000 rpm would be acceptable but a redline speed of 5800 rpm would not. The operator would regularly exceed the redline speed based on the "feel" of the engine. As an engine is tuned for higher speed in high performance or racing applications the need to simultaneously address structural considerations to increase redline speed becomes quickly apparent.

In many other applications a high-speed governor is included as part of the fuel metering system. The governor is designed to monitor engine speed and rapidly reduce fueling as the speed increases. The torque curve for such an engine will include a steeply dropping torque from speeds at or slightly above rated speed, to a zero-load, high idle speed. This is shown on the diesel engine operating map in Fig. 2.4. High-speed governors are almost universally used on diesel engines, and are included on spark-ignition engines in many industrial and non-automotive applications.

Also shown on these maps are lines of constant specific fuel consumption. Minimum specific fuel consumption will typically be seen at high loads, and at relatively low engine speeds. As engine speed increases the engine becomes less efficient due to the rapid rate of friction increase. At low speeds the specific fuel consumption will again increase because of increased heat transfer losses. This phenomenon is more pronounced in diesel engines where the air-fuel ratio drops with the increased torque. This results in increased flame temperatures during combustion, and minimum specific fuel consumption typically occurring between rated and peak torque speeds. As load is reduced at any speed the specific fuel consumption again increases. This is due to a combination of increased pumping work and the fact that friction losses stay nearly constant while the brake output is dropping.

Note that the specific fuel consumption increases more rapidly as load is reduced with the spark-ignition engine than with the diesel. This is due to the increased pumping work associated with the throttled intake on the spark-ignition engine.

At this point it is instructive to briefly review and compare the fundamental combustion characteristics of diesel and spark-ignition engines. The reader is referred to texts on combustion in engines for a more detailed treatment—the information presented here is merely an overview needed for engine design considerations. Figure 2.5 provides a general comparative look at automotive diesel versus spark-ignition engine operating maps for the same vehicle. Two important points should be emphasized. First, the spark-ignition engine inherently operates at higher speeds, and over a wider speed range. In part this is due to the necessarily heavier construction of the diesel engine. Faced with peak cylinder pressures two to three times those seen in spark-ignition engines, the diesel combustion chamber and power transfer components must be more robust. Higher rotating and reciprocating forces associated with these heavier components limits their maximum speed. However, a more importan speed limitation results directly from the combustion process of the diesel engine. While the flame speed in a spark-ignition engine scales very well with engine speed, diesel combustion does not scale well with speed. Diesel combustion

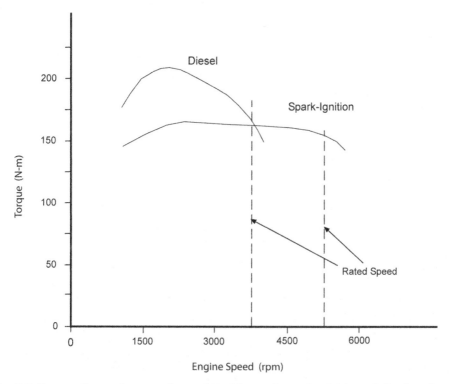

Fig. 2.5 Comparative engine maps for spark-ignition engine and turbocharged diesel engine of similar displacement

requires a sequence of processes—atomization of liquid fuel injected into the combustion chamber; vaporization of the fuel droplets; mixing of the vaporized fuel with the surrounding air; and finally combustion. Clearly these processes do not all scale with engine speed. The maximum speed of a diesel engine is thus limited by the time required to complete the combustion process before efficiency is penalized by extending combustion too far into the expansion stroke. A practical limit for a direct-injection diesel engine will be on the order of 3500–4500 rpm.

The second point to note from Fig. 2.5 is the significantly higher torque rise of the diesel engine. In the case of the spark-ignition engine, the mass ratio of air to fuel must remain approximately constant over the full-load speed range from peak torque to rated speed. With a constant air-to-fuel ratio the torque output can change only with changes in thermal efficiency (brake specific fuel consumption) or volumetric efficiency. The specific fuel consumption drop with reduced speed, seen previously in Fig. 2.3 provides a natural but relatively small increase in torque output. To a greater extent, the shape of the torque curve is dependent on how the volumetric efficiency changes with engine speed. The resulting change in torque with engine speed is significantly less than that which can be achieved in the diesel engine. In the case of the diesel engine a full charge of air is drawn into the cylinder under all conditions. The torque output at any given speed is governed by the amount of fuel injected, and again by the thermal efficiency that can be achieved at that speed. The result is generally a significantly higher rate of torque rise, as shown in Fig. 2.5.

Two final notes should be made about this comparison. First, the diesel may not always demonstrate this much higher rate of torque rise. Because the engine may be set up to operate in the same installation as a spark-ignition engine one may choose to inject more fuel at higher engine speeds to increase peak power. Also the capability of the drivetrain may preclude high torque rise and limit the fueling at low speed. Finally, if the spark-ignition engine is turbocharged boost control versus speed provides another parameter allowing the torque curve shape to be adjusted—in some cases this may allow the torque curve to approach the torque rise of a diesel engine.

2.2   Automobile, Motorcycle, and Light Truck Applications

Automobiles, motorcycles, and light trucks clearly encompass a wide range of vehicles. While detailed coverage of each of these markets cannot be provided in this book several general observations can be made to provide perspective for the designer. In each of these cases, with the exception only of racing engines, the duty cycles seen by the engines will be extremely low. That is, the vast majority of time is spent at very light loads and low engine speeds. While the steady-state loads are increasing as smaller displacement, boosted engines are adopted to increase efficiency, the duty cycles continue to be relatively low. Excursions to higher speeds and loads continue to be limited to brief transients, with many drivers never operating their engines at full load. This observation is of vital importance in two regards. First, it has a great impact on mechanical development of the engine for durability; it creates the difficulty of developing the engine for an expected duty cycle, but recognizing that there may be a wide range of driver operating practices, resulting in significant variation in expected engine life. Second, while the engine achieves its best thermal efficiency much nearer to full load, it spends very little operational time there. Conventional drivetrains are not well suited to providing maximum fuel efficiency, and a great deal of attention is now being given to alternatives such as hybrid electric drive.

Of primary importance to most customers is the drivability of the vehicle. While drivability is a term having many meanings, for these customers its primary meanings are those associated with rapid acceleration, smooth transitions between operating conditions, and minimal noise and vibration.

Most of these markets are quite cost sensitive, and any design features that add cost must be carefully studied for market acceptance. Is the resulting improvement something the customer will appreciate and be willing to pay for?

Fuel economy varies in importance, correlating most closely with fuel costs in the region or country in which the vehicle is to be operated. Weight is a criterion most customers would not identify as having importance, but its impact on package size, cost, and fuel economy make it an important consideration for the engine designer.

Reliability and durability will be more carefully defined in Chap. 3. For now it is important to say that most segments of these applications have expectations of surpassing some threshold engine life, or durability—some number of miles that can be expected before the engine would need to be replaced or rebuilt. Since most customers of these vehicles are not thinking about life to overhaul at the time of initial purchase design improvements that improve durability beyond the customers' threshold expectations can seldom be justified—especially if they add cost to the engine. Reliability refers also to minimizing or eliminating unexpected problems that force repairs over the useful life of the engine. As with virtually every engine application this aspect of reliability expectations are high.

An installed engine operating map for these applications is typically presented as shown in Fig. 2.6. The figure shows tractive force at the drive wheels versus vehicle speed. The individual curves each show the steady-state full-load engine torque curve with a different drive ratio between the engine and drive axle. The dashed lines show each of the five transmission ratios with a particular axle ratio, and the solid lines show the same five transmission ratios with another axle ratio. The right end of each curve is at the engine's redline speed in the particular gear, and thus the maximum vehicle speed in that gear. Also shown on the figure is the force required to move the vehicle on a level road at any given speed. The top speed of the vehicle is determined by either the intersection of this curve with the engine's torque curve corresponding to the highest transmission gear, or by the redline speed of the engine if the torque curve is above the vehicle load curve. A final limit on this diagram is the maximum tractive force or wheel spin limit. If the engine torque as multiplied through the transmission and drive axle gears exceeds this limiting force the tire will break free from the pavement. The force at which this occurs is dependent on the gripping force between the tire and the particular road surface.

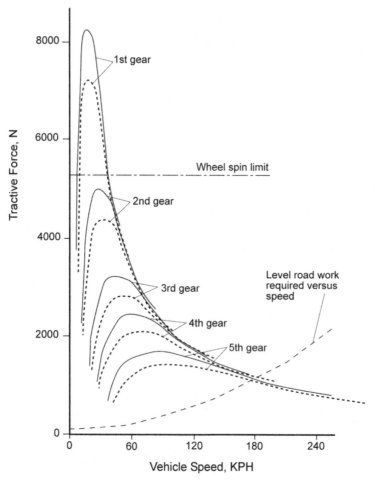

Fig. 2.6 Tractive force in each transmission gear, with two different axle ratios, and level road vehicle load versus engine speed

The plots presented in Fig. 2.6 cover steady-state operation. Also important in automobile applications is transient performance—specifically acceleration rate. The acceleration rate for the same vehicle, at the lower (numerically higher) axle ratio, is shown in Fig. 2.7. Also shown in the figure are the shift points for maximum acceleration. The lower axle ratio of Fig. 2.6 provides more rapid acceleration, and is thus favored in many high performance applications. The higher ratio (numerically lower) results in higher maximum vehicle speed, and of more importance in most applications, it results in lower engine speeds at any given vehicle speed. Because of the strong impact of friction on fuel economy the lower engine speeds provide better fuel economy while the engine will feel less responsive (lower acceleration rates) in the vehicle.

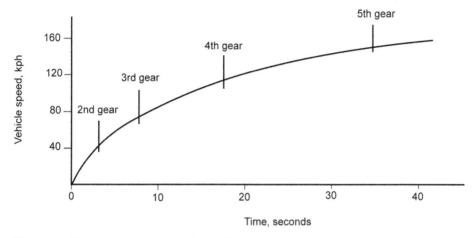

Fig. 2.7 Vehicle speed versus time under conditions of maximum acceleration, indicating points at which each gear shift takes place

2.3   Heavy Truck Applications

Whereas automobile and light truck duty cycles are almost invariably quite low those seen in heavy truck applications are very high, with significant percentages of operating time spent along the full-load torque curve. Combining this with the fact that mileage accumulation rates are quite high (200,000 km per year is typical), results in great emphasis on engine durability or useful life to overhaul. Fuel economy often receives great emphasis as the combination of high usage and high loads makes fuel costs significant. One important distinction within the heavy truck market is between weight-limited and volume-limited hauling. In weight-limited markets, where the maximum allowable vehicle weight is attained before the cargo volume is filled, engine weight is an important criterion—any reduction in engine weight translates directly into increased payload. Economics figures strongly in purchase decisions, and the cost of design improvements must be tied closely to resulting fuel economy, weight, reliability or durability improvements.

Figure 2.8 shows the engine operating map for a diesel engine used in a heavy truck. Overlaid on the engine map is the operating sequence seen when climbing a hill. While the exact power requirements will depend on the aerodynamic drag and rolling resistance of the particular vehicle, it will typically require a power of approximately 150 kW to propel a 36,300 kg truck along a level road at 100 km per hour. The mass chosen is the maximum gross vehicle weight allowed for an 18-wheeled tractor-trailer traveling the interstate highways in the United States (80,000 lbm GVW = 36,287 kg). If the particular engine has a rated power of 300 kW the engine will be operating at approximately half-load when traveling at freeway speed on a level road. This is depicted as point '1' in Fig. 2.8. The engine speed at point '1' will depend on the final drive ratio (the product of the transmission gear ratio in the highest gear and the differential gear ratio). Selecting a numerically higher ratio will increase the engine rpm, and a lower ratio will decrease rpm. The trade-off is pri-

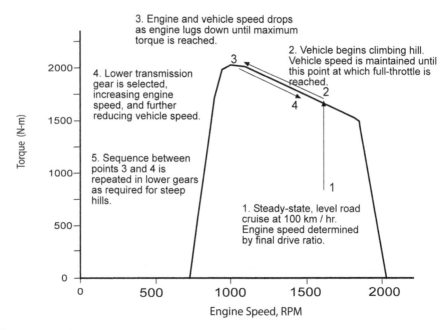

Fig. 2.8 Operating sequence for a heavy truck encountering an uphill grade

marily one between fuel economy and drivability. Increasing the engine speed will allow the operator a greater rpm range and require fewer gear shifts, but fuel economy rapidly deteriorates at high engine speed due to increased friction losses.

Returning now to the sequence of events in Fig. 2.8, if the operator encounters a hill she or he will depress the accelerator pedal further in order to maintain vehicle speed. At point '2' the engine reaches wide-open throttle. Vehicle speed has been maintained to this point, as engine speed has remained constant, and no gear shifts have occurred. Once point '2' has been reached the engine and vehicle will begin slowing down. The engine's torque rise allows increased work to be done to lift the vehicle up the hill, but it is now done with a steadily dropping speed. As point '3' is approached the operating conditions must be changed to avoid stalling the engine. Assuming the truck is still climbing the hill, the operator must select a lower gear (downshift). This brings the engine speed back up (as indicated by point '4') while vehicle speed continues to drop.

An alternative portrayal of heavy truck operation is given in Fig. 2.9. This figure is similar to that given earlier for the automobile application. Power required at the wheels is plotted versus vehicle speed for a level road, and 1 and 2% upgrades and downgrades. Also shown is the full throttle engine curve in high gear. The steeply downward sloped portion of this curve represents the high speed governor, and the transition point to the flatter portion occurs at rated power. Note that for this very typical vehicle the engine will reach full load when encountering a 1% uphill grade, and will be required to downshift if the grade reaches 2%. Similarly, at somewhere between a 1 and 2% downhill grade the throttle will be closed and the engine will be motored by the vehicle.

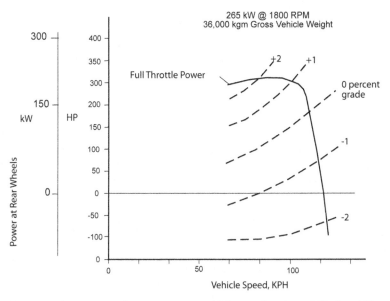

Fig. 2.9 Heavy truck power requirements versus vehicle speed under uphill, downhill, and level road conditions

Based on the discussion just presented it should be clear that engine speed range and torque rise will be critical design parameters for these customers. The more torque rise available to the operator the less rapidly the engine will slow down with increasing grade. The combination of torque rise, gearing, and speed range will dictate how often the operator must shift gears—an important consideration for cross-country driving.

2.4   Off-Highway Applications

Because of the high torque rise and durability requirements diesel engines are used in the vast majority of agricultural and construction applications. In almost all of these applications the operator sets but does not continuously adjust the throttle position. While the engine map requirements are generally very similar to those for heavy trucks, the fueling control strategy is quite different. This is depicted in Fig. 2.10. The two sets of parallel lines overlaid on the diesel engine map represent lines of constant throttle position for on-highway and off-highway applications. In the case of a heavy truck for example, the operator is controlling vehicle speed through continuous adjustment of accelerator pedal position. The desire is to maintain a constant vehicle speed over variable terrain. In any chosen gear constant vehicle speed translates directly to constant engine speed (assuming the manual transmission used in nearly 90 % of these applications). The shallow slope of the lines of constant throttle position results in the operator having full use of accelerator pedal position (0–100 % throttle) at any given speed.

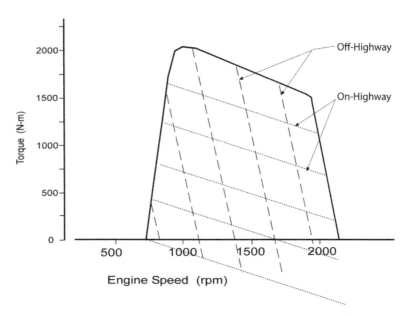

Fig. 2.10 Desired throttle control characteristics for on- and off-highway applications of diesel engines

In contrast, the objective in off-highway applications is to set the throttle position, and minimize the amount of engine speed variation as the load changes. The much steeper lines of constant throttle position ensure that the engine speed varies over a narrow band, even with large changes in load. If the on-highway control scheme were used a large change in engine speed would be seen even with very small changes in load. Using the off-highway control scheme in an on-highway engine would result in a very touchy accelerator pedal, with almost the entire range of fueling occurring over a small percentage of pedal travel. Whether it is done with a fully mechanical fuel injection system or through the use of electronic control, it is necessary to provide these very different throttle control characteristics for different applications.

As was previously discussed, the remaining customer expectations for these markets will be very similar to those for heavy trucks. The fuel economy versus engine speed range trade-off often plays out differently. The number of hours of use per year tends to be lower, and thus fuel economy plays a smaller role in equipment purchases. On the other hand, the equipment sees more excursions toward peak torque operation, making a wider speed range attractive. Similar engine designs to those for heavy trucks will be used, but often with higher rated speeds and equal torque peak speeds.

In off-highway applications such as railroad locomotives and some mining vehicles the engine is used to drive an electric generator and electric motors then drive the wheels. This approach further separates the torque characteristics of the engine from those required at the wheels. While a mechanical drive system can increase the force available at the wheels by going to higher and higher gear ratios the electric motor has the characteristic of maximum torque at zero rpm. To approach this characteristic with a mechanical transmission would require a very large, complex, and expensive drivetrain. The required engine characteristics for an electric drive system are very similar to those for other heavy-duty off-highway applications, but the engine can be optimized over a narrower speed range. This may allow higher peak work or power output, and an opportunity to optimize the engine for greater efficiency.

A unique area of IC engine application is in the wide array of marine applications—everything from small spark-ignition outboard engines to extremely large, low-speed diesel engines in huge ships. The power demand seen by engines in marine applications is summarized in Fig. 2.11. The application is unique in that power output follows a propeller curve, where the power requirement increases at a cubic rate (theoretically) with speed. This ensures that while the engine will see rated power at wide-open-throttle it will not see peak torque conditions. The engine is generally loaded along the prop curve shown in Fig. 2.11. There are a few deviations from this general statement. First, in smaller, high performance boats, the boat's position in the water will shift as it accelerates—this is typically referred to as the boat coming up "on plane." This has the effect of shifting the prop curve to the right. If the prop curve at speed intersects with the peak power point on the engine map, then while the boat is accelerating it will see lower speed, full-load operation. In large boats there may be significant auxiliary loads (electric power, hydraulic winches, etc.) that draw load while the boat is being propelled at high throttle positions. These too will cause the engine to see high loads at speeds below rated power.

The significance of the marine operating characteristics for engine development is that the engine can be developed with very low rates of torque rise. If the designer does not have to simultaneously accommodate high power and high rates of torque-rise the rated power of the engine can often be significantly increased. In some markets—those where

Fig. 2.11 Marine engine power requirements versus boat speed

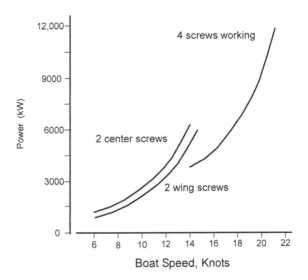

fuel economy is not important, such as high performance pleasure craft—this can be furthered by increasing rated speed. The net result is that marine engines using the same carcass as those for other applications will often be rated at nearly double the peak power output. Different ratings will be available depending on the hours expected to be spent under high power output conditions.

2.5 Recommendations for Further Reading

Engine testing in a dynamometer test cell is discussed in detail in this book by Plint and Martyr (see Plint and Martyr 1995).

The following papers provide examples of experimental and analytical work in matching the engine to a transmission and vehicle. The final reference discusses heavy truck applications (see Thring 1981; Barker and Ivens 1982; Ren and Zong Ying 1993; Jones 1992).

References

Barker, T., Ivens, J.: Powertrain Matching of a Small Automotive Diesel. I Mech E MEP-160, London (1982)

Jones, C.: Heavy-Duty Drivetrains—The System and Component Application. Truck Systems Design Handbook, SAE PT-41 (1992)

Plint, M., Martyr, A.: Engine Testing: Theory and Practice. Butterworth-Heinemann, Oxford (1995)

Ren, H., Zong Ying, G.: An Analytical System of the Automobile Powertrain Matching. SAE 931964 (1993)

Thring, R.: Engine Transmission Matching. SAE 810446 (1981)

Engine Validation and Reliability

3

3.1 Developing a Reliable and Durable Engine

An integral part of the engine design process is that of ensuring that the product has sufficient reliability and durability. As used in this book *durability* refers to the useful life of the engine. For the engine system this is the average life-to-overhaul. For most of the major engine components it includes an expectation of reuse when the engine is overhauled. *Reliability* includes also infant mortality and the unforeseen problems that require attention over the engine's operational life. Engine design and development must include validation processes to ensure that the durability and reliability expectations are met. The success of this endeavor is crucial to the success of any engine design.

As an example consider the need for cylinder head durability. The cylinder head is a major component of the engine, and must perform its functions without problem for the useful life of the engine. It is generally expected that if the engine is overhauled the cylinder head will not be replaced but will be reused. The cylinder head is a complex component simultaneously exposed to a variety of loads including cylinder pressure, combustion temperature, and high levels of clamping loads at the head bolts and press-fit loads at the valve seats. Any of these loads, singularly or in combination, may result in cylinder head cracking. If such cracking occurs even as infrequently as once per 1000 heads the engine will quickly develop a bad reputation among customers. A reputation for head cracking is extremely difficult to overcome and may result in an unsuccessful product—one that must be replaced or significantly redeveloped prior to recovering the tooling investment. As this example demonstrates it is absolutely imperative to ensure that the design validation process identifies all reasonable possibilities for inadequate durability or reliability prior to the engine's release to production.

In the example just presented the complexity of a given engine part and the combination of loads to which it is exposed were stated. The development challenge is made still

© Springer Vienna 2016
K. Hoag, B. Dondlinger, *Vehicular Engine Design*, Powertrain,
DOI 10.1007/978-3-7091-1859-7_3

more complicated by the combination of material property control, manufacturing process control, and the variety of customer abuses which may or may not have been fully anticipated. The structural analysis process is typically based on nominal, or expected material properties, and the assumption that the part will be manufactured as specified in the design drawings. Neither of these assumptions is completely accurate, and deviations from the nominal conditions must be included in the analysis process. This requires working closely with the manufacturing engineers and material suppliers throughout the development process to ensure that process and specification control limits are selected based on the true needs for durability and reliability. Over-specifying and under-specifying control limits both add cost to the product. Conservative engineering decisions often lead to over-specifying both material and manufacturing control limits. Because of remaining unknowns in the validation process this is often necessary, but has become an increasingly less affordable luxury. If the challenges just described are not enough the engineer remains faced with anticipating the creative ways in which the customer might use the product. This remains the biggest unknown, and often the source of greatest aggravation in designing for reliability and durability.

Many engine components require that every effort be made to eliminate any possibility of failures. The cylinder head example discussed earlier certainly falls into this category as do cylinder blocks, crankshafts, and flywheels. Pistons, connecting rods, camshafts and bearings are certainly components where every effort must be made to ensure extremely low failure rates. However, at some point the development costs and piece prices become prohibitive, and a trade-off must be recognized versus what the customer is willing to pay for the product and what level of product problems the customer deems to be acceptable. As an extreme example, consider the engine that is absolutely guaranteed to operate with only minimal care and no component failures for the life of the product. The engine can be sufficiently over-designed to make this guarantee, and it will sound very attractive to the customer until the price is revealed. Almost every customer will then opt for the far less expensive product offered by the competitor, and will assume the risk that some repairs will be required over the life of the vehicle. Finding the appropriate balance between acceptable initial price and acceptable repair rate is a difficult process requiring careful customer and market studies and statistical analysis of repair rates. The result of the process is an anticipated engine repair rate, and a warranty accrual that is factored into the initial selling price of the engine. In effect the customer is paying a lower price for the engine, and is then paying for an insurance policy (the warranty accrual) to help offset the risk of product problems. The "extended warranty" now offered at extra cost with many engines and vehicles furthers this concept. The repair rate is typically reported in "repairs per hundred"—the total number of repairs made per 100 engines. This number is often further broken down by individual engine components. An important aspect of the continued development of an engine after it goes into production is that of tracking repairs per hundred by individual component, and making design changes to improve components having high repair rates, while if possible not increasing (or better yet, further reducing) their costs.

In some engine applications the useful life of the engine is an important consideration used by the customer in making purchase decisions. This is certainly the case in many heavy trucks as well as for engines used in industrial and construction applications. It is typically not a significant criterion in automobile purchases although it may be important to anyone expecting to rapidly accumulate miles on their vehicle. The average life-to-overhaul, often referred to statistically as the "B50 life," becomes an important parameter for these applications. A related parameter, the B10 life, or the time at which the first 10% of the engines requires overhaul, is often tracked to ensure adequate life-to-overhaul for applications that do not accumulate miles as rapidly (most automobile and light truck applications). The continued development of production engines for increased B50 or B10 life is often important. In these cases analysis of data from the field is done to determine the engine system or components most often responsible for overhaul. Continued development focuses on these parts of the engine as a means of increasing the life-to-overhaul. For example, analysis may reveal that the B50 life of a given engine is 125,000 miles, and that the most common reason for overhaul is head gasket leakage. Further development of the head gasket results in increasing the B50 life to 140,000 miles. At this point the most common reason for overhaul is found to be low oil pressure due to main bearing wear, and development efforts shift to this component.

Various concepts pertaining to engine development for durability and reliability have been introduced. In later chapters these concepts will be applied to the discussion of specific components and sub-systems of the engine. The remaining sections of this chapter address the general mechanisms and principles that drive wear and failure of engine components. The concepts relevant to each engine component will then need to be applied in developing appropriate durability and reliability validation techniques.

3.2   Fatigue Analysis

The discussion of mechanisms that define the durability of an engine appropriately begins with fatigue. It is the primary failure mechanism for many components in engines, and the focus of the greatest fraction of the analysis and testing to which most engines and engine components are subjected. Of necessity this section provides only a brief overview of the science of fatigue, emphasizing its application to engine development.

The stage is set with a brief review of a typical stress-strain diagram depicted in Fig. 3.1. Most readers should be quite familiar with this diagram showing the deformation of a metal bar as it is loaded in tension (left portion of the figure), and a cycle of tension and compression (right portion). The strain is the ratio of the change in length to the original length of the bar, and the stress is the applied tensile or compressive load per unit area. As long as the applied load per unit area is below the yield strength of the material, the bar will return to its original length when the load is removed, following the linear portion of the plot shown in the figure. The deformation that occurs with load under these conditions is referred to as *elastic deformation*. Once the yield strength is exceeded the bar will be

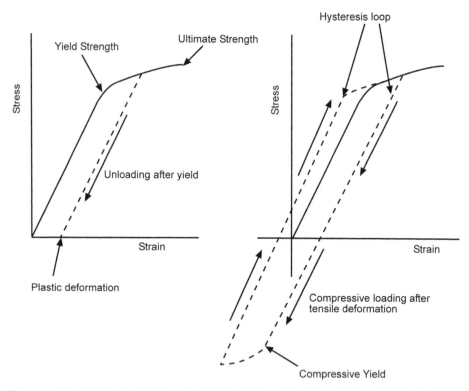

Fig. 3.1 The stress-strain diagram, showing simple tensile loading (*left*), and a load cycle of tension and compression (*right*)

permanently deformed. If the load is now removed the bar returns along a parallel line such as the example dashed lines shown in Fig. 3.1. When the load reaches zero the bar remains at a longer, deformed length. The loading has resulted in ***plastic deformation***. A similar loading excursion through compressive yield is also shown. The slope of the elastic (linear) portion of the plot provides a measure of the ***elasticity*** of the material. Materials with greater elasticity will have greater deflection per unit load, and thus a shallower slope. The ***modulus of elasticity*** is the ratio of stress to strain, and is thus the standard measure of elasticity. The difference in strain between that at the material's yield strength and that at its ultimate strength determines the ***ductility*** of the material. A brittle material will elongate very little before fracture, while a ductile material will demonstrate significant elongation. The typical sign convention is that tensile stress and strain are given positive values, and compressive stress and strain are given negative values. In the discussion of engine components compressive loading will often be seen.

When a material is repeatedly loaded and unloaded it will eventually fracture even if the loading is well below the ultimate strength. Although it will take many more cycles, it will also fracture under cyclic loading well below the yield strength. Such fractures are referred to as ***fatigue*** fractures, and are the subject of this section. The fatigue properties of a material are most often presented graphically in the form shown in Fig. 3.2. The plot

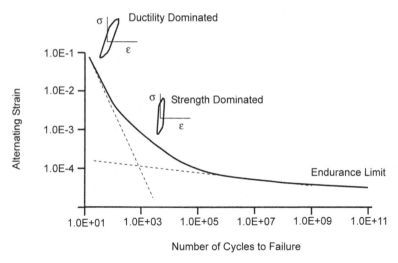

Fig. 3.2 Characterization of fatigue life using an S-N diagram

shown is typically termed an *S-N diagram*, where 'S' is the alternating cyclic stress in each direction about a mean of zero, and 'N' is the number of cycles resulting in a fracture. At the right portion of the figure the alternating stress remains significantly below the yield strength of the material, and the loading remains entirely elastic. An *endurance limit* is identified in this portion of the figure. When working with ferrous materials the endurance limit is often defined as an alternating stress magnitude which if not exceeded will result in *infinite life*. This remains debated terminology, as some fatigue experts will argue that if enough cycles are applied a fracture will still be seen even at alternating stress levels below the endurance limit. From a practical standpoint the concepts of an endurance limit and infinite life remain important as the alternating stress must be kept below that which would result in failure during the life of the product. For example, a crankshaft may literally be exposed to 3 billion alternating load cycles during a useful life that includes its use in two or three rebuilds. Even if that crankshaft would have failed if exposed to 10 billion alternating load cycles for all practical purposes it has not exceeded its endurance limit, and it has thus been designed for infinite life. In the case of non-ferrous materials the S-N diagram does not level off as the alternating load is reduced, and the concept of an endurance limit cannot be applied.

Returning to Fig. 3.2, as the alternating load is increased the material will fracture under fewer load cycles. As the curve is traversed toward the left it deviates from the purely elastic loading line when the material begins to yield, and plastically deform. The load cycle appears as shown in the inset stress-strain diagram. As the yield point is first exceeded the loading is referred to as *strength dominated*. Further alternating load increases result in greater plastic deformation with each load cycle. This is further depicted in the stress-strain inset termed *ductility dominated* loading. The region of fatigue loading relevant to most engine component development is purely elastic loading. A few components, such as the cylinder head firedeck and exhaust manifolds may experience some plastic deformation as a result of high thermal loading, but even these remain in the strength dominated regime.

One of the first challenges of fatigue analysis is that of obtaining accurate fatigue properties for the material being used. Many materials are not well characterized, and even among those that have been extensively studied, such as gray iron and various aluminum alloys, considerable variability is found. One explanation for this variability can be found in the fact that the material's response to fatigue loading is dependent not only on the alloy composition but on the processing to which the material was subjected and the resulting microstructure. For geometrically complex castings and forgings this means that the fatigue properties will vary even from one location to another within a given part. Another important variable is that of temperature. Especially in the case of aluminum alloys the fatigue strength is strongly temperature dependent, and this must be taken into consideration in the analysis. The fatigue strength of an aluminum piston at full-load operating crown temperatures is on the order of half that at ambient temperature.

Whenever possible it is recommended that fatigue specimens be taken from the particular regions of interest in sample parts, and that fatigue data (an S-N diagram) be generated as part of the analysis process. Figure 3.3 shows several commonly used fatigue specimen test fixtures. Also indicated in the figure are the types of stresses imposed by the test. The fatigue test should be chosen to best replicate the types of stresses that will be seen by the part being analyzed. For example, if a cylinder head is being analyzed, and the engineer is especially concerned about the fatigue life in the valve bridge regions, test specimens should be taken from the valve bridges of the head castings (or in the case of a new design, a similar casting using the same alloy and casting process). Because the fatigue loading of the valve bridge has a highly compressive peak as well as a smaller tensile peak the four-point bending test shown in Fig. 3.3 would be most appropriate for generating an S-N diagram.

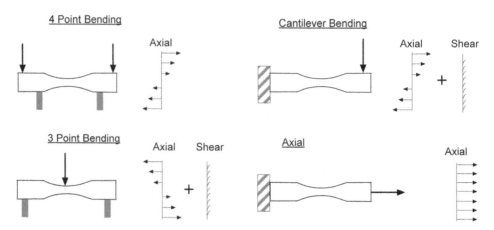

Fig. 3.3 Example specimen tests for fatigue life characterization

The next challenge faced in engine development is the variety of loads to which engine components are simultaneously exposed. The part may experience stresses imparted through the manufacturing processes or during assembly before one even considers the operating stresses. In many engine components there are then two distinct alternating stress cycles that must be considered—changes in loading that occur with changes in operating conditions, and superimposed on these loads, the alternating loads that are experienced with each engine cycle. As an example, consider the cylinder head valve bridge depicted in Fig. 3.4. Because of the complexity of the cylinder head casting significant stresses will be induced during the casting solidification process. The casting shrinks as it cools, and so the last regions to cool impart loads on regions that had cooled more rapidly. These casting *residual stresses* are significantly reduced or eliminated during an annealing process following the initial casting cool-down. Thus they can generally be ignored, but the possibility of remaining residual stress should never be forgotten. The next loads seen by the valve bridge occur during assembly, when the valve seats are pressed into the heads, imparting a tensile assembly stress to the valve bridge, as shown in Fig. 3.4b. The clamping loads resulting from assembling the cylinder head to the cylinder block at the head gasket joint typically have minimal direct impact on valve bridge loads. However, the combination of cylinder head stiffness and clamping load significantly constrain the head during operation. The valve bridges are directly exposed to the combustion process, and as the combustion chamber surface heats up the casting attempts to expand. However, the constraints just identified prevent expansion from occurring, resulting in the highly compressive thermal loads shown in Fig. 3.4c. Finally, the combustion chamber is exposed to cylinder pressure that changes rapidly throughout each engine operating cycle. This *high-cycle* alternating load (depicted in Fig. 3.4d) is superimposed on the *low-cycle* thermal load. For this particular component, and this location within the component, it is generally found that the highest load magnitude is that induced by the thermal gradients. However, the high cycle rate of the cylinder pressure loading makes the greatest con-

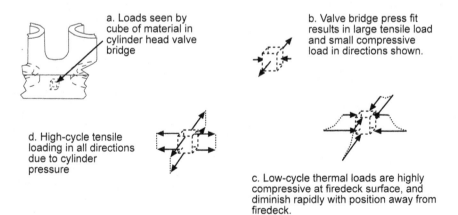

a. Loads seen by cube of material in cylinder head valve bridge

b. Valve bridge press fit results in large tensile load and small compressive load in directions shown.

d. High-cycle tensile loading in all directions due to cylinder pressure

c. Low-cycle thermal loads are highly compressive at firedeck surface, and diminish rapidly with position away from firedeck.

Fig. 3.4 Example showing the complex loading experienced in the valve bridge of a cylinder head

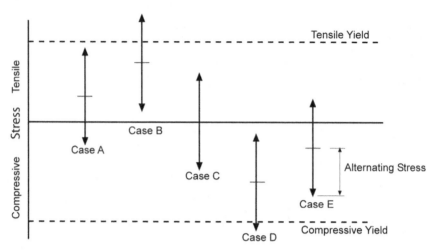

Fig. 3.5 Several cases having the same alternating load superimposed on different mean loads

tribution to fatigue. Along with the thermal loads the assembly stress contributes to the mean loading, and cannot be ignored. While it is prohibitive to fully consider every load impacting the fatigue life of a given component proper analysis requires that the dominant loads, the directions and paths over which they act, and the constraints defined by the assembly be correctly identified. Discussion of analytical and experimental techniques for identifying the relevant loads and constraints will be presented in the context of various component discussions in later chapters.

The cylinder head valve bridge example introduced the importance of understanding the manufacturing and assembly loads before considering the cyclical loads that might lead to fatigue fractures. The reason for this is explained with reference to Fig. 3.5, where the same alternating load is shown applied in cases having different mean loads. The mean load variation results in the alternating load excursions approaching (or exceeding) either the compressive or tensile yield strengths of the material. For most materials the magnitude of the compressive yield strength will be different from that in tension. The chosen material's strength in compression and tension, the mean load, and the alternating load all have important impacts on the fatigue life of the component. For example, the fillet radii of a crankshaft are often cold rolled or shot peened to impart a compressive residual stress to the surface. For a given alternating load these processes reduce the peak tensile stress and increase the compressive peak. The net result is a fatigue life improvement.

The importance of mean loads as well as alternating loads in defining the fatigue life of a material suggest that different S-N diagrams will be required for each possible mean load. This would be quite impractical, and the need for multiple S-N diagrams is eliminated by defining an *equivalent fully-reversed stress (EFR stress)*. The EFR stress is defined as the alternating stress that with zero mean stress would produce a fatigue fracture in the same number of cycles as the actual combination of mean and alternating stress. The EFR stress allows all of the various combinations of mean and alternating stress to be collapsed

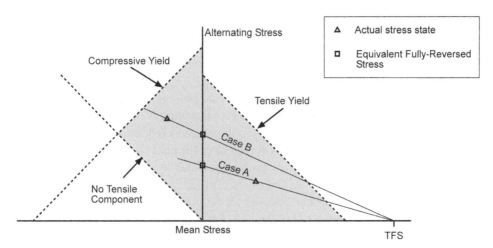

Fig. 3.6 Goodman model for determining equivalent fully-reversed (EFR) stress

onto a single S-N diagram where the plotted alternating stress is that for conditions having zero mean stress.

The remaining problem is that of identifying a model that accurately identifies the EFR stress. Various models have been developed based on theoretical considerations and experimental validation. The Goodman diagram shown in Fig. 3.6 depicts one such model well-developed under the stress conditions typically seen in engine development. In the Goodman diagram mean stress is plotted on the horizontal axis, and alternating stress on the vertical. The mean stress may be positive or negative, depending on whether the stress is tensile or compressive respectively. The alternating stress is always a positive number indicating the absolute value of the difference between the maximum and minimum stress through the load cycle. Limiting lines are indicated on the diagram based on combinations of mean and alternating load that would result in tensile or compressive yielding of the material. These limits along with one for cases where the entire load cycle remains compressive define the shaded region of Fig. 3.6 for which the Goodman model can be applied. Application of the Goodman diagram involves drawing a line from the true fracture strength through the plotted actual combination of mean and alternating stress. The point at which this line intersects the alternating stress coordinate is identified as the EFR stress. The relationship between true fracture strength and ultimate strength should be noted as one often sees ultimate strength used in Goodman diagrams. The ultimate strength is the load per unit area under which a test specimen fractures under tensile loading. However, it is based on the initial cross-sectional area of the test specimen, and most metals "neck down" under tensile loading. The true fracture strength takes into account the cross-sectional area change, and is the correct value to use in the Goodman plot.

For case 'A' shown in the figure the mean stress is tensile, and the EFR stress is greater than the actual alternating stress. For case 'B' the compressive mean stress results in a reduced EFR stress. This result further demonstrates the impact of compressive residual stress on fatigue life.

It must be emphasized that the Goodman diagram is a model allowing one to approximate an equivalent fully-reversed stress. Nevertheless it has been found to provide good agreement with measured results over the realms of conditions for which it has been developed.

The tools are now in place for quantitative fatigue analysis. For many engine components the dominant loading is that induced with each operating cycle. Examples of components for which these high-cycle loads are most important include crankshafts, connecting rods, piston pins and the pin boss and skirt regions of the piston, main and rod bearings, cam lobes and tappets or cam followers, push rods, rocker levers, and valve springs and retainers. These components may accumulate as many as several billion load cycles over their useful lives, and hence must be designed for "infinite life." The loads must be below the endurance limit first introduced in Fig. 3.2. For many of these components an S-N diagram for the specific part can be generated through rig testing. This can be done for components where the loads and constraints are well defined, and it allows the separate step of determining the EFR stress to be eliminated. The life of the part can be directly quantified, and the effects of design changes or manufacturing process variation can be accurately determined. Taking crankshaft web fillets as an example, a crankshaft section can be tested in a rig that reproduces the bending loads seen by the fillet. By varying the alternating load imposed on the specimen an S-N diagram can be generated. The diagram is specific not only to the component but to the particular location of fatigue fracture generated by the test. The plot says nothing about fatigue life at the nose or flywheel mounting flange of the same crankshaft.

For components where the primary loads are "low-cycle" the allowable alternating stresses are higher, perhaps approaching or even exceeding the material's yield strength. Examples of engine components falling into this category include the thermal loads encountered by cylinder heads, piston crowns, and exhaust manifolds, and the torque reaction forces experienced by the crankcase, engine mounts, and the flywheel and housing. The techniques used for fatigue analysis under these conditions are referred to as ***cumulative damage theory***. By this theory each load cycle contributes a fraction of the "damage" that will ultimately result in a fatigue fracture. Various load cycle counting techniques are used to identify complete load cycles from a continuously varying operating cycle. An EFR stress is identified for each individual load cycle, and is then used in conjunction with an S-N diagram to identify its fractional contribution to a fatigue fracture. When the sum of these fractional contributions reaches unity a fatigue fracture is expected.

As an example of cumulative damage theory, consider the case where engine mount failures are occurring prematurely in a particular construction vehicle application. Because failures have already occurred, parts can be observed to determine the exact failure location, and strain gauged parts can be installed in the application. Installed strain readings provide the mean load, and the load changes are then measured throughout the cycle under which the failures have been occurring. The resulting load versus time is shown in Fig. 3.7. The EFR stress is determined for the first load cycle shown in the figure, and

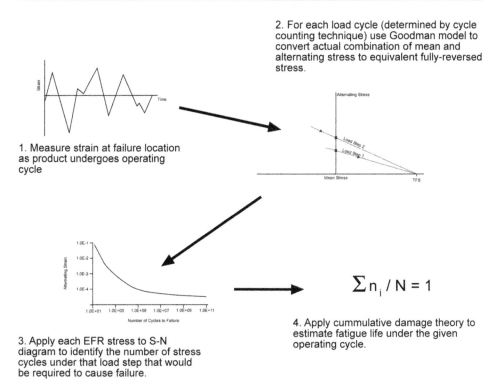

2. For each load cycle (determined by cycle counting technique) use Goodman model to convert actual combination of mean and alternating stress to equivalent fully-reversed stress.

1. Measure strain at failure location as product undergoes operating cycle

3. Apply each EFR stress to S-N diagram to identify the number of stress cycles under that load step that would be required to cause failure.

$$\sum n_i / N = 1$$

4. Apply cummulative damage theory to estimate fatigue life under the given operating cycle.

Fig. 3.7 Application of cumulative damage theory

from the S-N diagram for the engine mount material it is found that under this load cycle a fracture would be expected in 10,000 cycles. By cumulative damage theory it is postulated that operating the engine through this single cycle resulted in the damage equivalent to 1/10,000th of that resulting in fracture. Applying the same sequence to the next load cycle identifies it as contributing 1/5000th of the fracture. This value is summed to that from the first load cycle, and the process is continued throughout the duty cycle for which measurements have been obtained. If it is found that the resulting sum is 1/500th it would be predicted that a fracture would result if the duty cycle were repeated 500 times. If the duty cycle is sufficiently well known it might be possible to test for agreement between the predicted life and that seen in the field. The technique described is certainly inexact, but can nevertheless be used quite effectively, especially in assessing the relative or comparative impact of design changes.

The brief discussion of fatigue analysis described in the preceding paragraphs included examples of rig tests and measurements taken from an installation in the field. Rig testing is extremely effective in reducing test cost and complexity, and often allows significantly higher cycle accumulation frequencies than available through engine testing. Rig testing of specific components (like the crankshaft introduced earlier in this section) will be described throughout the book. Effective rig testing requires accurate simulation of the most relevant loads as well as appropriate fixture design—most of the engineering investment in rig testing is typically focused on developing the loads and constraints that best simulate engine operation.

Another important approach for fatigue analysis is the use of analytical methods, which may be as simple as hand calculations, but more often rely on finite element or boundary element computational techniques. These tools may be quite effective in setting initial design direction before hardware is available. The models are also useful in further optimizing a design once they have been validated through early prototype testing. Most of the time investment in analytical methods is typically in initial model construction and validation. Fatigue analysis requires each load step to be calculated only once, and the fatigue life then predicted using post-processing routines that calculate the EFR stress and predicted cycles to fracture for each node of the model. This approach allows very rapid solution. For detailed parts relatively coarse models of entire components or component groups may be constructed to define gross deflections and constraints. Detailed, fine mesh sub-models of particular regions of interest are then constructed, and run using the deflections obtained from the coarse models. An example of fatigue analysis applied to a cylinder head will be discussed in Chap. 10. A coarse model of the entire cylinder component or sub-assembly is constructed. This model is insufficiently accurate to predict stress profiles, but can accurately identify deflections when exposed to thermal and mechanical boundary conditions. A much finer sub-model is constructed containing the region around the particular location of interest, providing accurate results with acceptable run times. The same technique is applied in Chap. 11 to look at main bearing deflection.

Engine testing should finally be discussed. While it is generally far more expensive and time consuming than either rig testing or analytical methods it holds the distinct advantage of including all relevant loads—some of which may be quite difficult to properly simulate through other methods. The complementary use of engine testing with rig testing or analysis often provides the best solution to accurate fatigue life prediction.

3.3   Friction, Lubrication, and Wear

Engine operation requires rolling and sliding contact between metal components at a variety of locations. Each of these locations involves different materials, surface finishes, forces between the components, and lubrication environments. The topic of *tribology*— the various disciplines impacting friction, lubrication and wear—is thus of great importance to engine durability.

In order to better understand the mechanisms of friction and wear, and the role of the lubricant, it is helpful to begin with a brief, fundamental look at *adhesion* due to molecular bonding at a surface. Figure 3.8 is a simplified molecular representation of a metal surface. At an atomic level the electron shell structure of any given atom determines a number of bonding sites where the atom desires to share electrons with adjacent atoms in order to satisfy bonds and minimize its energy level. The bond sites may be satisfied through metallic,

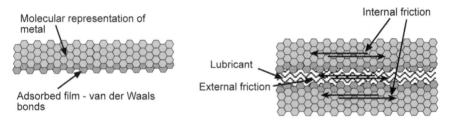

Fig. 3.8 Molecular portrayal of lubricated contact, depicting internal and external friction

ionic, covalent, or van der Waals bonds. The metal has a crystalline structure in which the interior atoms have had each of their bond sites satisfied through metallic bonding. However, at the metal surface it is not possible for all of the bonds to be satisfied by adjacent metallic atoms. The surface atoms with these unsatisfied bonds have a higher energy level, and a desire to reduce their energy level through contact with other atoms. Gases (typically oxygen) adjacent to the surface will initially adsorb into the surface through van der Waals bonding, and with time will form inter-metallic species such as aluminum oxide (Al_2O_3) and various iron oxides (Fe_3O_4, Fe_2O_3, FeO).

Whenever two solid materials come into sliding contact a friction force is required to maintain the sliding motion. The force is proportional to the product of adhesive bonding and the true surface area of contact based on the surface finishes of the two materials. Oxidation significantly reduces the adhesive forces since more of the surface bonds have already been satisfied. In some cases the adhesion forces occurring at the surface between two metals becomes greater than the bond forces within one of the materials (the "softer" material), and interior, metallic bonds are broken causing material transfer from one surface to the other. In the field of tribology this is referred to as ***internal friction***, while the breaking of adhesive bonds at the surface between the materials is termed ***external friction***. Placing a lubricant between the two surfaces separates the surfaces, thus minimizing or eliminating adhesive bonding directly between the surfaces. The adhesion between either surface and the lubricant, and molecular shear within the lubricant itself determine the magnitude of the friction forces at a lubricated joint.

With this very brief look at the fundamental molecular mechanisms attention now turns to its application throughout the engine. The mechanism of internal friction just described results in a progressive failure mechanism at various possible locations within the engine. Locally high friction resulting from insufficient lubrication causes a temperature increase, and may result in micro-welding and subsequent tearing. This is the mechanism referred to as ***scuffing*** in an engine. It may occur between bearings and shafts (cam bearings, main or rod bearings), or between the piston skirts and cylinder walls. In each case, metal has been transferred from the softer material to the harder material—bearing material transferred to the shaft, or piston skirt material transferred to the cylinder wall. The local temperature increase caused by the high friction may result in local melting and streaking of the softer material. This more aggressive material transfer is termed ***wiping***, and is seen in more severe cases on the same components. ***Galling*** on a cam lobe is a similar streaking of cam

Fig. 3.9 The dynamic friction coefficient versus Sommerfeld number, depicting the regimes of sliding contact

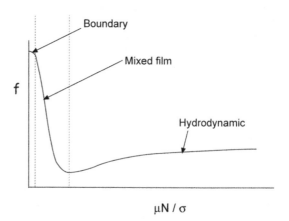

roller or tappet material transfer to the cam lobe. Finally, if the bond strength occurring due to friction between the two surfaces becomes great enough that continued engine operation results in breakage at another location, the failure is termed a *seizure*. A "spun" bearing is an example of seizure. The bearing surface has seized to the shaft, and the back surface of the bearing breaks free and spins in the journal. A seized piston typically results in connecting rod breakage and further progressive damage.

The friction coefficient is defined as the ratio of the force required to overcome friction to the load force acting over the particular sliding contact area. The friction coefficient is plotted versus a dimensionless group of terms known as the *Sommerfeld number* in Fig. 3.9. The Sommerfeld number includes lubricant viscosity, relative speed between the two surfaces, and the load applied per unit of contact area. If one assumes a given joint (with a given load and oil viscosity) the magnitude on the horizontal axis changes solely with the sliding velocity. When the parts are at rest the friction coefficient is high due to direct metal-to-metal contact between the surfaces and resulting high adhesive forces. There is little or no lubricant between the surfaces, and any reduction of the adhesive force is due to the absorbed gases and oxidized surfaces discussed at the beginning of this section. The situation is referred to as *boundary lubrication*. As motion is initiated and lubricant is fed to the joint the surfaces begin to separate, reducing the adhesive forces and the friction coefficient during this *mixed film* lubrication regime. As the name implies, portions of the contact area are separated by lubricant, while portions remain in direct, metal-to-metal contact under mixed film lubrication. Finally, as speed continues to increase a sufficient oil film is created to completely separate the two surfaces. There is no further metal-to-metal contact, and the friction coefficient reaches a minimum. The lubrication regime is now *hydrodynamic*. With further speed increases in the fully hydrodynamic regime the friction coefficient again climbs due to increasing shear forces within the lubricant and between the lubricant and the surfaces.

The goal in any engine design is to ensure hydrodynamic lubrication is achieved wherever possible, and as rapidly as possible after engine start-up. Practical limitations

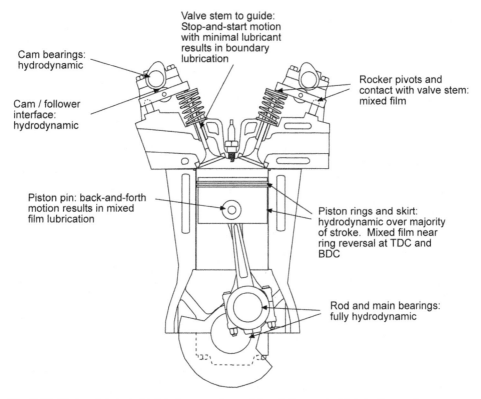

Cam bearings:
hydrodynamic

Valve stem to guide:
Stop-and-start motion
with minimal lubricant
results in boundary
lubrication

Rocker pivots and
contact with valve stem:
mixed film

Cam / follower
interface:
hydrodynamic

Piston pin: back-and-forth
motion results in mixed
film lubrication

Piston rings and skirt:
hydrodynamic over majority
of stroke. Mixed film near
ring reversal at TDC and
BDC

Rod and main bearings:
fully hydrodynamic

Fig. 3.10 Various lubricated joints in an engine and the sliding contact lubrication regimes

preclude hydrodynamic lubrication at some surfaces within the engine. In Fig. 3.10 the various lubricated joints within the engine are shown, and the lubrication regimes experienced by those joints are identified. There are various places within the engine where shafts spin in journal bearings and hydrodynamic lubrication can be maintained under all but start-up conditions. These include the crankshaft main and rod bearings, the camshaft spinning in its bearings in the block or head, and roller pins on valve train systems having roller followers in contact with the cam lobes. Other fully hydrodynamic joints include that between the cam lobe and tappet or roller, and the contact between teeth in a gear-drive systems.

The contact between the rings and the cylinder wall would be expected to be fully hydrodynamic through the majority of each piston stroke. However, as the piston approaches the top (TDC) and bottom (BDC) of each stroke the relative velocity between the rings and wall drops to zero. Based on Fig. 3.9 one would conclude that the rings would experience boundary lubrication, but here a further concept must be introduced. When a hydrodynamic lubrication film was present and the velocity suddenly drops to zero a finite amount of time is required for the lubricant to "squeeze" out of the joint under a given applied load. The time required is based on the applied load, the lubricant viscosity, and the available area and material geometry. The theoretical film thickness decay rate is given by the following equation:

$$\frac{1}{h^2} = \frac{1}{h_o^2} + \frac{1}{\pi} \frac{2W(a^2 + b^2)\,t}{3a^3b^3\eta}$$

In this equation 'h' is the oil film thickness versus time, 't.' The applied load is given by 'W' and the contact area is assumed rectangular, of dimensions 'a' and 'b.' Oil viscosity is given by 'η.' It should be noted that this equation is theoretical, and it is based on a constant film thickness over the entire area. In the case of piston rings the rate of film thickness decay can be controlled with ring surface geometry. This will be discussed further in Chap. 15. The point to be made here is that the piston rings see mixed film lubrication as they approach each end of the piston stroke, but the oil film does not completely disappear in the short time during which the piston is not moving. This remains an important design issue because piston side loading is high at precisely the same time, resulting in the problem of bore polishing due to direct metal-to-metal contact and thus cylinder wear.

There are many joints within the engine where the continual stop-and-start motion resulting from cyclic operation entirely precludes hydrodynamic operation. Such joints include the piston pin joint at the piston (and at the connecting rod in full-floating designs), and the rocker levers at both their pivots and their contact with the valve stems. In each of these cases lubricant is supplied to the joints when they are unloaded (or continuously upstream of the load), but because of the low relative velocities and relatively high loading significant metal-to-metal contact occurs throughout engine operation. These joints must be designed by careful characterization of the wear rates, ensuring that the wear occurs slowly enough for adequate performance over the useful engine life.

Another joint experiencing stop-and-start motion is that between the valve stems and their guides in the cylinder head. This joint has the further difficulties of minimal oil supply, and in the case of the exhaust valves, high temperature. Supplying any more oil to this joint than the minimal amount required contributes directly to oil consumption and exhaust emissions. Therefore, the design objectives are to minimize the amount of oil supplied to the joint, and then try to retain any oil that enters the joint. Once again, wear rates must be characterized to ensure that clearances do not become excessive over the life of the engine.

The role of component surface finish has already been mentioned, and is of critical importance to lubricated joint design. As surface finish is improved a thinner oil film is required for hydrodynamic lubrication. This holds the advantage of allowing a given joint to achieve hydrodynamic location more quickly and thus with less wear. It also allows greater loads to be applied over a given bearing area—reducing required bearing area or allowing higher engine loads to be successfully applied. Joint alignment is another important variable affecting the minimum required oil film thickness. Any misalignment results in a portion of the contact area carrying more of the load, and thus seeing greater loading

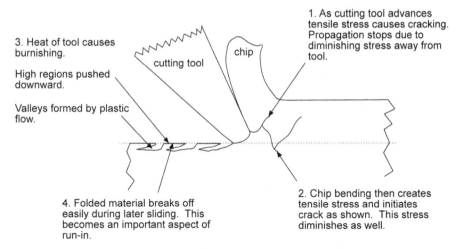

3. Heat of tool causes burnishing.

High regions pushed downward.

Valleys formed by plastic flow.

cutting tool

chip

1. As cutting tool advances tensile stress causes cracking. Propagation stops due to diminishing stress away from tool.

4. Folded material breaks off easily during later sliding. This becomes an important aspect of run-in.

2. Chip bending then creates tensile stress and initiates crack as shown. This stress diminishes as well.

Fig. 3.11 The metal cutting process and the resulting surface characteristics

per unit area than anticipated. This may result in metal-to-metal contact in the overloaded region of a joint designed for hydrodynamic lubrication.

The process of metal cutting is summarized with reference to Fig. 3.11. The tool depicted may be a lathe or mill cutter, or on a smaller scale, a grinding stone. As the tool passes over the surface a tensile force is applied to the material ahead of the tool. The tensile force subsides with distance from the tool, resulting in a non-propagating crack. Bending of the chip creates another tensile force acting approximately 90° from the first. It too is non-propagating. This sequence of tensile cracking is the process by which material removal occurs. In ductile materials the cracks tend to wander along grain boundaries, and the resulting surface finish is dependent on grain size. Behind the cutting tool remaining fragments are bent towards the surface. The plastic deformation creates valleys, and the now folded fragments will later break off easily during engine run-in. Manufacturing process selection and the capability of a process to provide a desired surface finish are important aspects of engine design for minimizing wear. The surface finishes typically resulting from various manufacturing processes are summarized as follows:

Milling or turning	1–10 micron
Boring	0.5–5 micron
Drilling	2–5 micron
Broaching	0.5–2.5 micron
Grinding	0.5–5 micron
Honing	0.05–1 micron
Lapping and superfinishing	0.025–0.1 micron

As-cast or forged surfaces:

Sand casting	5–25 micron
Forging	2–5 micron
Die casting	1–2 micron

3.4   Further Wear and Failure Mechanisms

While the preceding sections describe the primary mechanisms determining an engine's durability there are several other mechanisms that affect specific components. These further mechanisms are identified and briefly discussed in the following paragraphs.

Fretting results from vibrating motion between two or more solid surfaces. The vibration results in abrasive wear and build-up of metal and oxide particles. In the engine fretting wear occurs when this vibrating motion occurs between two components due to insufficient clamping load. Two critical locations where fretting wear is problematic in engines are depicted in Fig. 3.12. The split bearing shells used at the crankshaft main and rod bearings are clamped in place using bearing caps as shown. The clamping load provided by the bolts must be sufficient to ensure that the cap does not separate due to loading of the bearing shell in the cap. If the clamping load is insufficient the cap is pulled away from its mating surface, causing abrasive sliding between the surfaces. As the sliding continues the build-up of wear particles aggravates the problem, and the wear rate increases. Fretting at this joint results in progressive loss of clearance between the bearing and shaft, and ultimately may lead to scuffed or seized bearings. The second region of fretting wear depicted in the figure is at the mating surface between the back of the bearing shell and the journal in which it is installed. The same clamping load holding the bearing cap in place is intended to keep the bearing shell from moving as well. The bearing shell is designed slightly oversized relative to the journal diameter. As the cap clamps the bearing shell

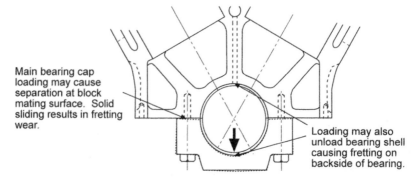

Main bearing cap loading may cause separation at block mating surface. Solid sliding results in fretting wear.

Loading may also unload bearing shell causing fretting on backside of bearing.

Fig. 3.12 Fretting wear locations in a split bearing assembly

in place the oversized shell results in bearing crush, applying an outward force from the bearing shell to the journal around its entire perimeter. If the bearing crush is insufficient the bearing shell may move as the cap is loaded. The movement again results in fretting, in this case wearing material away from the backside of the bearing shell. The result is a loss of contact area through which bearing loads can be transferred. The load is carried over a smaller area, resulting in bearing overload and increased fatigue of the bearing. The problems described here will be further discussed in Chap. 11 where bearing design is discussed in greater detail.

Impact loads are especially important throughout the engine's valve train. The first, and in most cases most important, location to be taken up is that between the valves and seats. Among the design goals for optimal engine performance is to open and close the intake and exhaust valves as rapidly as possible in order to minimize flow restriction. Among the design goals are to maintain high valve lift as long as possible, and then close the valves as rapidly as possible. The valves are not allowed to slam shut, but the closing velocities are defined by the closing ramps on the cam lobes. The rate at which the valves can be closed is determined to a great extent by durability of the contact region between the valves and seats. As a valve contacts its seat there is abrasive sliding without lubricant. Especially in the case of the exhaust valves this contact wear occurs at very high temperature. The rate of resulting valve guttering or beat-in must be acceptably low for the desired engine life. High temperature, high hardness alloys are chosen for the valve and seat materials, and steps are taken to control the valve temperatures to the extent possible. While the valve and seat interface surely sees the most severe impact loads the remaining components of the valve train cannot be ignored. In engines using "solid" valve trains, and thus requiring valve lash there are impact loads throughout the system when the valves are rapidly opened. Such impact loads are minimal but may still exist in systems using hydraulic lifters to continuously take up the valve train lash. These topics will be further discussed in Chap. 17 where valve train design is addressed. From the standpoint of durability validation the results of impact loading are addressed through careful characterization of wear rates at each interface throughout the valve train to ensure that the wear rates remain low enough to meet life expectations.

Cavitation results from the implosion of vapor bubbles in a fluid due to a sudden increase of pressure. The mechanism is important in both the cooling and lubrication systems of engines. In diesel engines it is also important in the nozzles of the high pressure fuel injectors.

Taking as an example the rod bearings in an engine, the cyclic operation may result in a sudden pressure drop in the lubricant film. A portion of the lubricant vaporizes due to the pressure drop, and bubbles are formed. If this is followed by a rapid pressure increase the droplets implode, and high-energy micro-jets travel through the fluid. These jets contain sufficient energy to erode surfaces with which they come into contact. In the case of the bearing example cavitation results in tiny holes drilled into the bearing surface. With time the number and size of these drilled holes results in overloading of the remaining bearing material, and progressive failure through fatigue cracking and further material removal.

Cavitation is also common on the impellers of oil pumps and water pumps as these components also see rapid pressure drops and increases. Finally, the rapid movement of a cylinder wall due to the impact of piston side-loading may result in cavitation damage of the cylinder walls. This is especially prevalent in engines having removable cylinder liners in direct contact with the coolant (wet liners), and is sometimes severe enough to erode all the way through the cylinder wall. It should be noted that combustion leakage into the engine coolant is an additional source of bubbles, requiring only a rapid pressure increase to result in cavitation. The phenomenon of cavitation is extremely difficult to predict, and must be recognized and addressed through engine testing. Specific cavitation problems will be further discussed in the chapters covering cooling jacket design and engine bearings.

3.5   Recommendations for Further Reading

There are several current textbooks covering the general science of fatigue. A recommended text for readers desiring a greater fundamental understanding of fatigue is (see Schijve 2010).

For general reference on engine component testing the following text provides a good overview (see Wright 1993).

The *Handbook of Tribology and Lubrication, Volume II, Theory and Design,* edited by Robert W. Bruce, and published by CRC in 2012 contains several articles addressing tribology, wear mechanisms, and application to engines.

A classic work on internal combustion engine durability, originally published in German, has recently been translated to English (see Greuter and Zima 2012).

References

Greuter, E., Zima, S.: Engine Failure Analysis—Internal Combustion Engine Failures and their Causes. SAE International, Warrendale (2012)
Schijve, J.: Fatigue of Structures and Materials, 2nd edn. Springer, Vienna (2010)
Wright, D.H.: Testing Automotive Materials and Components. Society of Automotive Engineers, Warrendale (1993)

The Engine Development Process

4

The intent of this chapter is to provide an overview of the processes involved in developing a new engine from the initial need identification to production release. A flowchart of the processes has been developed by AVL List GmbH, and is presented in the Figure. For ease of reference, the paragraph headings in this chapter correspond to encircled numbers in the figure.

1. Concept Selection and Customer Needs It is essential to gain a solid understanding of the needs of the customer and market for which the new engine is being designed prior to beginning the hardware design process.

- What is the application for which the engine is being designed? Is the same engine being planned for more than one application? (either different vehicles such as different model automobiles, or different applications such as automobiles and boats).
- What are the performance expectations of the customer? The customers often cannot articulate their expectations in engineering terms, but they will certainly know whether or not they have been met! Performance expectations will include rated power; the speed range between rated power and peak torque; the torque rise or torque curve shape over this speed range; the match between the torque curve and the vehicle drivetrain (gear ratios, torque converter slip, etc.); and transient response.
- What are the further customer needs, expectations, and priorities? Examples of items that may be directly or indirectly valued by the customer include durability (engine life to overhaul), fuel consumption, engine weight, service intervals and service requirements. Which of these items are most important to the customer? What are their expectations regarding each item? What compromises are they willing to make if the compromise allows another item to be better addressed or makes the product less expensive?

© Springer Vienna 2016
K. Hoag, B. Dondlinger, *Vehicular Engine Design*, Powertrain,
DOI 10.1007/978-3-7091-1859-7_4

- What cost targets must be met in order to make the product attractive? How will this product compare with competitors' products in terms of the items listed in the previous paragraphs, and the cost of the engine or vehicle?
- What are the market trends that might affect future power requirements, future design feature needs, and future pricing expectations?

2. Initial Thermodynamic Calculation Depending on answers obtained to the questions in the preceding section, as well as fuel prices and availability and applicable emission regulations the combustion system must be selected. Should the engine utilize compression-ignition or spark-ignition? The compression-ignition engine is attractive in markets for which fuel economy, torque rise, and durability are especially important. Spark-ignition engines are more attractive when high power and light weight are required, or when engine cost must be minimized. Stringent emission standards are also more easily met with spark-ignition engines.

Based on the performance expectations identified in the customer and market studies an initial performance analysis is conducted to determine the required displacement to produce the needed combination of speed and power. Displacement is determined by the air flow requirements for complete combustion, and requires an initial assessment of the fuel efficiency and volumetric efficiency of the engine. Comparisons between naturally aspirated configurations and those utilizing smaller displacement in conjunction with a supercharger or turbocharger will need to be done at this time. Estimating the total displacement based on initial performance estimates and combustion system decisions is covered in Chap. 5.

The engineers conducting the initial performance analysis must work closely with the engineers doing the initial engine layout analysis. In addition to the total displacement, the number of cylinders over which the displacement is distributed, and the bore and stroke of each cylinder must be determined. These critical early design decisions impact both the mechanical design and the performance characteristics of the engine. They are covered further in Chap. 6.

3. Initial Mechanical Calculation The previous paragraph introduces the partnership between layout design and performance analysis in determining the number of cylinders and the bore and stroke of each cylinder. These decisions along with the configuration of the cylinders determine the layout of the engine. Much of the effort in determining engine configuration focuses on the application. What will be the dimensions of the engine compartment? Are all of these dimensions fixed, as in the case where a new engine is being developed for an existing vehicle, or can some of the dimensions be modified as needed to optimize the design? If the engine compartment dimensions are still being determined which dimensions will it be most attractive to minimize? For example, if vehicle aerodynamics is to receive high priority it might be especially important to minimize engine height. In motorcycle applications width might be especially critical. These questions along with engine balance, cost, complexity, engine speed requirements, and serviceability must be considered in determining the number and configuration of the cylinders.

4. Fix Displacement and Configuration An important early milestone is that of fixing the displacement and configuration. This is done based on the initial performance and design layout analysis, and provides a basis for furthering the design layout and beginning tooling procurement. It must be emphasized that these design decisions must be solidified for much of the further work. Modifying these decisions will become prohibitively expensive soon after this milestone is reached. The only exception is the possibility of allowing margin for a displacement change in the next stages of the layout design process. Allowing margin for later increasing the bore or stroke is discussed further in Chap. 8.

5. Concept Design Several critical dimensions such as the cylinder-to-cylinder spacing and the deck height are determined at this early stage of the engine design process. These dimensions are explained in detail in Chap. 8, and they in turn impact many further aspects of the engine design and durability.

Critical dimensions throughout the engine are now determined with an eye to optimizing the design. These dimensions have important impacts on operating loads and the ability of the engine to adequately distribute loads and maximize durability. For example, the cylinder-to-cylinder spacing impacts crankshaft length and stiffness, and main and rod bearing area. It also impacts head bolt spacing and gasket sealing. It impacts the available block structure and the load distribution between the main bearing bolts and cylinder head bolts. Finally it impact cooling jacket design and available flow area between the cylinders and coolant transfer between the block and head at each cylinder. Cylinder-to-cylinder spacing is only one example. Similar lists can be made of the systems impacted by deck height, camshaft placement, and many other design decisions. As a result the optimization process requires the designer to work closely with those conducting the initial structural and noise and vibration analysis.

Material selection for the cylinder block and head must be done at this time. This in turn affects further design decisions. For example, selecting aluminum for the crankcase may drive the decision to use a die cast aluminum oil pan for the further structural rigidity it provides.

Another aspect of layout design is that of determining the resulting outer package dimensions. This was introduced during the initial layout analysis, but must now be more closely determined because of its possible impacts on vehicle or engine compartment design.

6. Thermodynamic Layout Fundamental to controlling costs and minimizing the time required for engine development is the use of computational tools prior to constructing hardware. It is desirable to utilize the available analytical tools to optimize both the combustion chamber and the air handling system as much as possible. Zero-dimensional engine cycle simulation is used to begin developing the intake and exhaust system. Valve lobe design and valve opening and closing events can be defined quite closely based on the current capability of engine system models, used in conjunction with prior experience. Runner length and cylinder-to-cylinder interaction can also be accurately determined. If the engine is to be supercharged or turbocharged the performance analysis can include the simulation of these devices. Multi-dimensional fluid flow modeling can be used to develop the compressor (and turbine for the turbocharged engine) during this stage as well.

Increasingly valuable tools are those used for multi-dimensional modeling of the combustion chamber and the combustion process itself. The direction may be set for combustion chamber geometry. The location of the sparkplug, or the location and spray geometry of the diesel fuel injector can be studied. Piston and cylinder head geometry can also be assessed. Finally, the bulk air motion generated during the intake process, and the resulting turbulence can be studied. At the time of this writing the results of these analyses remain mixed but are rapidly improving. The tools can be best used when they have been previously carefully validated for similar engines, and when the engineers conducting the study can supplement the study with strong practical experience with both the tools and the engine concepts being simulated. In other words it remains easy to be misled since many parameters are available and the models use a mixture of empirical and physically based models.

It should be noted from Fig. 4.1 that the efforts in performance analysis are shown as continuing from this point throughout the engine development process. By continuing to utilize the performance analysis tools, and validating them with experimental data as it becomes available, the tools can be used to rapidly assess design modifications and aid in setting direction for needed design changes. Further modification for specific application requirements or changes once the engine is in production can also be supported with these tools.

7. Mechanical Analysis As with performance analysis the various structural analysis tools are valuable in setting design direction and identifying potential development challenges before constructing hardware. Finite element and boundary element techniques are used to begin assessing fatigue limits and component durability under mechanical and thermal loading. Clamping loads and critical joints such as main and rod bearing caps and the head gasket seal can also be assessed with these tools. Dynamic simulations are used to begin developing the camshaft and valve train and to assess torsional vibration in the crankshaft system and in some cases camshaft torsionals. Computational fluids analysis can be used to begin optimizing coolant flow paths and the details of the cooling jackets. Fluid network calculations can be used to begin developing the lubrication system.

Although detailed structural analysis adds time and resource commitments early in the engine development process the objective almost invariably realized is to reduce the overall development time and expense. The cost of making design changes goes up with time. It is certainly less expensive to make design modifications before any engines have been constructed. Once tooling construction has begun the cost of design changes escalates rapidly.

Once again, as with the performance analysis tools, it is important to maintain the structural models throughout the engine development process. As design changes are made during successive hardware iterations the modifications should be incorporated in the models. Measurements taken during rig and engine testing should be used to verify and update the boundary conditions and improve the modeling accuracy. This effort pays off in having tools that can then be used to rapidly evaluate further design modifications throughout the engine development process and even after the engine has been released to production.

It should be noted that structural analysis as well as detail design and tooling construction actually include several parallel paths as the efforts are applied to each engine component and sub-system.

8. Acoustic Analysis Of increasing importance in many vehicular engine applications are the related considerations of noise, vibration, and harshness (NVH). Many factors in both the design of the engine and its installation impact the resulting noise and vibration signatures. The engine configuration determines which of the mechanical forces within the engine can be balanced and the magnitude and frequency of remaining unbalanced forces that are transmitted to the vehicle through the engine mounts.

The cylinder block and head structure determine their roles in noise transmission to other engine components as well as direct noise radiation from their own surfaces. Modal analysis is done to identify the critical frequencies at which the noise transmission and radiation will be amplified. Design changes to reduce the vibration amplitude, and to shift the resonant frequencies away from desired engine operating speeds can be identified through early analysis. The models required for NVH analysis are to a great extent the same models required for structural analysis thus reducing the resource requirements.

9. Construct Tooling (Not Shown in Figure) As the critical dimensions of the engine are established through the layout design process construction of the tooling required for production casting, forging, and machining can begin. Assembly line design and construction must also be done. Quality control procedures and production test requirements must be determined, and the necessary equipment procured and installed.

Earlier market studies will have defined the anticipated production volume and design life that in turn set the direction for tooling design. In all but the very lowest production engines the major components will be constructed from fixed tooling; this significantly increases the production rate and significantly reduces the machining costs per part. However it also increases the tooling investment, requires many of the dimensions and design features to be finalized early in the tooling construction process, and makes later design changes extremely expensive.

In Fig. 4.1 tooling construction is identified as the critical path. In most cases the engine development timeline is governed by the time required to procure, install, and qualify the machine tools for major components such as the cylinder block and head(s), the crankshaft, and the connecting rods. Input from the detail design process feeds into the tool development, but the amount of design modification must diminish with time as indicated in Fig. 4.1 or project delays and cost overrun will be significant.

10. Detail Design and Costing The detail design process works from the design layout and finalizes the dimensions and tolerances required for design integrity. Manufacturing, material, and process specifications are defined, as well as procedures for quality control.

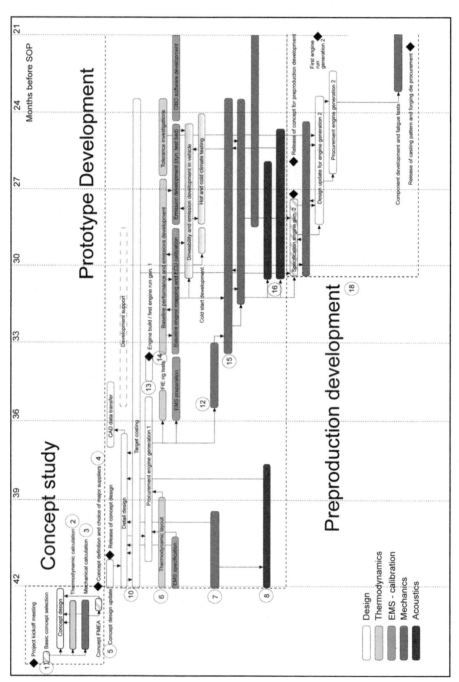

Fig. 4.1 Flowchart of engine development process

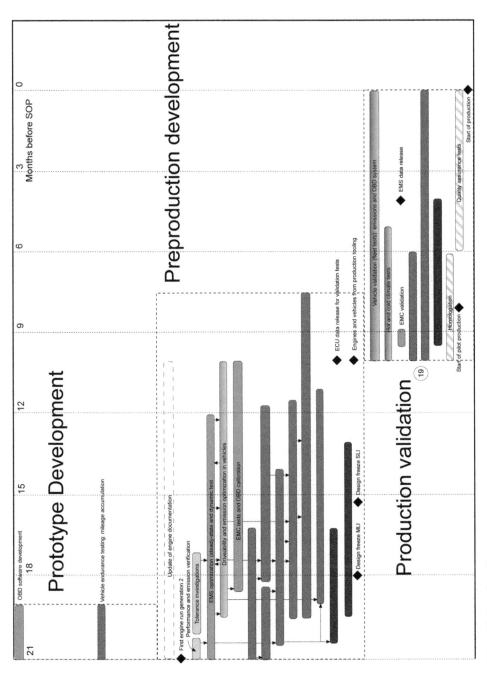

Fig. 4.1 (continued)

Another important aspect of detail design is the documentation, tracking, and production release of each part. Each company will have their own part numbering and tracking system. The new engine may utilize some existing parts but the vast majority will undoubtedly be new. Each part required for the engine will need to be tracked in a variety of ways. These include cost tracking, inventory and supply to the assembly line, and aftermarket or service parts supply.

11.   Breadboard Testing (Not Shown in Figure) While there is considerable effort throughout the industry to continue to improve the performance analysis capabilities it remains important to supplement the analytical studies with experimental validation and further optimization. In order to begin the experimental development efforts before production-like prototypes are available breadboard testing is often done. Such testing involves modifying an existing engine to include prototype hardware that begins to represent the planned engine. While the crankcase from an existing engine is used the piston, cylinder head, and air handling systems are replaced with prototypes that represent the planned engine as closely as possible. In this way the air handling and combustion system development can proceed to the hardware development phase. As the analytical tools continue to be further developed the need for breadboard testing is reduced, thus reducing development costs.

12.   Rig Testing The next stage in the hardware development and durability validation processes involves rig testing of individual components and sub-systems as parts become available. While such testing is more expensive than analysis it is far less expensive than engine testing. At the heart of rig testing is the development of tests that accurately simulate the most important loads faced by each component. Once the correct fixturing and boundary conditions have been identified the tests often allow load cycles to be accumulated at a much higher frequency than can be done in engines. Many of the components can be very well optimized prior to testing them in complete engines.

13.   Prototype Builds Upon completion of the layout design, and based on initial analysis efforts one or more groups of prototype engines will be built. Castings will be made from temporary molds created from the computer-generated component layouts. Components such as the crankshaft, connecting rods and camshafts will be machined from billet steel until the casting molds or forging dies are available. Most of these engines will be used for performance and mechanical development. Some may be supplied for vehicle application engineering. In many cases several prototype builds will be done as the development process continues, with each new build reflecting further design optimization and each build including more parts manufactured with production tooling. In some cases the later prototype builds will include field test engines given to selected customers for evaluation.

The prototype engine builds—especially the later builds—also provide the opportunity to study the manufacturing and assembly processes. Design features that are difficult to assemble can be identified and the appropriate modifications made to avoid later engine production bottlenecks.

14.   Performance and Emission Development Testing The process that began with performance analysis and breadboard testing is now applied to the prototype engines. The degree of effort expended at this point is dependent on the accuracy and detail of the earlier performance analysis, and the ability of the breadboard engines to characterize the intended engine design. A common objective in engine development is to continue to improve the upstream analysis capability and minimize the efforts at this stage. Final combustion and performance recipes are optimized at this point. Emission certification is a necessary part of this effort.

Testing and optimization will be conducted both in dynamometer test cells and in vehicles on chasis dynamometers or on the road. Especially in automobile engines the performance development and optimization includes optimizing the combination of engine and transmission with the vehicle control system. This includes such parameters as gear ratios, automatic transmission shift points, and lock-up torque converter controls.

15.   Function and Durability Testing As with the performance testing the engine validation testing follows on the work previously done through structural analysis and rig testing. Depending upon the accuracy and detail of the earlier efforts reduced resource expenditures may be possible in the validation testing phase. Most engine manufacturers have specified sequences of qualification testing that includes various steady-state and cyclical tests designed based on their history to accurately assess the durability of particular components. While recent trends are to do more validation with combinations of analysis and rig testing some durability considerations continue to require engine testing. A portion of this testing may also be done in vehicles.

16.   NVH Testing Another important aspect of the prototype engine testing and development is that pertaining to NVH. Here too the testing follows on earlier modeling and analysis, and as the modeling capability grows the resources required for testing are reduced.

The NVH testing will include the engine by itself, typically tested in a semi-anechoic chamber where sound power measurements are made at various locations around the engine. Modal analysis testing of the engine structure and its impact on other engine components will be conducted in dynamometer test cells. Much of the testing will also be done in vehicles, where drive-by and free field noise measurements are made.

17.   Tooling Complete A critical milestone is the completion of the tooling construction and qualification. At this point the manufacturing lines are ready for production release. Remaining design changes must be minimal, or large cost overruns result.

18.   Pre-Production Builds It is typical to build a relatively small number of further engines from the production tooling before full production release. This ensures no remaining problems in the production engine assembly processes, and provides additional engines for field testing, customer evaluation, and remaining certifications for exhaust emissions and noise. If problems do arise in the field, this small initial production run limits the costs and exposure associated with addressing the problems.

19. Production Validation and Release The final milestone is the release to production. In many cases the build rates are ramped up slowly as initial warranty information is carefully tracked and necessary customer support is provided.

From this point the development of the engine will continue, with emphasis on further product improvement (based on warranty information) and cost reduction. Further engineering efforts may also address application of the engine into additional vehicles, or rating and performance modifications for further markets. For example an automobile engine might be uprated and aesthetic modifications might be made for a sports or high performance version of the car.

Determining Displacement

5

5.1   The Engine as an Air Pump

Once a market need has been identified the first step in designing a new engine is to determine its required displacement. In this context the engine should be viewed as a positive displacement air pump. In order to produce the required work one must burn sufficient fuel. In order to completely react the fuel to products one must supply sufficient air. Determining the displacement required for a given engine is thus a matter of working backwards from the desired work output (or power at a given shaft speed), as further depicted in Fig. 5.1. Each of the steps in this process will now be covered in greater detail.

The first step in setting forth the expectations of a new engine is to determine the desired operating map. The characteristic operating maps for diesel and spark-ignition engines were discussed in Chap. 2, and various vehicular applications were reviewed. The needs of a particular application will define the maximum power required, and the rpm at which it should occur; the rate of torque rise as the rpm decreases from that of rated power; and the speed range between that of peak torque and that of peak power.

Once the full-load torque curve requirements have been determined the expected specific fuel consumption must be estimated. At the selected speed, the product of the specific fuel consumption, the desired power output at that speed, and the mass ratio of air to fuel provides the required air flow rate as shown in Eq. 5.1a and b.

$$\dot{m}_{fuel}\left[gm/hr\right] = \left(BSFC\left[gm/kW - hr\right]\right)\left(Power\left[kW\right]\right) \tag{5.1a}$$

$$\dot{m}_{air}\left[gm/hr\right] = \left(\dot{m}_{fuel}\left[gm/hr\right]\right)\left(Air/Fuel\right) \tag{5.1b}$$

© Springer Vienna 2016
K. Hoag, B. Dondlinger, *Vehicular Engine Design*, Powertrain,
DOI 10.1007/978-3-7091-1859-7_5

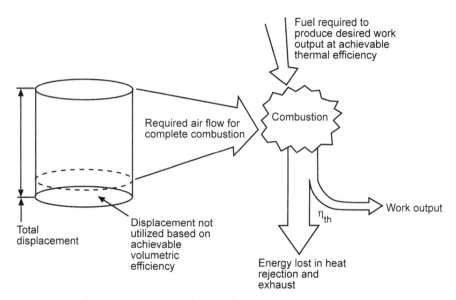

Fig. 5.1 Determining displacement based on combustion and output requirements

It will be important to ensure that both the rated and the peak torque conditions can be met, so the goal will be to determine air flow needs over the desired engine speed range.

As indicated in Fig. 5.2 the work output of the spark-ignition engine is limited by its displacement over its entire speed range. In contrast, the diesel engine is displacement-limited only at maximum torque, and may have an excess of air at higher engine speeds. Approximate air-to-fuel ratios for each engine are also indicated in Fig. 5.2. For the spark-ignition engine the chosen air-to-fuel ratio will typically be that resulting in maximum power or work output. This maximum occurs at an air-to-fuel ratio richer than that required for complete combustion—on the order of 11.5 to 12.5 to one for most liquid fuels. It should be briefly noted that for exhaust emission control (three-way catalyst operation) the air-to-fuel ratio in most engines is very closely controlled to its chemically correct mixture of approximately 15 to one under almost all operating conditions. Because the excursions to full load tend to be very brief the air-to-fuel ratio is almost invariably allowed to decrease in order to achieve maximum power, and in some cases to reduce full-load combustion temperatures.

The practical minimum air-to-fuel ratio attainable in diesel engines is leaner than that theoretically required for complete combustion. A minimum air-to-fuel ratio on the order of 20 to one is typical at maximum torque. The air-to-fuel ratio increase required at higher speeds is dependent on the desired torque curve and the turbocharger boost controls versus engine speed.

The next step in determining the displacement is to estimate the expected specific fuel consumption. If a new engine is being developed the engineer is not in a position to obtain measurements, but several resources are available to support this estimate. The first is to collect data from similar engines. Measurements could be made, or values could be

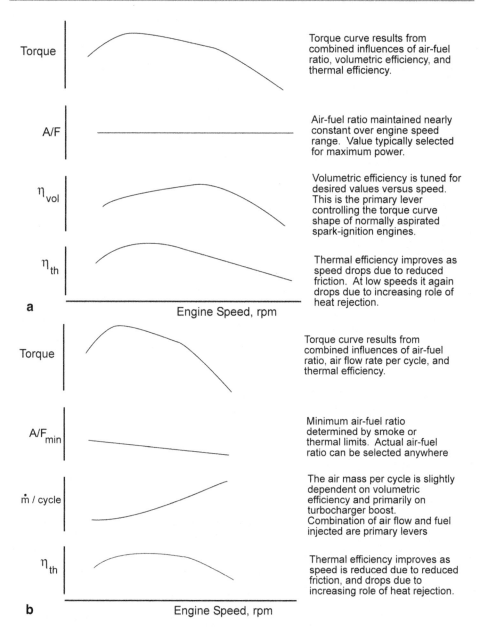

Torque curve results from combined influences of air-fuel ratio, volumetric efficiency, and thermal efficiency.

Air-fuel ratio maintained nearly constant over engine speed range. Value typically selected for maximum power.

Volumetric efficiency is tuned for desired values versus speed. This is the primary lever controlling the torque curve shape of normally aspirated spark-ignition engines.

Thermal efficiency improves as speed drops due to reduced friction. At low speeds it again drops due to increasing role of heat rejection.

Torque curve results from combined influences of air-fuel ratio, air flow rate per cycle, and thermal efficiency.

Minimum air-fuel ratio determined by smoke or thermal limits. Actual air-fuel ratio can be selected anywhere

The air mass per cycle is slightly dependent on volumetric efficiency and primarily on turbocharger boost. Combination of air flow and fuel injected are primary levers

Thermal efficiency improves as speed is reduced due to reduced friction, and drops due to increasing role of heat rejection.

Fig. 5.2 a Variables defining the torque curve in the naturally-aspirated spark-ignition engine. **b** Variables defining the torque curve in the turbocharged diesel engine

obtained from the literature—technical publications will often be more reliable than marketing brochures for reported specific fuel consumption data. Another very good approach is to use an engine cycle simulation to model the planned engine. Several very well developed commercial codes are available that will provide close estimates. Beginning with validated models of similar production engines will add confidence to this calculation.

5.2   Estimating Displacement

Once the required air flow rate is known the displacement can be determined directly by looking at the engine as an air pump of fixed displacement operating at a particular speed. At any given rpm the displacement volume is filled with air every revolution in a two-stroke engine, or every second revolution in a four-stroke engine. The volumetric efficiency was defined in Chap. 1, and provides a measure of how well the displacement is utilized. Since the volumetric efficiency is the ratio of the actual mass flow of air inducted into the engine to the ideal rate based on completely filling the displacement volume, the actual displacement volume required in order to supply the needed air will be the ideal volume divided by the volumetric efficiency. A displacement rate can now be calculated by multiplying an engine's displacement by one half the engine speed (for a four-stroke engine), and then multiplying by volumetric efficiency:

$$Displ.\ Rate[liter/min] = (Displacement[liter]) * \left(\frac{N[rev/min]}{2\ rev/cycle} \right) * \eta_{vol} \qquad (5.2)$$

By rearranging this equation and introducing the density to determine the mass per unit volume of air supplied to the engine one can calculate the displacement required to supply the necessary air flow rate at the given engine speed:

$$D[liter] = \left(\frac{\dot{m}_{air}\ kg}{min} \right) * \left(\frac{m^3}{\rho\ kg} \right) * \left(\frac{1000\ liter}{m^3} \right) * \left(\frac{min}{N\ rev} \right) * \left(\frac{2\ rev}{cycle} \right) * \left(\frac{1}{\eta_{vol}} \right) \qquad (5.3)$$

Several examples now demonstrate how the steps just covered can be used to estimate the displacement that will be required of a new engine. Attention turns first to a typical spark-ignition passenger car engine. The same approach will then be applied to assess the competitiveness of a diesel for the same application. Finally, the different assumptions required to size a heavy-duty diesel truck engine will be shown.

Spark-Ignition Automobile Engine   The goal in this example is to design a new spark-ignition engine for an automobile application. The engine is to have a rated power of 100 kW at 5500 rpm. A literature study has suggested that the specific fuel consumption at rated power will be approximately 300 gm/kW-hr. Finally, a volumetric efficiency of 86% at rated speed is assumed. The estimated displacement is calculated for these assumptions as:

$$\dot{m}_{fuel} = (100\ kW)*\left(\frac{300\,gm}{kW\text{-}hr}\right)*\left(\frac{kg}{1000\,gm}\right)*\left(\frac{hr}{60\ min}\right) = 0.5\ kg/min$$

$$\dot{m}_{air} = \dot{m}_{fuel}*A/F = (0.5\ kg/min)*(12.5) = 6.25\ kg/min$$

$$D[liter] = \left(\frac{6.25\ kg}{min}\right)*\left(\frac{m^3}{1.18\ kg}\right)*\left(\frac{1000\ liter}{m^3}\right)*\left(\frac{min}{5500\ rev}\right)*\left(\frac{2\ rev}{cycle}\right)*\left(\frac{1}{0.86}\right)$$

Displacement = 2.2 *liters*

The predicted displacement of 2.2 L is quite reasonable for such an engine. Let us now as-sume that an engine is needed for a small sports car. The same rated power is required, but the engine will be designed to deliver this power at 7200 rpm. While the higher rpm will lead to an increase in specific fuel consumption this has been partially offset by designing the engine to run on a higher octane fuel, and increasing the compression ratio accord-ingly. The estimated specific fuel consumption is 320 gm/kW-hr. A tuned intake system and high performance camshaft are being designed, and the estimated resulting volumetric efficiency is 90%. The predicted displacement is now:

$$\dot{m}_{air} = (100\ kW)*\left(\frac{320\,gm}{kW\text{-}hr}\right)*\left(\frac{kg}{1000\,gm}\right)*\left(\frac{hr}{60\ min}\right)*(12.5) = 6.67\ kg/min$$

$$D[liter] = \left(\frac{6.67\ kg}{min}\right)*\left(\frac{m^3}{1.18\ kg}\right)*\left(\frac{1000\ liter}{m^3}\right)*\left(\frac{min}{7200\ rev}\right)*\left(\frac{2\ rev}{cycle}\right)*\left(\frac{1}{0.90}\right)$$

Displacement = 1.7 *liters*

It should be noted that especially in high performance applications the same calculations are often performed in the opposite direction. An engine of a certain displacement is given, and the objective is to maximize the power output. What rpm can be safely achieved? What can be done to the intake and exhaust systems to maximize the volumetric efficiency at that rpm? What other modifications can be made to minimize the specific fuel consump-tion (maximize the power output for a given quantity of fuel injected)?

Diesel Passenger Car Engine A diesel engine option is now to be offered in the same car considered in the first part of the preceding example. The same power is to be produced, but the final drive ratio must be selected differently, and the engine developed for a rated speed of 4200 rpm. The diesel engine will provide an inherent fuel economy advantage, so a specific fuel consumption of 260 gm/kW-hr is estimated. The lower engine speed of the diesel can be offset by its higher torque rise capabilities, and a 23 to one air-to-fuel ratio is selected at rated speed to allow margin for the desired torque rise at lower speeds. The slower speed should allow a higher volumetric efficiency to be achieved so a value of 91% is chosen. The estimated displacement is:

$$\dot{m}_{fuel} = (100\ kW) * \left(\frac{260\,gm}{kW\text{-hr}}\right) * \left(\frac{kg}{1000\,gm}\right) * \left(\frac{hr}{60\ min}\right) = 0.43\,kg/min$$

$$\dot{m}_{air} = \dot{m}_{fuel} * A/F = (0.43\,kg/min) * (23) = 9.97\ kg/min$$

$$D\,[liter] = \left(\frac{9.97\ kg}{min}\right) * \left(\frac{m^3}{1.18\ kg}\right) * \left(\frac{1000\ liter}{m^3}\right) * \left(\frac{min}{4200\ rev}\right) * \left(\frac{2\ rev}{cycle}\right) * \left(\frac{1}{0.91}\right)$$

Displacement = 4.4 liters

Unfortunately, under these assumptions the diesel engine will be twice as large as the gasoline engine designed for the same vehicle. This would certainly be unacceptable, and the power output would have to be reduced considerably (as was often the case in older, diesel powered automobiles). However, a very attractive option for the diesel engine is to add a turbocharger. If it is assumed that the incoming air can be compressed to 1.8 atm., and cooled back to 47 °C a further fuel efficiency improvement will be achieved as well. Assuming the specific fuel consumption can be reduced to 235 gm/kW-hr, and that all other parameters remain the same, the predicted displacement is now calculated as:

$$\dot{m}_{air} = (100\ kW) * \left(\frac{235\,gm}{kW\text{-hr}}\right) * \left(\frac{kg}{1000\,gm}\right) * \left(\frac{hr}{60\ min}\right) * (23) = 9.00\,kg/min$$

$$P_{port} = (1.8\ atm) * \left(\frac{1.01 \times 10^5\ N}{m^2\,atm}\right) * \left(\frac{K}{287\ N\text{-}m}\right) * \left(\frac{1}{320\ K}\right) = 1.98\,kg/min^3$$

$$D\,[liter] = \left(\frac{9.00\ kg}{min}\right) * \left(\frac{m^3}{1.98\ kg}\right) * \left(\frac{1000\ liter}{m^3}\right) * \left(\frac{min}{4200\ rev}\right) * \left(\frac{2\ rev}{cycle}\right) * \left(\frac{1}{0.91}\right)$$

Displacement = 2.4 liters

With this modest level (for a diesel) of turbocharging a displacement almost identical to that of the spark-ignition engine can be achieved. Several options may be available to further optimize this engine; recognizing as well the greater torque rise capability of the diesel the resulting package will be quite attractive.

Heavy Truck Engine The diesel engine used in a heavy truck application will be optimized very differently. The desired rated power in this engine will be 300 kW. The combination of high load requirements and very long life expectations in a much heavier engine dictate a lower rated speed—1800 rpm is very typical. The high torque rise requirements will result in the need for greater excess air at rated power, so an air-to-fuel ratio of 32 to one is chosen. In order to provide the required air flow, the turbocharger will be set up to provide 3 atm. of boost pressure. A large charge air cooler will still allow a charge temperature of 47 °C (probably necessary in order to meet exhaust emission requirements). Finally, the combination of lower speed and higher boost will result in lower specific fuel consumption. 210 gm/kW-hr will be assumed. The resulting displacement is calculated as:

$$\dot{m}_{air} = \left(300\ kW\right)*\left(\frac{210gm}{kW\text{-}hr}\right)*\left(\frac{kg}{1000gm}\right)*\left(\frac{hr}{60\ min}\right)*\left(32\right) = 33.6\ kg/min$$

$$\rho_{port} = \left(3.0\ atm\right)*\left(\frac{1.01\times10^5\ N}{m^2\ atm}\right)*\left(\frac{K}{287\ N\text{-}m}\right)*\left(\frac{1}{320\ K}\right) = 3.3\ kg/min^3$$

$$D\left[liter\right] = \left(\frac{33.6\ kg}{min}\right)*\left(\frac{m^3}{3.3\ kg}\right)*\left(\frac{1000\ liter}{m^3}\right)*\left(\frac{min}{1800\ rev}\right)*\left(\frac{2\ rev}{cycle}\right)*\left(\frac{1}{0.91}\right)$$

Displacement $= 12.4\ liters$

This again is very typical of the conditions and engine displacements seen in heavy trucks and construction equipment.

It must be noted that recent advances in turbocharger controls, while beyond the scope of this book, allow compressor outlet pressure to be more closely controlled versus engine speed. As a result, it may not be necessary to provide as much excess air at rated power as used in these examples. Repeating the calculations with rated power air to fuel ratios on the order of 20 to one allow specific power output to be substantially increased. Examples are seen in many of the recently introduced high-performance diesel engines.

5.3   Engine Up-rating and Critical Dimensions

Perhaps one of the most common challenges faced by engineers working with internal combustion engines is that of increasing the output of a production engine. The engine was developed at a displacement determined by the rationale described in the preceding section. Now that the engine is in production market demands indicate the need to increase the power output by perhaps 15 or 20 %.

From the discussion just presented several options for increasing the engine's power output can be readily identified. Small increases might be achievable through fine-tuning—perhaps improving the volumetric efficiency or reducing specific fuel consumption. However while a few percent power increase might be achieved the results are seldom sufficient to fully provide the required additional power.

Another approach that might be taken with spark-ignition engines is that of increasing the engine speed. The leverage this provides is significant, but invariably it results in fuel economy penalties due to increased friction and pumping work. In many cases it cannot be considered at all due to increased inertia forces and stress in the reciprocating components.

The next option might be to add a supercharger or turbocharger. For a spark-ignition engine either option might be considered although they both add cost. The engine must be significantly modified to ensure that the combustion process does not violate knock limits, and that durability is not sacrificed. In the case of modern diesel engines a turbocharger has already been fitted, so the turbocharger would be resized for increased intake manifold pressure. This too requires re-optimizing the combustion system such that peak cylinder pressure limits are not exceeded, but it is often the most attractive approach for diesel engine up-rating.

Especially in the spark-ignition engine, increasing the displacement is often the only viable approach. On older engine designs this was often easily done, as engine dimensions were quite generous. This is no longer the case since considerable attention is now paid to minimizing weight and package dimensions. The challenges of increasing the bore or stroke of an engine are detailed in Chap. 8 where the critical dimensions and resulting design trade-offs are identified.

Engine Configuration and Balance

6

6.1 Determining the Number and Layout of Cylinders

Once the fuel type, engine operating cycle, total displacement, and supercharging decisions have been made for a new engine, the next tasks will be to decide upon the number of cylinders over which the displacement will be divided, and the orientation of the cylinders. The factors that must be considered include cost and complexity, reciprocating mass and required engine speed, surface-to-volume ratio, pumping losses, packaging, and the balancing of mechanical forces. The majority of this chapter will address the mechanical forces and engine balancing as this is a key factor explaining why particular numbers and orientations of cylinders are repeatedly chosen.

6.2 Determining the Number of Cylinders

Having determined the total engine displacement, the number of cylinders into which that displacement is divided must be selected. Table 6.1 shows common cylinder volumes for different applications. These volumes vary depending on engine speed, duty cycle, and fuel choice.

Cost and complexity certainly lead the engineer to the lowest practical number of cylinders for any given engine. The quantities of many parts in the engine are directly multiplied by the number of cylinders, so the fewer the cylinders, the lower the cost of materials and assembly. On a cost-per-unit power basis, supercharging an engine will require less displacement and fewer cylinders, which reduces base engine cost. However, adding a compressor, intercooler, and additional intake plumbing to enable supercharging increases the cost of the total engine assembly. For a given cylinder size and power output, it is less expensive to continue to add more cylinders, until it starts to become cost neutral to add

© Springer Vienna 2016
K. Hoag, B. Dondlinger, *Vehicular Engine Design*, Powertrain,
DOI 10.1007/978-3-7091-1859-7_6

Table 6.1 Typical ranges for cylinder volume

Engine type	Cylinder volume (cc)
Small utility	140–500
Motorcycle (sport)	100–350
Motorcycle (cruiser)	375–1000
Automotive gasoline	350–850
Automotive turbo-diesel	475–840
Over-the-highway truck turbo-diesel	840–2660

supercharging above approximately seven cylinders. In other words, it may cost less to build a four-cylinder, super-charged engine of a given power output than a comparable eight-cylinder engine of the same power output.

Conflicting with the requirement to reduce cost and complexity, *reciprocating mass* is reduced by increasing the number of cylinders for a given engine displacement. As cylinder diameter is reduced the mass of the piston, rings, piston pin, and connecting rod diminish as well. At any instant in time the force transmitted through a connecting rod, and seen by the rod and main bearings, crankshaft, cylinder block and main bearing caps is the product of the mass of the reciprocating components and the instantaneous rate of acceleration or deceleration. Higher engine speed requirements result in greater acceleration rates, and in order to keep the reciprocating forces manageable the reciprocating mass must be reduced. It follows directly that increasing the number of cylinders is attractive for high speed engines, as it allows the mass to be distributed, reducing the force transmitted at each cylinder.

Next, the engine designer must choose the *cylinder arrangement*: in-line, vee, W, horizontally opposed, radial, or Wankel, as shown in Figs. 6.1, 6.2, 6.3, 6.4 and 6.5. Packaging

Fig. 6.1 Inline configuration

Fig. 6.2 Vee configuration

Fig. 6.3 Horizontally opposed
configuration

Fig. 6.4 Radial configuration

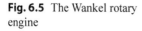

Fig. 6.5 The Wankel rotary engine

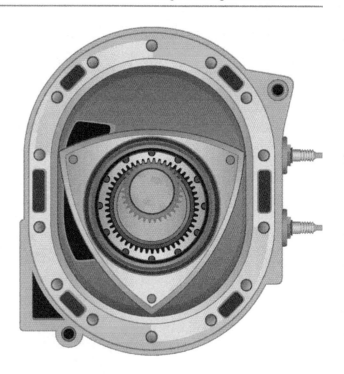

considerations include not only the length, width, and height of the engine, but installation requirements such as those pertaining to intake and exhaust fittings and locations, and cooling system connections. The need for duplication of some components with certain engine configurations—cylinder heads or exhaust piping in vee engines for example—must also be considered. The in-line engine, with a single bank of cylinders provides the simplest configuration. Because only one bank of cylinders must be served, the intake, exhaust, and cooling systems are all easily configured. However, the engine height creates packaging challenges in some applications, and as the number of cylinders increases, length may become prohibitive. The height is sometimes addressed by tipping the engine in a "slant" mounting, or by mounting it horizontally.

The reduced height of alternative configurations comes at the price of increased width. The vee configuration allows a reduction in both height and length at the expense of width, but now requires two banks of cylinders to be supported with cooling, intake, and exhaust systems. A further potential disadvantage is that of reduced crankshaft and connecting rod bearing area. Whereas the in-line engine allowed a main bearing to be placed between each cylinder, the vee configuration results in a pair of cylinders between each pair of main bearings. With only slightly increased spacing between the main bearings sufficient bearing area must be supplied for two rod bearings. These challenges are met without difficulty in most automobile engines, but are very difficult to meet in heavy-duty applications. The compact 'W' configuration can be thought of as two vee engines placed side-by-side with a single crankshaft. The 'W' engine further demonstrates the trade-off between engine width and length. While there are now four banks of cylinders their close proximity reduces (but does not eliminate) the challenges of intake, exhaust, and cooling system packaging.

The horizontally opposed engine provides minimum height, but a very wide package. The length of the engine is similar to that of the vee engine, and it faces similar bearing area challenges. The support system challenges are made greater by the greater physical distance between cylinder banks.

Beginning from an in-line engine changing the configuration to a vee significantly reduces the length of the engine along the crankshaft. To exaggerate that line of thought, a radial engine can be though of as only one cylinder long, with many cylinders arranged around the crankshaft. This dramatically reduces the length of the engine, but significantly complicates the connecting rod to crankshaft packaging. A special type of articulated connecting rod is required. This arrangement was typically used in air-cooled aircraft engines, where each exhaust port needed to share the same amount of cooling airflow for maximum power density.

A different type of internal combustion engine is the rotary piston engine, or Wankel engine named after its creator. In this arrangement, a triangular rotor moves around the crankshaft, creating increasing and decreasing volumes within its housing. Several combustion events per rotation per working cylinder are possible, which enables a high power density for this engine. The rotary motion of the working piston produces much less vibration than a conventional reciprocating piston engine. Unfortunately, an inherently poor combustion chamber shape and difficulty sealing the combustion chamber lead to fuel consumption and exhaust emission challenges that have limited the Wankel engine's use.

6.3   Determining the Cylinder Bore-to-Stroke Ratio

Once the number and configuration of cylinders has been determined the final remaining basic layout question is that of bore-to-stroke ratio. For any given cylinder displacement volume a theoretically infinite range of cylinder bore-to-stroke ratios could be chosen. Choosing the bore-to-stroke ratio will have a fundamental impact on the overall engine size. As the bore increases the block length will grow; as the stroke increases the deck height will raise. Choosing the bore-to-stroke ratio is also one of the key drivers to how the engine will perform and has many effects on engine power output, valvetrain architecture, thermal efficiency, combustion efficiency, pumping losses, and mechanical friction. The bore-to-stroke ratio is calculated in Eq. 6.1, and comparison data is presented in Fig. 6.6.

$$\text{Bore-to-Stroke Ratio} = \frac{\text{Cylinder Bore Diameter or Piston Diameter}}{\text{Piston Stroke Length}} \qquad (6.1)$$

The first thing to consider when choosing bore-to-stroke ratio is the affect the ratio will have on rated or peak power speed. Valve diameter and timing events are set to optimize

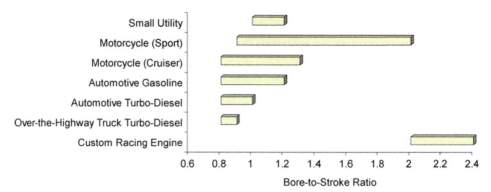

Fig. 6.6 Bore-to-stroke ratio comparison

intake air velocity at the desired peak power speed. The bore-to-stroke ratio and valvetrain must be matched to maximize the output of a given engine. A large bore-to-stroke ratio will enable higher engine speeds, and due to the larger packaging size will enable larger valves, which will allow the intake air velocity to be optimized at high engine speeds, resulting in an engine with high power density. If a small bore-to-stroke ratio is chosen, the long stroke enables the peak power speed to occur at lower engine speed. Comparison data is shown in Fig. 6.7.

The reciprocating piston has continually changing speed and acceleration rates through-out its stroke. At TDC, the piston speed is zero, and it must accelerate to maximum piston speed near mid-stroke, and then decelerate to a full stop at BDC for each stroke. The maximum piston speed and acceleration are related to the piston stroke and the engine speed. The higher the acceleration, the higher the mechanical stress on reciprocating en-

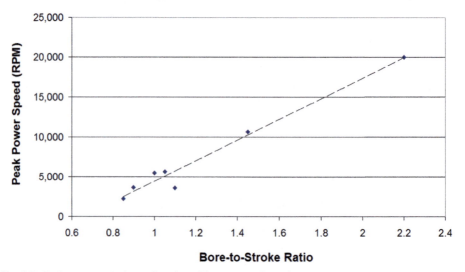

Fig. 6.7 Peak power speed as a function of bore-to-stroke ratio

gine parts. A useful way to describe the design limit is with the ***mean piston speed*** shown in Eq. 6.2. Note the tradeoff that increasing the engine stroke has on engine speed for a given mean piston speed. A high mean piston speed is one of the structural limits imposed on any engine design.

$$\text{Mean Piston Speed, } V_p = 2 \cdot S \cdot N \qquad (6.2)$$

Where:

S =   Piston stroke
N =   Engine Speed

The role of the valvetrain is to allow sufficient air exchange capacity (breathing) and to control the timing of the intake and exhaust events, while not exceeding the limits of contact stresses on the camshaft or the stress limits of other valvetrain components. An approximation of valve inertia forces is illustrated in Eq. 6.3. As the engine's bore is increased, the diameter of the intake valve must also increase to allow sufficient breathing capacity at higher engine speeds. As the valve diameter increases, so does its mass. At low engine speeds, a pushrod or overhead valve (OHV) arrangement may be sufficient to maintain valve timing and meet component stress requirements. As engine speed increases, a stiffer valve spring is required to keep the valve following the cam profile. As engine speed increases beyond a certain point, valve accelerations increase to the point where the valve spring will no longer be able to control the valvetrain due to the large mass, and an overhead cam (OHC) arrangement may be required to reduce valvetrain mass and maintain valve timing. An alternate approach to achieve these same combinations of breathing, timing and stress requirements is to split the single valve mass into two or more valves. This will enable sufficient valve area for breathing while reducing the individual valve mass and inertia.

$$F_v \approx m \cdot h \cdot N^2 \qquad (6.3)$$

Where:

F_v = inertia force on valve
m = mass of the valve
h   = max lift of the valve

General industry values are shown in Table 6.2. As regulatory pressure increases to improve fuel consumption and minimize exhaust emissions, the industry will continue to respond with engines of higher specific output (down-sizing and down-speeding). These general industry values will continue to rise with further development.

Table 6.2 Typical performance ranges for engine types

Engine type	Bore-to-stroke ratio (B/S)	Peak power speed (RPM)	Mean piston speed (m/s)	BMEP at peak torque (MPa)
Small utility	1.0–1.2	3600	8–10	0.8–1.0
Motorcycle (Sport)	0.9–2.0	6800–14,500	16–24	1.0–1.5
Motorcycle (Cruiser)	0.8–1.3	4250–7000	13–23	0.9–1.2
Automotive gasoline	0.8–1.2	4000–7000	13–24	1.1–1.3
Automotive turbo-diesel	0.8–1.0	3300–4000	9–13	1.3–2.1
Over-the-highway truck turbo-diesel	0.8–0.9	1800–2600	10–14	1.9–2.3
Custom racing engine	2.0–2.4	20,000	22–33	1.6

Surface-to-Volume ratio refers to the ratio of combustion chamber surface area to combustion chamber volume, and has a significant effect on heat transfer and combustion efficiency. The actual value of surface-to-volume ratio changes as cylinder volume changes due to piston movement throughout the operating cycle. While surface-to-volume ratio is nearly independent of bore-to-stroke ratio when considering the BDC volume, it is strongly dependent on bore-to-stroke ratio at or near the TDC cylinder volume. For the same cylinder displacement, a larger bore and shorter stroke engine has a higher TDC surface-to-volume ratio and thus greater heat rejection from the cylinder. The surface-to-volume ratio is of greatest importance near TDC, when energy is rapidly released during combustion, and heat rejection to the combustion chamber walls is to be minimized.

The surface-to-volume ratio is impacted by both the number of cylinders and the bore-to-stroke ratio of each cylinder. As can be seen from Eqs. 6.4 to 6.6 below, the larger the cylinder, the lower the surface-to-volume ratio. Increasing or decreasing the number of cylinders for a given engine displacement correspondingly increases or decreases the surface-to-volume ratio.

$$\text{Surface Area of a Sphere} = A_S = 4\pi\, r^2 \tag{6.4}$$

$$\text{Voume of a Sphere} = V_S = \frac{4}{3}\pi r^3 \tag{6.5}$$

$$\text{Surface-to-Volume Ratio (S/V)} = \frac{A_S}{V_S} = \frac{4\pi r^2}{\frac{4}{3}\pi r^3} = \frac{3}{r} \approx \frac{1}{r} \tag{6.6}$$

Because of the effect bore-to-stroke ratio has on heat transfer, it will also have a significant effect on *combustion efficiency*. For different reasons the combustion chambers of both diesel and spark-ignition engines become more difficult to optimize as bore diameter increases. For a given cylinder size, the bore-to-stroke ratio will have a significant effect on the shape of the combustion chamber. As the bore diameter increases, the aspect ratio of the combustion chamber will flatten out to a shallow disc.

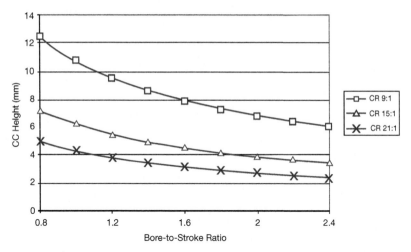

Fig. 6.8 Effect of bore-to-stroke ratio on combustion chamber height

In spark-ignition engines increasing the bore-to-stroke ratio increases the flame travel distance, slowing the energy release rate and increasing the potential for early flame quench and cycle-to-cycle variability. As bore diameter is increased above approximately 105 mm a second sparkplug will be required to reduce flame travel distance and time. A larger bore also increases the top piston ring crevice volume and correspondingly increases un-burnt hydrocarbons. In the diesel engine the larger bore results in a shallower combustion chamber to maintain compression ratio, and leads to a greater propensity for fuel injection spray impingement on either the cylinder head or piston surface.

The bore-to-stroke ratio will also affect the maximum *compression ratio* (CR) possible, as defined in Eq. 6.7. A high compression ratio is desirable because it improves the thermal efficiency and generally the fuel efficiency of the engine. For a fixed cylinder volume, as the bore increases and the stroke decreases, it is difficult to achieve a high compression ratio as the valves begin to interfere with the piston crown at TDC-overlap as shown in Fig. 6.8. The plots were generated assuming the combustion chamber is a cylindrical disc of equal diameter to the piston bore.

$$\text{Compression Ratio} = \frac{\text{Volume max}}{\text{Volume min}} = \frac{V_{cc} + V_{cyl}}{V_{cc}} \tag{6.7}$$

Where:

V_{cc} = Combustion Chamber Volume
V_{cyl} = Swept Volume of Cylinder

Pumping losses increase as the length of intake and exhaust ducts increase, and as their diameters decrease. As the number of cylinders increase additional duct length is often required. The adverse effects can often be countered through tuning—selecting duct lengths

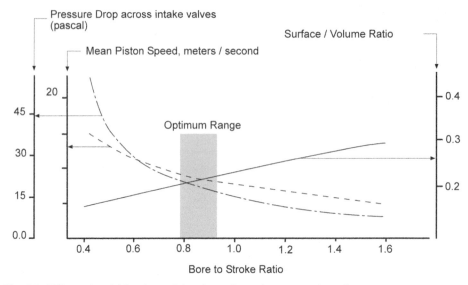

Fig. 6.9 Effects of variables determining the optimum bore-to-stroke ratio

that result in dynamic pressure pulses aiding the cylinder filling and emptying processes at chosen engine speeds. An additional challenge is that of duct diameter, and specifically valve diameters as the number of cylinders is increased, and the resulting bore diameter of each cylinder becomes smaller.

Frictional losses are also modestly affected by bore-to-stroke ratio. For large bore-to-stroke ratios the same combustion pressure is applied over a greater piston area, leading to higher unit loads on bearings. The higher unit loading requires larger bearings and increases hydrodynamic losses. Piston weight increases as a function of diameter and the resulting higher inertial forces will also increase the unit loads on the bearings. At the other extreme, the long stroke engine will see higher mean piston speeds which will increase rubbing friction between the piston and rings and the cylinder bore.

In summary, trade-offs associated with the parameters just described are presented for a given engine in Fig. 6.9. In one sense this plot is misleading, as each of the parameters is plotted on a different dependent axis. However it is important to consider the combined effect of each parameter. The bore must be made sufficiently large to keep the mean piston speed below its design target and to minimize pressure drop across the valves (since the pressure drop increases sharply as the valve area is reduced). As a general rule the bore is made no larger than is necessary to fulfill these two requirements—this minimizes heat rejection to the coolant. The resulting bore-to-stroke ratio will differ depending on the design criteria for the specific engine. In automotive applications, fuel efficiency and emission requirements dominate, so the bore-to-stroke ratio is typically just below unity to give the best balance of performance. In high performance applications where the engine displacement is typically limited by racing rules, and a high specific power is required, the bore-to-stroke ratio tends to be larger in order to maximize breathing and provide accept-able piston speeds at high engine speed.

6.4   Vibration Fundamentals Reviewed

An important factor that must be considered in determining the number and configura-tion of cylinders is that of controlling, and as much as possible balancing, the mechanical forces generated within the engine. In this section the basic concepts of mechanical vibra-tion are reviewed, and the forces generated within the engine are identified. Rotating and reciprocating forces are then separately discussed in the next two sections.

Vibration is defined as the response resulting from any force repeatedly applied to a body. The repeated force may be random, or a force of a given magnitude may be applied at some constant frequency. In the case of forces generated within an engine the mag-nitudes will be constant at any given speed and load, and will change with either speed (rotating and reciprocating forces) or speed and load (gas pressure forces). Since the fre-quency of the vibration forces is determined by engine speed it is convenient to define the vibration *order* as the vibration frequency relative to shaft speed. A *first-order* vibration is generated by forces that are applied over a cycle that occurs once every crankshaft revolu-tion. A *half-order* vibration occurs once every second crankshaft revolution. Half-orders may be important in engines operating on a four-stroke cycle—for example, the gas pres-sure force applied at one cylinder is repeated over a cycle of two crankshaft revolutions. A *second-order* vibration occurs twice every crankshaft revolution, and so on.

In assessing engine vibration it is convenient to use a Cartesian coordinate system. The forces acting on the engine can be thought of as attempting to translate the engine along, or rotate it around, each of the three axes. The engine assembly (typically engine and transmission or transaxle) is constrained by the engine mounts, and any forces or moments not counteracted within the engine are transmitted through the mounts to the vehicle. The forces acting along each axis must sum to zero, and the moments acting around each axis must also sum to zero or a mechanical vibration is transmitted to the vehicle.

The engine simultaneously experiences a variety of internal forces. First, at each cylin-der some portion of the crankshaft and connecting rod must be centered at some distance away from the crankshaft centerline as determined by the engine's stroke. As the crank-shaft spins this mass gives rise to a centrifugal force of a magnitude dependent on engine speed, and acting outward from the crankshaft at each instant in time along the centerline of the particular crankshaft throw. The reaction force is transmitted to the block through the main bearings—it is split between the main bearings surrounding the particular cylin-der, and further distributed along the crankshaft to the remaining main bearings in quanti-ties dependent on the length and stiffness of the crankshaft, and any block deflection that may be occurring. In a multi-cylinder engine such forces are simultaneously applied at each crankshaft throw. This results in various moments whose magnitude and direction are determined by the orientation of the throws along the crankshaft. Further mechanical forces are transmitted to the crankshaft at each cylinder due to the continuous acceleration and deceleration of the pistons. These reciprocating forces vary in magnitude with crank angle, and act along the axis of each cylinder centerline. Finally, gas pressure forces, varying continuously throughout the engine operating cycle, are transmitted to the crankshaft at each cylinder.

Two further observations must be made before proceeding into a detailed look at the calculation and management of these forces. First, although what the engine "sees" is the instantaneous summation of each of the forces just described, calculations will be facilitated by recognizing the powerful tool of superposition. If the forces are broken down as described in the previous paragraph each effect can be separately calculated. The resultant seen by the engine is simply the sum of each of the contributions along any given axis. Second, the reciprocating piston engine operates on the kinematic principles of converting reciprocating motion at the piston to rotating motion at the crankshaft. The conversion occurs across the connecting rod, with everything connected at its small end experiencing purely reciprocating motion, and everything at its large end experiencing purely rotational motion. The connecting rod itself experiences complex motion that would be extremely difficult to calculate. Fortunately such calculations are not necessary. A very close approximation can be achieved simply by weighing each end of the connecting rod. The mass is split about its center of gravity, between that in reciprocating and that in rotating motion.

6.5 Rotating Forces and Dynamic Couples

The centrifugal force generated at each crankshaft throw is calculated as depicted in Fig. 6.10. As the crankshaft spins each element of mass located at a distance away from the crankshaft centerline generates an outward centrifugal force calculated as

$$F_{Rotational} = \frac{M \cdot r \cdot \omega^2}{g_c} \tag{6.8}$$

Where:

M = mass
r = radial distance from shaft centerline (stroke/2)
ω = angular velocity, in radians per unit time
g_c = gravitational constant

The mass that must be considered in this calculation includes that of the crankshaft itself as well as that of the portion of the connecting rod affixed to the crankshaft at the given location. The resulting mass and its center of gravity are used to determine the values to be used in Eq. 6.8. Such calculations have traditionally been made using graphical methods

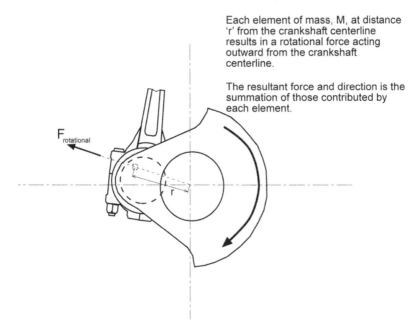

Each element of mass, M, at distance 'r' from the crankshaft centerline results in a rotational force acting outward from the crankshaft centerline.

The resultant force and direction is the summation of those contributed by each element.

$F_{rotational}$

Fig. 6.10 Summary of centrifugal force calculation, showing contributing element and variables pertaining to that element. Each element of mass m at distance r from the crankshaft centerline results in a rotational force acting outward from the crankshaft centerline. The resulting force and direction is the summation of those contributed by each element

in which the crankshaft section was laid out on a fine grid, and the contribution of each region was calculated to determine the resultant mass and location of the center of gravity. Today such calculations are included in the computer aided design (CAD) software used in most engine design work.

If the centrifugal force generated at any given cylinder or crankshaft throw is considered it should be apparent that mass can be added to the crankshaft opposite of the throw in order to completely balance the force. The added masses are referred to as the crankshaft *counterweights*. However, in a multi-cylinder engine it may be possible to balance the centrifugal force generated at one crankshaft throw with that generated at another throw. Referring to Fig. 6.11a a simplified crankshaft section is represented in which two adjacent throws are configured 180° apart. The centrifugal force generated at one throw is exactly counteracted by that generated at the adjacent throw, and the sum of forces along the plane of the crankshaft is zero. However, because the throws are located at different positions along the length of the crankshaft the moment generated by this pair of throws is not zero. As shown in the figure, a moment will be centered at the midpoint between the throws. This is referred to as a *dynamic couple*, and its magnitude is calculated as

$$\sum F_{Rotational} \cdot L \qquad\qquad (6.9)$$

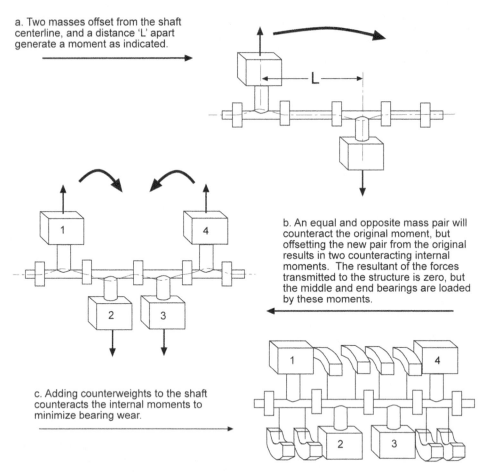

a. Two masses offset from the shaft centerline, and a distance 'L' apart generate a moment as indicated.

b. An equal and opposite mass pair will counteract the original moment, but offsetting the new pair from the original results in two counteracting internal moments. The resultant of the forces transmitted to the structure is zero, but the middle and end bearings are loaded by these moments.

c. Adding counterweights to the shaft counteracts the internal moments to minimize bearing wear.

Fig. 6.11 Dynamic couples and their balance, leading to the example of a four-cylinder crankshaft

Where:

L = distance along shaft axis where force is applied, to the neutral axis about which the
     moment occurs

In Fig. 6.11b the crankshaft representation is extended to that for a four-cylinder engine having the typical firing order of 1-3-4-2. The throws for cylinders two and three are 180° apart from those for cylinders one and four. The centrifugal forces all act in the same plane and again balance one another out. If the moments are now considered across the entire length of the crankshaft it is found that these too are completely balanced. The dynamic couple identified earlier (cylinders one and three) has been balanced by an equal and opposite couple (cylinders two and four) as depicted in the figure. However, because the centrifugal forces are applied at various points over the length of the crankshaft it is also

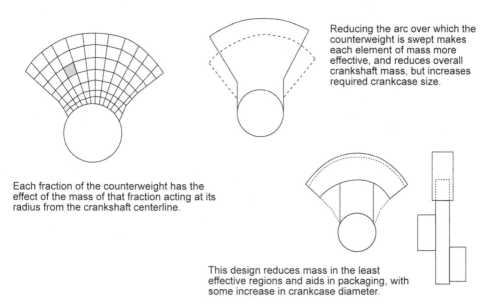

Reducing the arc over which the counterweight is swept makes each element of mass more effective, and reduces overall crankshaft mass, but increases required crankcase size.

Each fraction of the counterweight has the effect of the mass of that fraction acting at its radius from the crankshaft centerline.

This design reduces mass in the least effective regions and aids in packaging, with some increase in crankcase diameter.

Fig. 6.12 Counterweight design considerations

important to consider the force being transmitted from the crankshaft to the block at each bearing. While the balanced forces and moments result in zero net force attempting to translate or rotate the block (and thus having to be countered by the engine mounts) the two internal moments result in crankshaft deflection and additional main bearing loads at certain bearings. The axes about which the moments rotate are centered at the number two and four main bearings and these two bearings will not see any loads induced by the moments. Bearings one, three, and five see additional loads as a result of the moments. Counterweights, as shown in Fig. 6.11c, are therefore added to this four-cylinder crankshaft, not to achieve rotational balance but to reduce crankshaft deflection and main bearing loads.

It is important to emphasize again that the discussion so far has considered only rotational forces. For the simple case of the four-cylinder engine it was demonstrated that each centrifugal force or dynamic couple that attempts to translate or rotate the engine is completely balanced, and no net force remains that would be transmitted to the vehicle at the engine mounts. This is not to say that no vibration forces whatsoever are transmitted to the engine mounts—only that none resulting from rotational forces are transmitted. Reciprocating and gas pressure forces have not been introduced in this discussion and will in fact transmit forces through the engine mounts.

Several design trade-offs pertaining to the counterweights are summarized in Fig. 6.12. While the counterweights too can each be represented as a single mass acting at some local center of gravity, the actual counterweights consists of elements of mass at various distances from the crankshaft centerline and from the plane about which the mass is to be centered. Mass located at positions furthest from the shaft centerline, and closest to the center plane is most effective. Locating more mass in these most effective regions allows the total crankshaft mass to be reduced, but at the penalty of larger crankcase size. In some engines, especially those designed to operate at high speed it may be advantageous to devote attention to the aerodynamics of the counterweight. This too may result in the need for greater total crankshaft mass.

The four-cylinder engine taken as the example for the initial discussion of rotating balance and couples has a "planer" or "flat" crankshaft—the crankshaft is centered about a single plane, and all of the forces and moments act in that plane. This is not the case with various other engine configurations of interest for automotive applications. The calculations become more complicated, but the methodology remains the same. As an example, consider the inline six-cylinder engine whose crankshaft is represented in Fig. 6.13. The typical firing order is 1-5-3-6-2-4. The crank throws are each 120° apart, with the throws for cylinders one and six in the same plane, five and two another plane, and three and four in a third plane. Equal, outward centrifugal forces are acting at 120° intervals, and this

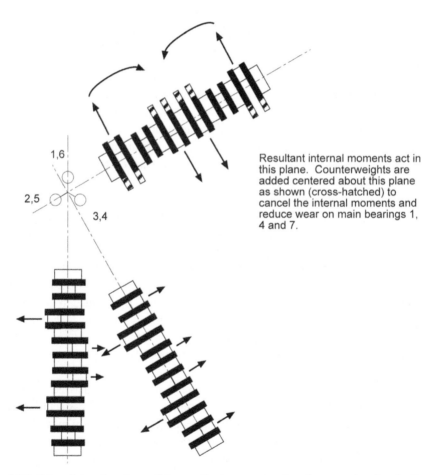

Resultant internal moments act in this plane. Counterweights are added centered about this plane as shown (cross-hatched) to cancel the internal moments and reduce wear on main bearings 1, 4 and 7.

Fig. 6.13 Determining the plane of the resultant moment, using the example of an in-line six-cylinder crankshaft

too results in balanced forces. The reader is encouraged to verify this fact by selecting a plane through any of the pairs of cylinders just identified, and calculating all of the force components acting in that plane. They will be found to sum to zero.

Again selecting any plane through the crankshaft and assessing the force components in that plane allows the moments to be calculated. It is found for this engine configuration that the moments balance across the engine as a whole, but that the front three cylinders and the back three cylinders each generate an equal and opposite internal moment. The counterweights required to counteract these moments must be centered about the plane in which the moments are maximized. The moments can be assessed over a variety of planes in order to determine the plane in which to center the counterweights, as further depicted in Fig. 6.13. In the case of the in-line six-cylinder engine the counterweights are centered in the plane 90° from that of cylinders two and five. Similar exercises can be done for any cylinder configuration.

Finally, it should be noted that an alternative to the approach just discussed is to counterweight each cylinder individually with mass opposite to each throw. For example, instead of the counterweights shown in Fig. 6.13 all of the cylinders would be equally counterweighted, with mass placed on each web and centered 180° from the centerline of the particular crank throw. Because each cylinder is balanced individually the resulting mass at each cylinder (12 counterweights in the in-line 6-cylinder example) is less than that shown at the 8 locations in the figure. This provides the advantage of a smaller crankcase, and both the advantages and disadvantages of a stiffer and heavier crankshaft. This approach can be taken relatively easily with a cast crankshaft, but if the crankshaft is to be forged the additional counterweights would have a detrimental impact on forging process complexity and cost.

The variety of engine configurations conceivable for automotive applications is too great to allow detailed discussion of each configuration. The principles discussed here can be applied for any configuration of interest. A summary of the results for a number of popular configurations is provided in Sect. 6.6.

6.6   Reciprocating Forces

The reader is reminded of the discussion in Sect. 6.4, where it was stated that while the engine simultaneously experiences rotating and reciprocating forces it is convenient to separate these for calculation purposes. Rotating forces and moments and their balance were the topic of Sect. 6.5. In this section reciprocating forces and balance will be taken up.

The piston, piston pin, rings, and the upper portion of the connecting rod are subjected to reciprocating motion, and thus repeated acceleration and deceleration. The reciprocating forces act along each cylinder centerline, and their magnitude varies continuously with crank angle. The magnitude of the reciprocating forces at any given crank angle position can be determined from Newton's Second Law as the product of the reciprocating mass and the instantaneous acceleration at that crank angle.

Piston velocity reaches zero at both the TDC and BDC positions, following a non-symmetric trace determined by the slider crank geometry between those positions. The peak velocity does not occur midway between TDC and BDC, but is skewed toward TDC. If the connecting rod were infinitely long the velocity trace would be sinusoidal, but as the connecting rod is made shorter the velocity trace becomes increasingly skewed. The resulting velocity versus crank angle is the following series expression:

$$V = -\omega^2 r[\sin\theta + 2a_2 \sin 2\theta + 4a_4 \sin 4\theta + ...] \tag{6.10}$$

Where:

ω = Angular velocity of crankshaft, in radians per unit time
r = Radial distance of rod bearing axis from crankshaft centerline (stroke/2)
θ = Crank angle relative to TDC
L = Connecting rod length

$$a_2 = \frac{L}{r}\left[\frac{1}{4}\left(\frac{r}{L}\right)^2 + \frac{1}{16}\left(\frac{r}{L}\right)^4 + \frac{15}{512}\left(\frac{r}{L}\right)^6 + ...\right] \tag{6.11}$$

$$a_4 = -\frac{L}{r}\left[\frac{1}{64}\left(\frac{r}{L}\right)^4 + \frac{3}{256}\left(\frac{r}{L}\right)^6 + ...\right] \tag{6.12}$$

Of current interest is the instantaneous acceleration, which is simply the velocity derivative with respect to time (in this case crank angle):

$$\frac{dV}{d\theta} = -\omega^2 r[\cos\theta + 4a_2 \cos 2\theta + 16a_4 \cos 4\theta + ...] \tag{6.13}$$

The instantaneous reciprocating force at any given cylinder can now be calculated as:

$$F_{Reciprocating} = M_{Reciprocating}\frac{dV}{d\theta}$$

$$F_{Reciprocating} = -M_{Reciprocating}\omega^2 r[\cos\theta + 4a_2 \cos 2\theta + 16a_4 \cos 4\theta + ...] \tag{6.14}$$

Equation 6.14 can be greatly simplified by making two observations. First, unless the connecting rods are very short it is observed that the higher order terms are quite small and only the first two must be considered. Second, the term '$4a_2$' is observed to be almost identically '$r/_L$.' These observations result in the simplified acceleration expression,

$$F_{Reciprocating} = -M_{Reciprocating}\omega^2 r\left[\cos\theta + \frac{r}{L}\cos 2\theta\right] \tag{6.15}$$

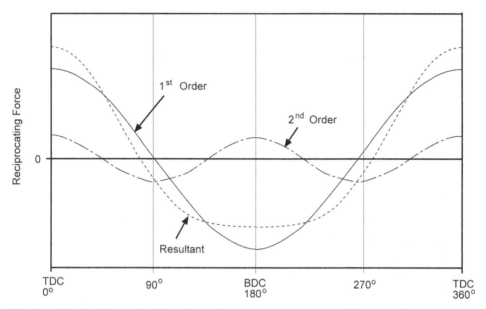

Fig. 6.14 First- and second-order reciprocating forces and the resultant force versus crank angle for an individual cylinder

It may now be recognized that of the two remaining crank angle terms the first varies at the rate of once per crankshaft revolution (first order), and the second varies at the rate of twice per crankshaft revolution (second order). While the reciprocating force seen by the engine at each cylinder is that calculated by this entire equation it is helpful to break the equation down into first- and second-order terms, and consider each separately. The first and second order terms and the resultant reciprocating force are each plotted in Fig. 6.14.

A physical understanding of the first and second order terms can be gained by returning to the in-line four-cylinder engine. If a second horizontal axis is added to Fig. 6.14, as shown in Fig. 6.15a, the net reciprocating forces from a bank of cylinders can readily be seen. On the second axis the first cylinder is located at its TDC position. Each remaining cylinder is located at the position it occupies when the first cylinder is at TDC. For the case of the four-cylinder engine the standard firing order, 1-3-4-2, results in pistons two and three at BDC, and piston four along with piston one at TDC. The reciprocating forces generated by each piston assembly at this moment in time are readily seen. For this engine, the two cylinders (one and four) decelerating to zero velocity as they approach TDC transmit their maximum first order reciprocating force upward on the cylinder block. At this same moment in time, cylinders two and three are decelerating to zero velocity as they approach BDC, and exert the same reciprocating force downward into the main bearing caps. The net result is zero, and the engine has balanced primary reciprocating forces. Looking now at the second order forces it can be seen from Fig. 6.15a that all four cylinders simultaneously reach their maximum second order force and this engine thus has a second

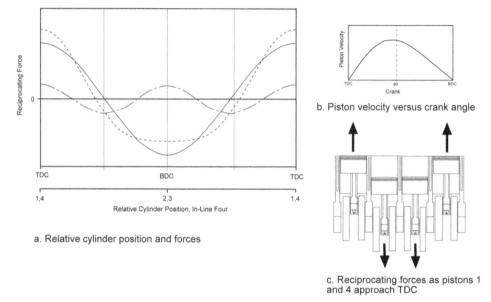

a. Relative cylinder position and forces

b. Piston velocity versus crank angle

c. Reciprocating forces as pistons 1 and 4 approach TDC

Fig. 6.15 Reciprocating forces in an in-line four-cylinder engine

order imbalance. The physical explanation can be seen with reference to Fig. 6.15b, c. In Fig. 6.15b the velocity profile discussed earlier is plotted. Since the maximum velocity occurred nearer to TDC than to BDC the two pistons approaching TDC were decelerated at a greater rate than those approaching BDC. This resulted in a greater upward reciprocating force than downward. One half revolution later cylinders two and three approach TDC as cylinders one and four approach BDC, and the same force imbalance occurs.

Finally, it should be recognized that the reciprocating forces may result in first- and second-order moments within a bank of cylinders. Depending on the relationship between cylinders within the bank these moments may be counteracted or remain unbalanced.

6.7   Balancing the Forces in Multi-Cylinder Engines

The calculation of rotating and reciprocating forces and resulting moments was summarized in the two preceding sections. These equations can now be applied to consider the balancing of any reciprocating piston engine configuration. Several common examples will be discussed further in this section. The discussion will not attempt to include all configurations that may be of interest for automotive engines. The reader is encouraged to apply the equations to look at configurations that have not been discussed.

The *in-line four-cylinder engine* has been considered in previous examples concerning both rotating and reciprocating balance. It was shown that the rotating forces create opposing dynamic couples that cancel one another out but result in internal crankshaft deflection and bearing wear. Counterweights are used as were shown in Fig. 6.11 to eliminate the

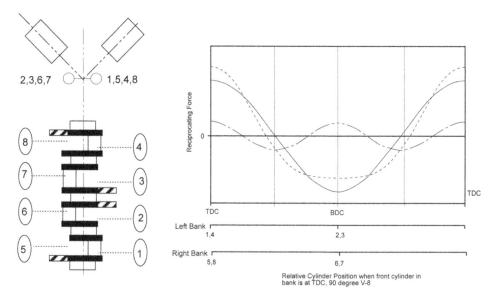

Fig. 6.16 Forces and balance in a 90° V-8 engine using a planar crankshaft

internal couples. The primary reciprocating forces are balanced, but there is an inherent second order imbalance that was explained with reference to Fig. 6.15. The symmetric relationship between the cylinders results in balanced reciprocating moments—both first-order and second-order. In most smaller four-cylinder engines the second-order force imbalance is accepted, and the mounts are designed to minimize its transmission to the vehicle. In order to eliminate the second order imbalance counter-rotating balance shafts are typically used in larger displacement (above two liter) engines. Two shafts are located at the same height in the block, and are driven such that they spin in opposite directions at twice engine speed. Each shaft contains an equal off-centered mass that generates a centrifugal force. Because the shafts are counter-rotating the horizontal components cancel one another out and the vertical components add. They are sized such that the vertical forces add to exactly cancel the second order forces generated by the piston assemblies.

Because the *V-8 engine* consists of two banks of four cylinders each it will be reviewed next. At first glance the forces acting on vee engines would seem far more complicated to analyze, but this is not necessarily the case. While there are now reciprocating forces acting in two different planes each bank of cylinders can be treated individually. If the reciprocating forces can be balanced within a bank they need not be considered further. The V-8 engine almost invariably uses a 90° angle between the two banks of cylinders, allowing even spacing between cylinder events and shared rod bearing throws. Two different crankshaft configurations can be used. The first is a planar crankshaft similar in appearance to that of a four-cylinder engine, but with two connecting rods sharing each throw. This configuration is summarized in Fig. 6.16. Each bank of cylinders appears exactly like the inline four-cylinder engine, and as a result the primary reciprocating forces are balanced within each bank. The two banks cancel the horizontal component of the second

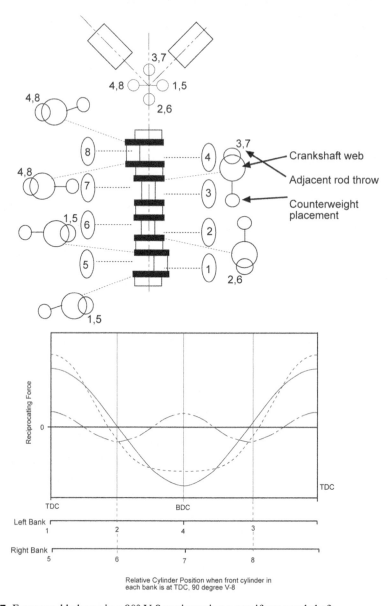

Fig. 6.17 Forces and balance in a 90° V-8 engine using a cruciform crankshaft

order reciprocating force, while the resulting vertical component is 1.414 times that of a single bank. Rotational forces are identical to those of an in-line four-cylinder engine, and the resulting counterweight placement is also identical.

Although it adds crankshaft mass and complexity the more common crankshaft configuration is a two-plane design generally termed "cruciform." This arrangement is shown in Fig. 6.17. The reciprocating force diagram for each bank of cylinders in this engine is

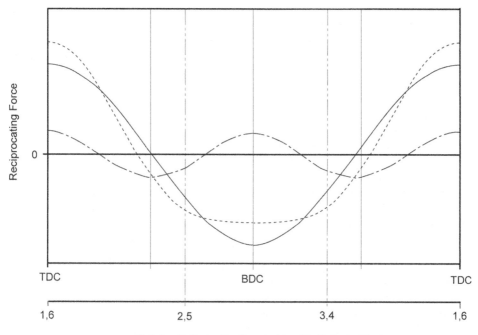

Relative Cylinder Position, In-Line Six-Cylinder Engine

Fig. 6.18 Reciprocating forces in an in-line six-cylinder engine

shown in the figure, and demonstrates both primary and secondary balance. Note that the horizontal axes show each bank at a moment in time when one piston is at TDC, and thus the position of maximum reciprocating force. These positions do not occur simultaneously on each bank, but are shifted in phase by 90°. The rotational forces result in a moment acting over the length of the crankshaft between the first and fourth crankshaft throw and a smaller moment acting across the second and third throw in a plane 90° from the larger moment. The plane in which the resultant moment acts is approximately 18° from that of the first and fourth throw. In order to avoid prohibitively large counterweights at either end of the crankshaft it is typical to place counterweights at each of the throws shown in the figure, acting in the planes of the crankshaft throws. These counterweights are used to offset between one-half and two-thirds of the moment. Balance is then completed with counterweights offset by 18° from the front and rear throws as shown.

Turning now to the *in-line six-cylinder engine* the reciprocating forces can be reviewed with reference to Fig. 6.18. The conventional firing order of 1-5-3-6-2-4 places the throws for cylinders one and six in the same plane. The lower axis in Fig. 6.18 depicts both of these cylinders at TDC. When these two cylinders are at TDC cylinders two and five will be at the position of 120 crank angle degrees, and cylinders three and four will be at 240°. The reciprocating force at cylinders one and six will be at its maximum in the positive direction, while the reciprocating force at each of cylinders two, three, four, and five will be at one-half their maximum in the negative direction ($\cos[120] = \cos[240] = -0.5$). The net first order reciprocating force is thus zero. The net second order force is also found to be zero, since $\cos2[120]=\cos2[240]=-0.5$. The rotational forces for this engine were discussed previously, and were shown to result in counteracting internal couples balanced with couterweights centered in a plane 90° from the crank throws for cylinders two and five. This was depicted previously in Fig. 6.13. The symmetric placement of the cylinders about the fore and aft center of the crankshaft results in complete balance of any moments resulting from the reciprocating forces.

The *V-12 engine* will be discussed next because of its similarity to the in-line six. The V-12 consists of two banks of six cylinders. The crankshaft looks exactly like that of the six-cylinder engine, but with each connecting rod throw made to accept two connecting rods. It follows that the engine has first- and second-order reciprocating balance within each bank. The rotational forces generate opposing internal moments identical to those of the in-line six-cylinder engine, again requiring counterweight placement identical to the in-line six. Note that the discussion did not mention the vee angle between the cylinder banks. Since the reciprocating forces in each bank are independently balanced the vee angle has no impact on mechanical balance. Vee angle does impact the firing forces, and this effect will be discussed further in the next section.

Several firing orders are seen with the *V-6 engine* but in each case it is important that the three cylinders within each bank are spaced such that the crankshaft throws are 120° apart. This results in first- and second-order reciprocating balance within each bank. In most V-6 engines the crankshaft has six separate throws, supported on four main bearings, with two throws between each bearing. The front and rear throw are in the same plane, 180° apart, with the remaining throws spaced symmetrically about this plane. The resultant moment is thus in the plane of the front and rear throws, and the counterweights are placed opposite these throws as shown in Fig. 6.19.

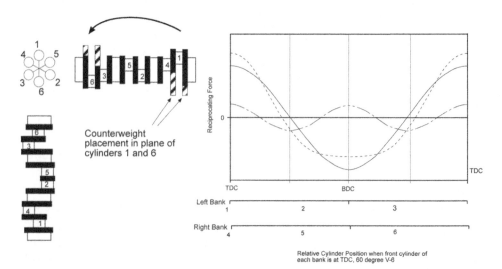

Fig. 6.19 Forces and balance in a V-6 engine

Table 6.3 Balance of reciprocating forces in in-line 5-cylinder engine

Cylinder	First-order	Second-order
1	cos[0]=1.0	cos2[0]=1.0
2	cos[72]=0.309	cos2[72]=−0.809
4	cos[144]=−0.809	cos2[144]=0.309
5	cos[216]=−0.809	cos2[216]=0.309
3	cos[288]=0.309	cos2[288]=−0.809
Sum	0.0	0.0

There are now several examples of both *in-line five-cylinder and V-10 engines*. The firing order commonly used with the inline five-cylinder engine is 1-2-4-5-3, and the cylinders are equally spaced at 72° intervals around the crankshaft. Both the first- and second-order reciprocating forces balance as shown by the following calculations in Table 6.3:

A first-order moment remains. Rotating balance is achieved by equally counterweighting each cylinder to individually balance it along the crankshaft centerline. This results in a relatively stiff, heavy crankshaft. The V-10 engine consists of two five-cylinder banks placed at either a 72 or 90° angle from one another. The 72° angle provides even spacing between firing pulses, but the 90° angle is more common since most V-10 engines are built from modified V-8 block tooling.

The discussion of multi-cylinder engine mechanical balance concludes with a look at *horizontally-opposed engines*. Beginning with the four-cylinder engine, a single plane crankshaft is used as depicted in Fig. 6.20. The throws are positioned identically to an in-line four cylinder engine, and therefore rotational balance is achieved with the same counterweight positions described for that engine. The peak reciprocating forces seen as the pistons approach TDC and BDC are also shown in Fig. 6.20. The two pistons approaching TDC (one in each bank) provide counteracting forces, as do the two pistons approaching

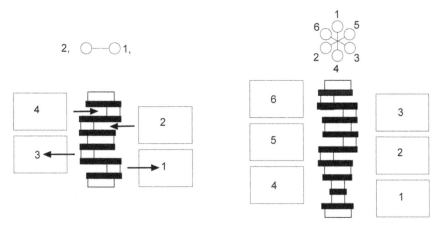

Fig. 6.20 Forces and balance in four- and six-cylinder horizontally opposed engines

BDC. Thus both first- and second-order reciprocating forces are balanced. Because the pistons approaching BDC generate a lower reciprocating force than those approaching TDC a second order moment results.

The horizontally-opposed six-cylinder engine is also depicted in Fig. 6.20. Like the V-6 reciprocating forces within each bank are completely balanced; although the typical firing order is different the throws within each bank are equally spaced at 120° intervals. First- and second-order couples are also completely balanced with this configuration.

6.8   Gas Pressure Forces

Another important category of forces acting internally on the engine is that due to gas pressure within each cylinder. The in-cylinder pressure varies continuously with crank angle, reaching its peak shortly after TDC during combustion. These forces cannot be balanced, and the engine mounts are designed to minimize their transmission into the vehicle. In a four-cycle engine the force generated at each cylinder goes through a complete cycle every two revolutions, and is thus half-order. If all of the cylinders in a multi-cylinder engine contribute equally, and the spacing between combustion events is equal, the resulting vibration order generated by the pressure force is one half the number of cylinders.

The angle between the banks of cylinders in vee engines becomes important when one considers the gas pressure forces. Since one cylinder from each bank shares each crankshaft throw the crank angle spacing between TDC positions in each cylinder pair is the same as the vee angle. Unless further design modifications are made the product of the vee angle and the total number of cylinders must equal 360 or 720° if an even spacing between the gas pressure forces is to be achieved. A 90° vee angle results in evenly spaced firing pulses for a V-4 or V-8 engine. A 60° vee angle provides evenly spaced pulses in a V-6 or V-12 engine. A V-10 engine would require a 72° vee angle.

Since the gas pressure forces cannot be balanced it is legitimate to question whether evenly spaced firing forces are required, and in fact they are not. There are many production examples with uneven firing pulse spacing. In some cases the decision to use a different vee angle is driven by manufacturing costs—a 90° V-6 or V-10 engine is often made based on modifications of an existing V-8 design. In other cases the decision might be driven by package dimensions—a narrow vee angle may be chosen to reduce engine width. In these cases it is most common to accept the uneven firing pulse spacing. Another alternative is to modify the crankshaft throws, offsetting the two shared bearings to maintain even firing. As an example consider the V-6 engine discussed with reference to Fig. 6.19. Even firing requires each firing pulse to be placed 120° apart. Sharing crank throws would require a 120° vee, resulting in a very wide engine. In Fig. 6.19, a web is shown between the two cylinders in each crankcase section, or bay. If the vee angle of the engine is 60° the two throws must be placed 60° apart and an unsupported web, or "flying web" is used to link the throws. On a 90° V-6 the two throws in a given bay are 30° apart. The crankshaft can be manufactured with an offset between the throws, and the flying web is not needed.

6.9 Recommendations for Further Reading

Engine balancing and configuration calculations were established many years ago, and few recent publications cover these concepts. For a recent detailed explanation of these concepts, using a helpful graphical analysis technique, the reader is referred to the following text (see Heisler 1995).

References

Heisler, H.: Advanced Engine Technology. SAE Press, Warrendale (1995)

Cylinder Block and Head Materials and Manufacturing

7

Before addressing cylinder block and head layout design it is important to understand the restrictions imposed by material and casting process selection. This chapter begins with a brief look at the aluminum and gray iron alloys typically used for cylinder blocks and heads. Magnesium alloys, and composite blocks with magnesium portions are receiving increased attention for weight reduction, and will also be briefly covered. Many of the design constraints are imposed by the capabilities of the chosen casting process, so the commonly used casting processes will next be introduced. The chapter concludes with an overview of the machining lines used for block and head production.

7.1 Cylinder Block and Head Materials

For many years the vast majority of automobile, truck, and agricultural and construction engines used cylinder heads and blocks that were sand cast from gray iron. While sand cast gray iron components remain important, automobile applications are seeing increased use of aluminum, and a variety of casting processes. Most new automobile engines use aluminum cylinder heads, while at this writing new engine block designs are closely split between aluminum and cast iron. This section begins with a look at gray iron and related alloys. Aluminum alloys are then discussed.

7.1.1 Gray Cast Iron

A variety of irons and steels including gray and ductile iron are alloyed from iron, carbon and silicon. Further alloying elements may be added to provide specific properties desired for a particular application. Some of these elements will be discussed later in this section.

© Springer Vienna 2016
K. Hoag, B. Dondlinger, *Vehicular Engine Design,* Powertrain,
DOI 10.1007/978-3-7091-1859-7_7

Both gray and ductile iron alloys have relatively high carbon and silicon content as compared to steel—between 2 and 4% carbon is typical, whereas 1% is considered "high carbon" in steel. Even with its much higher carbon content gray iron is relatively soft and easily machined. Its properties also differ considerably from those of ductile irons having very similar carbon content. These facts suggest that there is much more that distinguishes these materials than carbon and silicon content alone. The differences can be far better explained by examining the micro-structure of each.

In both gray and ductile iron the carbon precipitates out of the molten metal as graphite (as opposed to carbide in the case of steel). The gray iron alloys of interest for cylinder blocks and heads consist of mixtures of ferritic and pearlitic iron phases from which the carbon has precipitated out as graphite flakes as shown in the left photograph of Fig. 7.1. The silicon in the alloy creates precipitation sites controlling the size and distribution of the graphite flakes—increased silicon results in a finer distribution of smaller graphite flakes, generally resulting in increased strength. The resulting gray iron alloy is a low cost material that is relatively easy to cast and machine. The graphite flakes are resistant to shear between the iron crystals, and result in very high compressive strength. However these same graphite flakes provide crack initiation sites, reducing the tensile strength of the material. This will be an important consideration in designing engine components from gray iron.

By using a special ladle to add controlled amounts of magnesium to the melt as a casting is being poured the carbon can be made to precipitate out of the iron in the form of graphite spheres. This is shown in the photo at right in Fig. 7.1. The resulting alloy is ductile iron. The spherical graphite results in a considerably higher tensile strength than that of gray iron. However the material is more difficult to cast and machine and is considerably more expensive than gray iron. In engines ductile iron is often used for piston rings, exhaust manifolds, and main bearing caps. In every case the material is chosen for its increased tensile strength; for example, the firing forces result in high tensile loads in

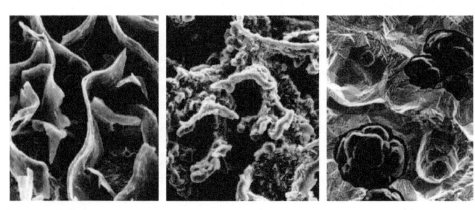

Fig. 7.1 Representations of *grey* iron (*left*), ductile iron (*right*), and compacted graphite (*center*) micro-structure

the main bearing caps, so while the block is cast from gray iron, the caps may be cast from ductile iron.

Returning briefly to gray iron casting and alloying, as the melt temperature drops after the casting is poured the carbon begins precipitating out when the **graphite eutectic temperature** is reached. The solidification process must then be completed before the temperature drops below a lower temperature termed the **carbide eutectic temperature.** If any carbon remains in the liquid phase when this lower temperature is reached it will solidify as carbide. An important aspect of gray iron castability is completing the solidification process within this temperature window. In complex castings thin sections may cool too quickly, rapidly dropping below the carbide eutectic temperature and thus forming carbide. Thick sections may solidify so slowly that carbides appear side by side with the graphite. Several alloying elements—copper, nickel, and cobalt—can be added to increase this window and improve castability. Copper is often especially favored, and its content must be carefully monitored as high copper content hurts fatigue strength.

Other alloying elements of interest include chromium and molybdenum. Chromium is sometimes added to gray iron to increase its strength. However it makes the material more difficult to cast as it reduces the temperature window discussed in the preceding paragraph, and it makes machining more difficult. Molybdenum is now quite often added to gray iron cylinder heads to improve high temperature fatigue life; it too reduces machinability.

Another iron alloy rapidly growing in usage for cylinder blocks is compacted graphite. While new to engine blocks compacted graphite was patented at the same time, by the same metallurgists as ductile iron. The year was 1948, and both materials were developed by introducing controlled quantities of magnesium to the melt. The middle photograph in Fig. 7.1 shows the intermediate graphite structure of compacted graphite. Where gray iron has a flake structure, and ductile iron a nodular structure, compacted graphite is described as a worm, or noodle structure. It has been kept from production applications until recently primarily due to casting process control. It should not be surprising that the structure is achieved through controlled magnesium introduction, but the problem lies in its close control. Too much and the nodular structure appears; too little and it reverts to the flake structure. Compacted graphite iron is specified as having no graphite flakes, and under 20 % nodularity.

The primary attractions for cylinder blocks are a higher tensile strength and greater resistance to wear of the cylinder wall surfaces. It should be noted that while compacted graphite has nearly twice the strength of gray iron, ANY presence of graphite flakes causes the tensile strength to plummet. This should not be surprising as one thinks of the role of graphite flakes in tensile failure as discussed relative to gray iron.

The next challenge is that of machining. The low tensile strength of gray iron makes it easy to machine. However, there is another important factor. All iron ores contain sulfur, and in the melt the sulfur is free and easily combines with oxygen and manganese. The iron ore contains manganese as well, and the resulting manganese sulfide coats the cutting tool. In compacted graphite two factors remove this mechanism. First, process control requires beginning with significantly lower sulfur ores. Second, the sulfur preferentially reacts with the magnesium introduced to the melt, thus taking away the role of manganese sulfide. The results are increased machining difficulties on several fronts. The approach that has initially been taken is to go to significantly higher feed rates to overcome machining resistance, and lower speeds to improve tool life. This combination allows volumes to be maintained, but requires very different fixture design, precluding running gray iron and compacted graphite parts on the same lines. The most recent approach has been to introduce tool heads with multiple cutting tools. By simultaneously cutting at several locations the feed rate can be reduced and the same overall metal removal rate can be achieved.

The increased strength, and the resulting ability to make much thinner, lower weight castings more than offsets the increased material, casting, and machining cost in many applications. Typical minimum wall section for cast iron is 4.5 mm thick, but as casting technology improves the material properties will allow 3 mm wall thicknesses, which open further possibilities for weight reduction. This weight reduction can be enabled because the minimum wall thickness is often limited by the casting process, while that actually required by the design, especially when using compacted graphite, might be significantly less.

7.1.2   Aluminum Alloys

The use of aluminum alloys has historically been driven by weight reduction, and it has been most often seen in high performance engines. Switching from cast iron to aluminum cylinder heads lowers the center of gravity of the engine in a vehicle, improving handling. More recently the combination of weight, thermal conductivity, and cost reduction has resulted in its use in many more automobile engines. An aluminum engine block will typically weigh between 40 and 55 % less than a comparable gray cast iron block. While the raw material cost of aluminum is higher than that of gray iron, and the energy required to produce aluminum from bauxite is high, this is offset by reduced final processing costs and a high degree of recycling. The lower casting temperature reduces the energy required to melt the material, and makes permanent mold and die casting processes possible, further reducing the cost of high volume parts. Finally the machining costs are reduced as tool life is increased relative to that for gray iron. The different thermal expansion rate of aluminum to cast iron for cylinder blocks is both a benefit to piston-to-cylinder fit, and a hindrance to maintaining crankshaft main bearing clearance. Aluminum alloys have three to four times greater thermal conductivity compared to gray iron, making it especially attractive for cylinder heads. The primary disadvantages of aluminum include its lower stiffness, high temperature creep relaxation, and poor wear characteristics relative to iron. Special design considerations must take these challenges into account.

Most aluminum alloys of interest for use in engines include copper (up to 5 %) and silicon (as much as 18 %). Slow cooling of aluminum and copper alloys results in the copper forming a separate "theta" phase along the "alpha" phase boundaries of the alloy, as

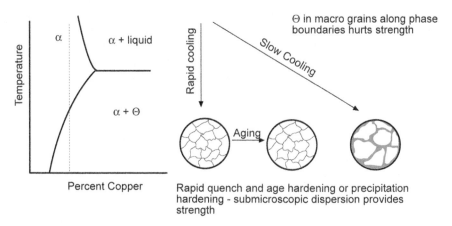

Fig. 7.2 Age hardening of aluminum-copper alloys

depicted in Fig. 7.2. This separate phase hurts casting strength, and can be minimized by rapid cooling or "quenching"– at least in the regions where strength is most critical. Under rapid cooling the copper forms a sub-microscopic dispersion. The alloy is subject to a further heat treatment (age hardening) in which the copper comes out of solution, but remains finely dispersed, thus minimizing the material's loss of strength. The silicon forms dendrites as it solidifies, and rapid cooling reduces the size and spacing of the dendrite arms; this too improves the alloy strength. The need for rapid cooling in critical regions, in order to achieve optimal strength, must be kept in mind as casting processes are discussed in the next section.

Because of the inherently poor wear characteristics of aluminum its use as a cylinder block material requires special cylinder wall considerations. In most cases this is addressed by casting iron alloy cylinder liners into the block. Aluminum 356 is the most typical alloy for such cylinder blocks. This alloy has no copper and only small amounts of silicon and magnesium, and is very easy to cast. Another approach increasingly seen for cylinder blocks is to use a high silicon alloy such as Aluminum 390. This alloy contains 17% silicon and 4.5% copper, in addition to small amounts of magnesium and manganese. The hyper-eutectic silicon content results in silicon particles that significantly increase hardness and provide an acceptable cylinder wall running surface. For scuff resistance the aluminum pistons are then coated with a thin layer of either iron or chromium. Another approach involves casting the cylinder walls from high silicon alloy aluminum and then surrounding them with a lower silicon alloy such as 356. This approach requires a casting process (typically die casting) that rapidly cools the molten block alloy before it can re-melt the cylinder walls.

An important requirement of the cylinder head casting is high temperature fatigue strength, and Aluminum 319 is commonly used. This alloy contains a small amount of silicon and about 4.5% copper. The casting process often includes the use of a water cooled "chill plate" to rapidly cool the casting at the firedeck surface, to disperse the alloying elements as shown in Fig. 7.2, reduce porosity, and increase density and material strength. Minimum wall thickness for cylinder head and engine block castings based on casting process capability is typically 2.5 mm; greater thicknesses are structurally required in many sections.

7.1.3   Magnesium Alloys

At this point in time production uses of magnesium have been limited to smaller engine components such as intake manifolds and covers, or have been limited to use in racing engine blocks where the higher cost of material and lower resistance to corrosion are less of a concern. It is beginning to generate interest as a cylinder block and head material for production engines. Its primary attraction is in providing the greatest potential for weight reduction while providing adequate strength. Estimates have suggested the resulting engine to be on the order of 25 % lighter than a comparable all aluminum engine, assuming the entire block and head could be made from magnesium. Design concerns include structural stiffness, cylinder wall characteristics, temperature limits, hot creep, corrosion resistance, casting control, and machining. Galvanic corrosion may occur when in contact with dissimilar materials, especially steel. One example engine has switched to aluminum fasteners to avoid this. Limited fatigue data, and an especially limited understanding of its high temperature capabilities add to the design challenge. Initial production use for cylinder blocks will be in conjunction with other materials to address the concerns just listed.

Material properties can vary widely based on alloy content and heat treat, some representative values are shown below in Table 7.1 for comparison purposes.

Table 7.1 Material property comparison

Material	Density (g/cm³)	Ultimate tensile strength (MPa)	Modulus of elasticity (GPa)	CTE, linear (µm/m-°C)	Thermal conductivity (W/m-K)	Relative material cost	Relative noise absorption
Gray cast iron	6.8–7.3	120–350	70–140	12–15	45–50	1.0	Best
Ductile cast iron	6.6–7.4	345–650	130–170	10–18	10–30	1.6	Good
Compacted Graphite Iron (CGI)	7.0-7.3	250–650	120–160	11–12	30–40	–	Good
Aluminum alloys	2.6–2.9	160–280	64–75	24–27	90–210	2.5–5.5	Poor
Magnesium alloys	1.8	90–230	40–50	26–30	160	3.6-6.0	Poor

7.2   Cylinder Block and Head Casting Processes

If it were possible to design the engine based solely on casting considerations each part would be designed such that reusable molds would define every feature. The exterior features of the part would be designed such that the molds could be pulled straight from the part in each direction. Even the number of directions from which the molds were to be pulled would be minimized. Clearly such requirements would compromise the exterior dimensions of parts such as cylinder blocks and heads, and they would further compromise or eliminate interior features. Before rejecting such requirements as totally unrealistic it should be noted that in small, high volume industrial engines such requirements are actually quite common, and do in fact constrain the design of the engine. However, in multi-cylinder automotive engines the compromises become unacceptable, and casting processes allowing the required design complexity must be found. The typical processes used for automotive cylinder blocks and heads are described in the paragraphs that follow.

7.2.1   Sand Casting

For many years this has been far and away the most commonly used production casting process for both aluminum and gray iron engine blocks and heads. The technology is very well developed, and the process allows intricate shapes, undercut geometry, and hollow cavities to be produced. The process involves creating a mold (negative of the part to be cast) made from a mixture of sand, bentonite clay, and water. The mold is packed around a steel pattern, and the pattern is then removed, creating the hollow cavity. Engine parts are typically created using a cope and drag mold, the features of which are shown in Fig. 7.3, where the cope and drag form the upper and lower portions respectively of the part to be cast. Interior cavities are created using cores molded from sand and an adhesive binder, or

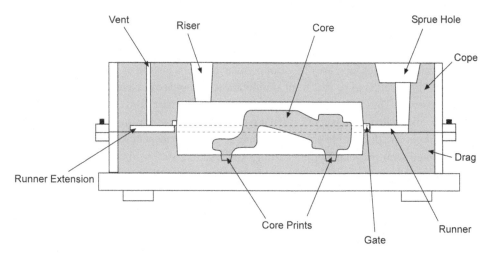

Fig. 7.3 Schematic of cope and drag mold for sand casting

are made from salt. The cores are held in place using locating tabs known as *core prints* as shown in Fig. 7.3. The core sand will later be removed from the casting at these print locations. The molten metal is poured into the cope and drag mold through the sprue hole. The metal flows through a series of runners around the perimeter of the mold, and is fed into the mold cavity through gates that have been placed and sized in such a way to optimize the fill rate and resulting properties of the casting. Runner extensions allow the initial portion of the melt to flow past the casting before the molten material flows through the gates into the casting. This is done because the melt from the top of the pouring ladle often contains impurities that would hurt the casting properties. Risers at various locations provide columns of material designed to compensate for shrinkage and resulting porosity as the casting cools. The sand itself must be porous to gas, allowing air from within the mold cavity, and gases from the core adhesives to escape from the mold.

Once the casting solidifies the sand is broken away from around the casting, and cleaned from the cores. The complex casting geometry results in various portions of the casting cooling and solidifying at different rates. This in turn creates residual stresses that are eliminated by annealing the casting. Because the sand surrounding the casting serves as an insulator the annealing process is often accomplished by allowing the casting to re-main surrounded by the sand for several hours after solidification. This is quite commonly done with cylinder block castings to reduce the casting process costs.

Because a new sand mold must be created for each part to be cast the sand casting process is relatively expensive, especially for high volume production. The resulting sur-face is rough as compared to that achieved with the other processes to be described, thus increasing machining requirements and further increasing costs. Further production chal-lenges include the gas and shrinkage porosity identified earlier. Core shift or core float—the movement of cores within the mold—sometimes causes problems as well. Finally, sand cleanout, especially from relatively small passages such as cylinder head cooling jackets may create production problems.

7.2.2   Permanent Mold Casting

As production volumes increase the use of permanent molds instead of sand molds becomes attractive. If the part is to be cast from aluminum, steel molds can be designed to pull apart from the casting. These molds can be used to create any surface from which the solid mold can be directly pulled, which restricts the use of undercut geometry. Interior cavities can be produced by using disposable cores in conjunction with permanent molds. This combina-tion is termed semi-permanent mold casting. The resulting surface finish is very good, and chill plates can be incorporated in the molds to improve strength in critical regions.

7.2.3   High Pressure Die Casting

Another form of permanent mold casting is die casting. Where the molten metal in permanent mold casting is gravity-fed, the metal is rapidly injected into the mold under high pressure in die casting. Wall thickness as little as 2.5 mm can be achieved. High Pressure Die Casting is attractive for further cost reduction in high volume production due to short cycle times, and is used for many aluminum engine parts. However, because of the rapid feed rate disposable cores are not typically used. During the high pressure die casting process a thin layer of metal quickly coats the mold surface and solidifies creating a dense skin. As the material beneath the surface solidifies, the resulting shrinkage creates porosity, and a distinct surface and sub-surface layer of markedly different material properties results. It is desirable to core oil and coolant passages and to avoid drilling into thick sections where porosity is present.

7.2.4   Lost Foam Casting

A relatively new process of rapidly increasing interest for cylinder blocks and heads is lost foam casting. This process can be used for both aluminum and gray iron casting although at this writing its production use for blocks and heads has been limited to aluminum.

Lost foam casting begins with an expandable polystyrene (EPS) model of the exact part to be cast. The EPS part includes all internal and external features of the final part. Its complexity can be increased by creating the EPS mold in several parts and then gluing them together. The EPS mold is then dipped in a refractory ceramic, covering the entire mold with a thin layer. The mold is then packed in loose sand. Finally, the molten metal is poured on the EPS mold causing it to vaporize, and replacing the mold with the cast metal part.

The resulting part has excellent dimensional stability, thus minimizing further machining. The part complexity that can be achieved often allows parts that would otherwise be made separately to be included as a single casting. This too reduces machining, as well as reducing assembly cost and eliminating gasketed joints. More intricate oil and coolant passages can be incorporated into the mold. A well-designed lost foam casting may result in reduced costs as compared to sand casting or permanent mold casting.

One disadvantage of lost foam casting is the inability to use cooling plates to improve the local properties of an aluminum casting. Another potential disadvantage results from the glue lines on the EPS mold and potential for core shift during gluing. These must be closely controlled to avoid detriments in the final part such as stress concentrations.

7.2.5 The Cosworth Casting Process

Originally developed for Formula One racing engines, the Cosworth process holds the attractions of very close dimensional control and the ability to cast very thin wall sections. The process is limited to non-ferrous alloys. The melt is supplied under controlled pressures to a zircon sand mold. The zircon mold is mixed with a binder and cured, resulting in the ability to control dimensions closely. The pressurized melt is supplied through gates to the base of the mold, and the controlled pressure and gate design results in a casting virtually free of porosity and inclusions. The process is now seeing some use for high volume production cylinder blocks.

Not common in the automotive industry, but occasionally used in aircraft and locomotives, are forged or welded engine blocks. For radial piston aircraft engines, the block halves are forged from steel for strength. For very large engines, blocks are sometimes fabricated by welding thick sheetmetal sections together. While the labor is time consuming and expensive, this typically lowers the weight of the engine block on the order of 10 % versus a cast block for the same application.

In concluding this section it must again be emphasized that many of the design features seen in cylinder blocks and heads are driven directly by the chosen casting process. Wall thicknesses and radii, and the placement of cooling jacket openings (core prints during casting, and pressed in "freeze plugs" to seal coolant passages on the final part) are examples. The shapes of passages and features from which permanent molds must be pulled are also examples. A comparison of casting methods is shown in Table 7.2.

Table 7.2 Comparison of casting methods

Process	Tooling	Labor	Piece cost	Typical economical quantities	Dimensional accuracy
Sand casting	Low	Med	Med	Small-Large	Low
Permanent mold casting	Med	Low	Med	Med-Large	Med
High pressure die casting	High	Low	Low	Very Large (> 10,000)	High
Lost-foam casting	Low	Med-High	Med	Small-Med	Med
Cosworth casting process	Med	Med	High	Small-Med	High

7.3   Cylinder Block and Head Casting Design Considerations

In general, it is important to minimize the weight of cylinder block and cylinder head castings, as these components are typically the heaviest in the engine assembly. It is good practice to avoid abrupt transitions from thick to thin sections and sharp corners in a casting, as these lead to difficulty flowing molten metal into the mold, and later lead to stress concentrations when the metal has cooled. For the cylinder block, it is important to avoid large flat panel sections, as these radiate noise. It is best to break these features up by curving the surface or adding ribs. *Casting draft* angle is a slight taper given to the mold surfaces perpendicular to the parting line to allow the easy removal of the casting, and is a key design consideration. The goal is to minimize machining on surfaces with a lot of casting draft, as this leads to excessive stock removal and could expose casting porosity.

During the design of the casting, it is common practice to employ casting simulation software. Casting simulation can predict flow patterns and velocities of metal filling the mold cavity, and can predict which areas of the casting will solidify first. This will often influence the final geometry of the part at an early design stage, and reduce the need for casting trials at later stages of development. Bearing bulkheads and cylinders in the block, the combustion chamber wall in the cylinder head, and the gates and runners in the die, are the main areas of focus (Fig. 7.5).

Figure 7.4 provides a simplified look at cylinder head casting for a water-cooled, dual overhead cam design seen in many automotive engines. The complexity of the needed port and cooling jacket geometry necessitates the use of non-permanent cores. The vast majority of engines in production today use sand cast cylinder heads. Semi-permanent molds—permanent dies for the outer dimensions, in combination with sand cores for the ports and cooling jackets—are used in a few applications. Lost foam is seen in a few

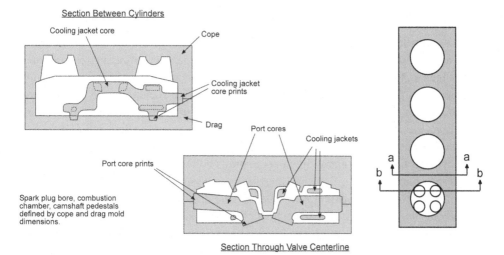

Fig. 7.4 Schematic depicting cylinder head casting considerations

Fig. 7.5 Cylinder head casting
cut-away

production applications, and might see increasing use. The figure shows a cope and drag mold, and depicts two sectional views—one through a port centerline and one between two cylinders. The as-cast dimensions of the outer perimeter, including the combustion chamber, spark plug bore, oil deck, and camshaft pedestals are determined by the cope and drag mold. The ports and cooling jackets are created with sand or salt cores. In the case of the ports the core print locations are quite straight-forward as the ports must be open to both the combustion chamber and the manifold flanges. The cooling jackets will be open to the cylinder block so in many engines core prints will be placed along the firedeck surface as shown in the figure. However, these will not be sufficient, either to hold the cores in position or for sand clean-out after casting. One additional core print will be located naturally by the coolant exit from the cylinder head to the thermostat housing. Further core prints are typically required at various locations along the sides of the head, an example of which is shown on the right side of the section between cylinders in Fig. 7.4. An example cylinder head casting cut-away The reader is encouraged to return to Figs. 1.6, 1.7 and 1.8

Fig. 7.6 A closed deck cylinder block

for examples of production cylinder heads; look particularly at the placement of the cooling jackets and ports. Cylinder head design will be covered in greater detail in Chap. 9.

A closed deck cylinder block is shown in Fig. 7.6. The geometry of the cylinder block
as shown suggests that either sand or semi-permanent mold casting will be required. If aluminum were selected as the block material, the crankcase, cylinders, and outside surfaces
could readily be designed such that permanent molds could be pulled directly away from
the casting. The cooling jackets of a closed deck engine require non-permanent mold cores
as further depicted in Fig. 7.7, increasing the cost of the casting. The primary benefit of the
closed deck engine design is to provide a structure to better support the top of the cylinder

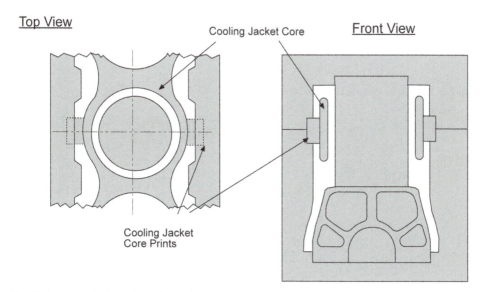

Fig. 7.7 Closed deck casting schematic

Fig. 7.8 Open deck cylinder
block

and head gasket. Due to combustion and inertial loads, the piston applies a radial thrust load on the cylinder, making the top of the cylinder deflect. This movement will fret the cylinder head gasket, which may lead to leaks if not limited by axial clamping load on the cylinder or radial support from the closed deck design. Additionally, a closed deck design will affect the cylindricity of the bore, due to radial growth from thermal loads. Thorough design is required to meet targets for cylindricity of the bore.

An open deck design is shown Fig. 7.8, and a casting schematic is shown in Fig. 7.9. Similar to the closed deck design, the outer surfaces are easily addressed with a permanent mold. The main difference is that the cylinder cooling passages are also created using permanent molds. No sand cores are required, which reduces tooling and piece costs. Production examples of each are shown in Figs. 1.6, 1.7 and 1.8. In addition to reducing costs relative to the closed deck design, the open deck allows improved cooling at the top part of the cylinder and the interface with the cylinder head. However, careful attention must be

Fig. 7.9 Open deck casting core schematic

Front View - Permanent Mold

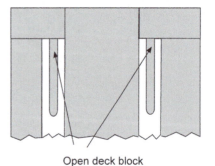

Open deck block

paid to the design of the head gasket to handle cylinder deflection due to piston thrust, and a multi layer steel (MLS) head gasket is typically used. Depending on the design of the cylinder block, it may decouple the deformation of the cylinder due to cylinder head bolt loads and thermal loads. In summary, an open deck design is less expensive and provides better cooling, but at the sacrifice of block stiffness. Cylinder block design will be covered further in Chap. 8.

7.4   Cylinder Block and Head Machining Processes

While advances continue to be made in flexible and computer controlled machining, engine parts are produced on fixed tooling for all but the lowest production volumes. It is important to involve manufacturing support as early as possible in the engine design process to balance the design requirement to the manufacturing capability, to enable the least expensive tooling and fixturing. Once the design is released for production, the majority of the cost is designed in and the cost of an operation rises exponentially with tighter tolerances. The type of fixturing and tooling will significantly influence the design, in order to achieve minimum cost. Frequently, a new engine design will be constrained to use existing tooling. While this drives compromise into the engine design, it reduces the overall cost to manufacture a new engine, which is important for the competitiveness of the business. The manufacturing lines for both cylinder blocks and heads are made up of a series of machines linked together in transfer lines. Each machine is constructed to repeatedly perform fixed operations specific to the particular part being produced. An example cylinder block transfer line is depicted in Fig. 7.10. The raw castings enter the "rough" end of the machining line, and the first machining operations establish the dimensional framework from which all further operations will be done. Locating "pads", or datums, on the casting are used to establish the reference position in 'x,' 'y,' and 'z' coordinates, and locating holes are drilled. It is important for these datums to be stiff, locate the part in all six degrees of freedom, and be as wide apart as possible. Each of the further machines in the transfer line will fixture the part by using pins fitted into these holes.

The transfer line begins with rough machining. A variety of operations are done at each machine, based on two criteria. The first is that of cycle time—the time required to complete all of the operations at each machine should be similar in order to avoid stacking parts up between machines or slowing the entire line down because of longer machining times required at a particular machine station. The second criterion is to minimize the number of times the part must be moved and again fixtured. In order to maintain the closest dimensional control it is desirable to maintain the part in a given fixture for as many operations as possible, as each time the part is re-fixtured it adds to the tolerance accumulation. While a given surface is exposed, conducting all of the operations on that surface is desirable. Where possible, it is desirable to dimension a part in whole degrees, as machining at fractional angles adds cost to the type of machining center required.

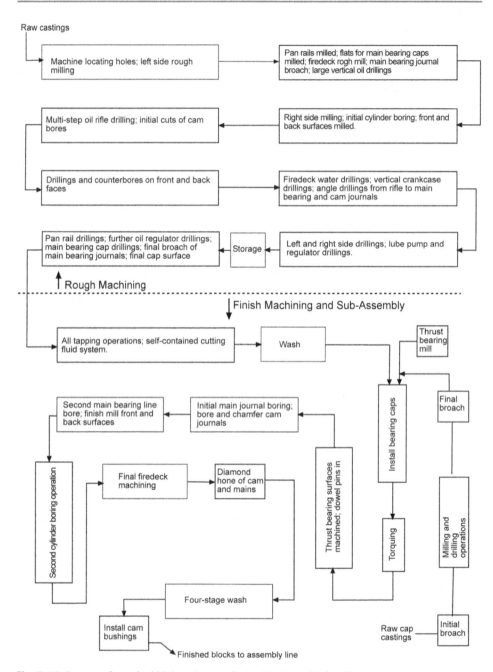

Fig. 7.10 Layout of a typical high-volume cylinder block machining line

Most machining lines begin with a series of rough operations, followed by an initial wash and then finish operations. Several of the finish operations are especially critical to the later performance of the engine. The effect of critical cylinder head dimensions include the concentricity of valve seats-to-guides on valve sealing; surface finish of firedeck machining on head gasket sealing; and the machined firedeck distance to the as-cast combustion chamber on the compression ratio from engine-to-engine. The effect of critical block dimensions include main bearing journal-to-journal alignment on bearing life; main bearing journals to cylinder bore alignment effect on cylinder and bearing wear; and cylinder bore surface finish effect on piston ring wear and sealing.

It is especially important for the engine designer to understand the limitations imposed on the design by the need to produce the engine on a transfer line using fixed tooling. While the fixed tooling is imperative for controlling costs and quality in high volume production it makes later design changes difficult and expensive. External dimensions such as the width of the cylinder block at the pan rails, or the deck height might be impossible to later increase as the revised part may not "fit" the machining line. As another example, moving a single drilling even a slight amount will be very expensive. The drilling is not made using an individual drill head but with a gear box holding many drill heads that simultaneously drill each of the holes on the particular surface of the part. Moving one drilling will involve modifying or replacing the entire gear box.

Certainly of interest to the engine designer is the ability of each machine to meet the tolerance specifications required for optimal engine performance and durability. There are key tradeoffs between what tolerance can be maintained in production, and the cost of the engine. Interchangeability of parts is key to mass production, and the design must be robust at the limits of the tolerance. A random sampling of the machined parts is used to determine the process standard deviation, and from that the process capability is calculated as the number of standard deviations that fit within the tolerance band of upper specification limit (USL) to lower specification limit (LSL), as shown in Fig. 7.11. The higher the number of standard deviations that can be fit within the tolerance band, the lower chance of a defect reaching a customer as illustrated in Table 7.3. Process capability is described by C_p (width of distribution with relation to the tolerance band), and C_{pk} (width, plus centering of the mean with relation to the tolerance band).

Fig. 7.11 Number of standard deviations between tolerance band

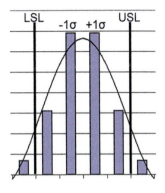

Table 7.3 Comparison of process capability (normally distributed)

Sigma level	Process capability index (Cp)	% of accurate parts produced by process	% of inaccurate parts produced by process	Defects per million parts produced
±1σ	0.33	31.7	68.3	690,000
±3σ	1.00	93.3	6.7	66,800
±6σ	2.00	99.99966	0.00034	3.4

7.5 Recommendations for Further Reading

The following reference provides an overview of casting technology, including design guidelines. It is not specific to engines, and does not cover processes such as lost foam or the Cosworth process, but provides coverage of sand casting, permanent mold and die casting (see Gervin 1995).

The papers listed below address the design of cylinder blocks for aluminum die casting (see Takami et al. 2000; Kurita et al. 2004; Yamazaki et al. 2004).

The following paper describes a composite block design using magnesium and aluminum for an in-line six cylinder automobile engine. It is followed by a study in which a production aluminum block is modified for magnesium casting, and a multi-company summary of magnesium block casting development (see Baril et al. 2004; Powell et al. 2004; Pedersen et al. 2006).

The following paper describes the use of compacted graphite for a bedplate. The second paper below presents a recent update on the use of compacted graphite for engines (see Warrick et al. 1999; Dawson 2011).

References

Baril, E., Labelle, P., Fischersworring-Bunk, A.: AJ (Mg-Al-Sr) Alloy System Used for New Engine Block. SAE 2004-01-0659 (2004)

Dawson, S.: Compacted Graphite Iron—A Material Solution for Modern Engine Design. SAE 2011-01-1083 (2011)

Gervin, S.J.: Cast Metals Technology. American Foundry Society, Des Plaines (1995)

Kurita, H., Yamagata, H., Arai, H., Nakamura, T.: Hypereutectic Al-20 %Si Alloy Engine Block Using High-Pressure Die-Casting. SAE 2004-01-1028 (2004)

Pedersen, A.S., Fischersworring-Bunk, A., Kunst, M., Bertilsson, I., de Lima, I., Smith, M.: Light Weight Engine Construction through Extended and Sustainable Use of Mg-Alloys. SAE 2006-01-0068 (2006)

Powell, B.R., Ouimet, L.J., Allison, J.E., Hines, J.A., Beals, R.S., Kopka, L., Ried, P.P.: Progress Toward a Magnesium-Intensive Engine: The USAMP Magnesium Powertrain Cast Components Project. SAE 2004-01-0654 (2004)

Takami, T., Fujine, M., Kato, S., Nagai, H., Tsujino, A., Masuda, Y.-h., Yamamoto, M.: MMC All Aluminum Cylinder Block for High Power SI Engines. SAE 2000-01-1231 (2000)

Warrick, R.J., Ellis, G.G., Grupke, C.C., Khamseh, A.R., McLachlan, T.H., Gerkits, C.: Development and Application of Enhanced Compacted Graphite Iron for the Bedplate of the New Chrysler 4.7 Liter V-8 Engine. SAE 1999-01-0325 (1999)

Yamazaki, M., Takai, A., Murakami, O., Kawabata, M.: Development of a High-Strength Aluminum Cylinder Block for Diesel Engine Employing a New Production Process. SAE 2004-01-1447 (2004)

Cylinder Block Layout and Design Decisions

8

8.1 Initial Block Layout, Function, and Terminology

In Chap. 5 the required engine displacement was calculated, and in Chap. 6 the number of cylinders, cylinder layout, and bore-to-stroke ratio were determined. These earlier decisions form the starting point for the discussion of this chapter. It is in this chapter more than any other that it will be necessary to limit the discussion to designs specific to automotive applications. Over the entire range of reciprocating piston internal combustion engines there is an extremely wide range of cylinder configurations and block layout and construction techniques. By limiting the discussion to automobile engines and heavy-duty engines in mobile installations, primary attention will be placed on in-line four, five and six cylinder engines, and vee six, eight, ten, and twelve cylinder engines. Casting the net a bit wider allows discussion of horizontally opposed four, six, and eight cylinder engines, and mention of the recently revived 'W-8' and 'W-12'. The cylinder block is the foundation of the engine, and supports the piston, cranktrain, cylinder head, and sometimes the valvetrain. It also houses the lubrication and cooling systems. It provides mounting points for the charging system, starting system, power take off (PTO), and typically has mounts which support the entire powertrain. The engine may be rigidly mounted as a structural member of the chassis, such as in a racecar or motorcycle. The cylinder block supports a variety of static, dynamic, and thermal loads, and must provide stiffness and alignment for many components. Because of the complexity of geometry, and complexity of loading, hand calculations are rarely used. Simplified finite element analysis (FEA) of a single power cylinder is usually the starting point, prior to analysis of the entire assembly.

Example cut-away illustrations of in-line and vee engines are shown in Figs. 8.1 and 8.2 respectively. In both of these engine configurations as used in automotive applications, the blocks are most often single-piece castings consisting of crankcase and cylinder sections. Engines in other industries are sometimes cast or fabricated with separate cyl-

© Springer Vienna 2016
K. Hoag, B. Dondlinger, *Vehicular Engine Design*, Powertrain,
DOI 10.1007/978-3-7091-1859-7_8

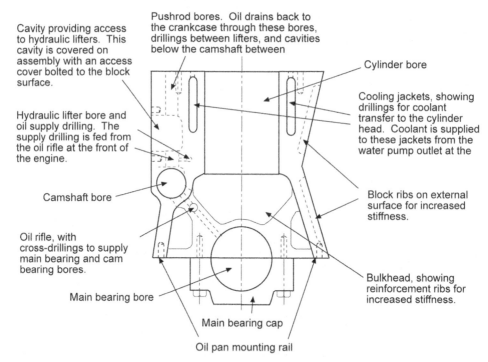

Cavity providing access to hydraulic lifters. This cavity is covered on assembly with an access cover bolted to the block surface.

Pushrod bores. Oil drains back to the crankcase through these bores, drillings between lifters, and cavities below the camshaft between

Cylinder bore

Cooling jackets, showing drillings for coolant transfer to the cylinder head. Coolant is supplied to these jackets from the water pump outlet at the

Hydraulic lifter bore and oil supply drilling. The supply drilling is fed from the oil rifle at the front of the engine.

Camshaft bore

Block ribs on external surface for increased stiffness.

Oil rifle, with cross-drillings to supply main bearing and cam bearing bores.

Bulkhead, showing reinforcement ribs for increased stiffness.

Main bearing bore

Main bearing cap

Oil pan mounting rail

Fig. 8.1 Front-view of in-line automotive cylinder block, identifying typical features

inder block and crankcase sections. The crankcase section consists of a series of parallel bulkheads containing the main bearing journals in which the crankshaft spins. The front and rear bulkheads form the front and rear outside mating surfaces of the cylinder block. The bulkheads are tied together by outside walls generally referred to as the block skirts. The oil pan is affixed to the pan rails along the base of the skirts. Much of the lubrication system, to be discussed further in Chap. 12, is incorporated in the crankcase. One or more banks of cylinders are cast above the crankcase portion. The cylinder walls may be cast directly into the block, or inserted liners may be used. The upper surface of the cylinder section is termed the firedeck, and serves as the mating surface for the cylinder head as shown in Fig. 8.3. The cylinder section includes cooling jackets and plays an integral role in the cooling system. This system will be covered further in Chap. 13. The cylinder block may also contain the camshaft as shown in Fig. 8.1. The engine in Fig. 8.2 has overhead cams, mounted in the cylinder heads. The remaining sections of this chapter will take up each portion of block layout in greater detail. Durability development will be addressed in Chap. 10.

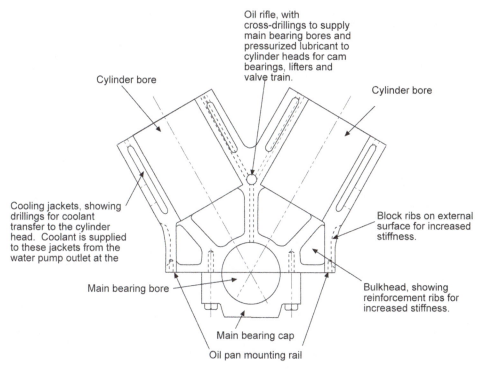

Oil rifle, with cross-drillings to supply main bearing bores and pressurized lubricant to cylinder heads for cam bearings, lifters and valve train.

Cylinder bore

Cylinder bore

Cooling jackets, showing drillings for coolant transfer to the cylinder head. Coolant is supplied to these jackets from the water pump outlet at the

Block ribs on external surface for increased stiffness.

Main bearing bore

Bulkhead, showing reinforcement ribs for increased stiffness.

Main bearing cap

Oil pan mounting rail

Fig. 8.2 Front-view of automotive cylinder block for vee engine, identifying typical features

Fig. 8.3 Example V6 cylinder block showing exposed firedeck surface where cylinder head will be affixed

8.2   Main Block Features

The crankcase portion of the block is further depicted in a simplified three-dimensional representation in Fig. 8.4. As was described in the previous section, the crankcase consists of a series of parallel bulkheads tied together with the block skirts which make up the outside surfaces of the block. There is one bulkhead for each crankshaft main bearing saddle. In most in-line engines there is a main bearing journal between each cylinder—five main bearings in each of the four-cylinder engines shown in Figs. 1.6a, b and 1.8b; six in a five-cylinder engine; and seven in a six-cylinder engine. In some cases the engine is made smaller and lighter by placing a main bearing only between every second cylinder, but this practice is not typically used as it decreases block and crankshaft stiffness and increases noise emissions. In vee engines there is a main bearing between each pair of cylinders across the vee—four main bearings in a V-6; five in a V-8 (as shown in Figs. 1.6c and 1.8a).

In a multi-cylinder engine it is necessary to split each main bearing journal radially in order to install the crankshaft. The caps are separately cast and bolted to the block to complete the bearing journal. Because the primary cap loading is tensile it is quite common for the caps to be made of a different alloy than the block to which they mate. For example, ductile iron caps are frequently used with gray iron blocks, especially in engines seeing high load factors and for which durability expectations are high. A recent exception to separately casting the main bearing caps is seen when the block is cast from compacted graphite. The much higher tensile strength allows the caps to be made from the same material as the block. This in turn allows the caps to be cast integral with the block, and then "cracked," using the same process that has become the norm with connecting rods

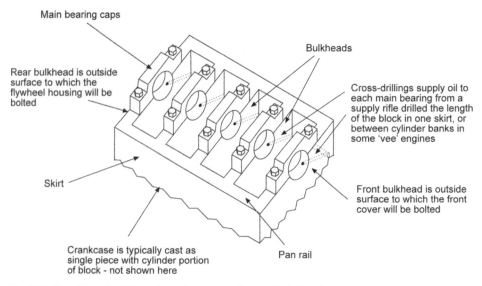

Fig. 8.4 Simplified depiction of crankcase, showing main design features

and their caps. The process is further discussed in Chap. 16 where connecting rods are discussed.

At the base of the skirts are the pan rails—pads of material running along the length of the skirts which are machined to mate with the oil pan, and which contain a series of tapped and drilled holes to secure the pan. Near the front or rear of the block an additional pad may be used immediately inboard from one of the pan rails to which the oil pump will be mounted. Alternately, the oil pump may be mounted co-axial with the crankshaft using a gerotor style pump, and with the housing integrated within the block or front cover. The oil pump, which will be discussed further in Chap. 12, is a positive displacement pump used in conjunction with a pressure relief valve and bypass passage back to the oil pan. In some cases the pressure relief valve is included in the lube pump housing, while in other cases it is incorporated into the block casting. An example of the latter approach is detailed in Fig. 8.5.

A supply drilling from the oil pump will intersect a horizontal oil rifle or gallery drilling running the length of the block to supply pressurized oil throughout the engine. The gallery drilling is cut through the entire length of the engine. Because of the length of the drilling, it is often made in progressively smaller steps from both the front and back of the block. The additional cost of the multi-step process is minimized by placing the successive operations at consecutive stations in the block machining line. Threaded plugs seal the rifle drilling at both the front and rear bulkheads. Cross-drillings through each bulkhead are then made to distribute pressurized oil to each main bearing. Further cross-drillings must be made at various locations in the crankcase to supply pressurized oil to the camshaft and valve train. If piston cooling nozzles are to be used in the engine the same gallery is often used. Other engines use a second gallery drilling devoted to these nozzles, also fed from the lube pump. Locations for the piston cooling nozzles must be designed into the block

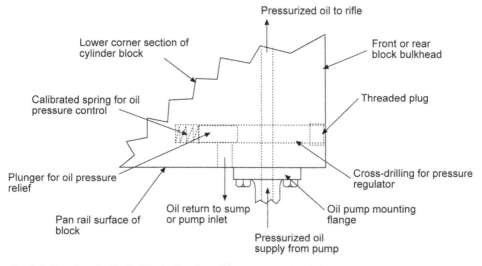

Fig. 8.5 Details of cylinder block showing oil pump mounting and pressure regulation

Fig. 8.6 Piston cooling nozzles

at each cylinder. These generally consist of machined holes cross-drilled from the gallery into which the nozzles are pressed, and flat surfaces on which the nozzles seat, and to which they are bolted as shown in Fig. 8.6. If the nozzles are designed to provide directed sprays, the location is critical and the machined pads may include dowel pins by which the nozzles are located. Piston cooling nozzles can also be seen in Figs. 1.6a, b, 1.8a and b.

The relatively open crankcase sections allow the casting cores to be held from the outside. In sand cast designs each cylinder in an in-line engine (or pair of cylinders in a vee engine) uses a single core around which the cylinder and crankcase bulkheads are cast. These open sections also lend themselves well to die casting, and this process has become increasingly popular with high-volume aluminum blocks.

As was stated in the opening section of this chapter, the front and rear bulkheads also serve as the front and rear surfaces of the block. In most automotive engines, the cam drive system is located immediately outside the front surface. This is the case for each of the engines of Chap. 1; the reader is referred especially to Fig. 1.7 where the cam drive system and its mounting to the front bulkhead can be clearly seen. The side-views of the engines in Figs. 1.6 and 1.8 also provide a look at front and rear bulkhead design. The majority of the front surface is milled flat, and the cam bearing plate, any auxiliary gears, and belt or chain tensioners are mounted to this surface as shown in Fig. 8.7. A stamped or cast cover then encloses this drive system, and is bolted to the block face. The cover must incorporate a gasket or seal against oil leakage, and in many engines it will also include sealed coolant transfer passages between the water pump and the cylinder section of the block. The crankshaft nose protrudes through the cover, and it is convenient to drive the water pump from the crankshaft nose, and therefore to mount it in front of the cam drive cover. The rear bulkhead is also milled, in this case mating with the flywheel housing or transmission casing as shown in Fig. 8.8. A semi-circular arc of bolts surrounds the upper half of the flywheel. Because of the criticality of shaft alignment between the engine and transmission this bolt pattern is usually supplemented by a pair of dowel pins on this mating surface,

Fig. 8.7 Cylinder block front bulkhead and timing cover mounting face

Fig. 8.8 Cylinder block rear bulkhead and bell housing mounting face

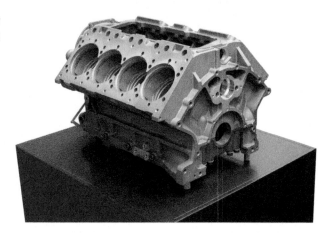

one on either side of the crankshaft. The dowel pins may not be necessary in transverse installations, where a chain drive replaces the direct alignment with the transmission shaft.

The brief discussion of this section introduces the main design features of the cylinder block. A discussion of block ribbing, primarily along the skirts and in some cases on the bulkheads will be deferred to Chap. 10, where block durability and noise are considered. The discussion now proceeds to critical crankcase layout dimensions. Various options for approaching further specific features of the design will then be covered.

8.3   Main Block Design Dimensions

A front-view layout showing critical crankcase dimensions for a vee engine is shown in the sketch in Fig. 8.9 and the example photograph of Fig. 8.10. While of necessity this book must cover engine design in a linear fashion it is important to recognize that much of the design work actually occurs along parallel paths. Evidence of these parallel paths can

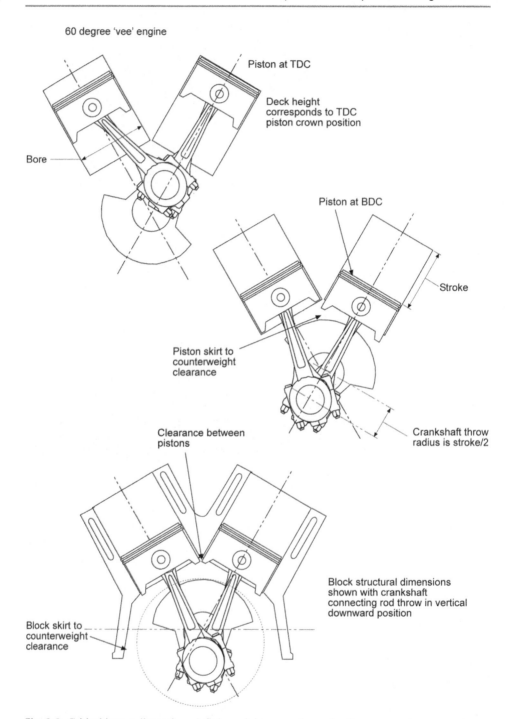

Fig. 8.9 Critical layout dimensions defining minimum package size for a vee engine

Fig. 8.10 Connecting rod
and counterweight to cylinder
block clearance

now be identified in Fig. 8.9. The main and rod bearing diameters will be determined as part of the crankshaft design and bearing sizing processes. Crankshaft counterweight diameter is an element of the engine balance and crankshaft design processes. The width of the crankcase at the pan rails will be further impacted by design features of the large end of the connecting rod. As the design layout moves from the crankcase to the cylinder section of the block critical dimensions will be determined by decisions regarding connecting rod length and piston design. The design begins by establishing deck height, then vee angle is considered, and finally cylinder spacing. This design loop is iterated as necessary.

8.3.1   Deck Height

A critical block layout dimension is the ***deck height***. The dimensional stack-up that defines deck height is illustrated in Fig. 8.11. The deck height is the distance from the crankshaft main bearing centerline to the firedeck surface, where the cylinder head mates to the block.

Combustion chamber design will be discussed further in the next chapter where cylinder head design is presented. It will become apparent in that discussion that for both diesel and spark-ignition engines the piston rim very closely approaches the firedeck surface as it reaches TDC as shown in Fig. 8.12. This being the case, the static deck height is determined by the sum of one-half the stroke, plus the connecting rod length from rod bearing centerline to piston pin centerline, plus the piston height from its pin centerline to rim (compression height) as shown in Eq. 8.1.

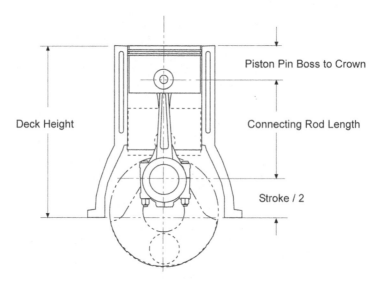

Fig. 8.11 Layout variables determining the deck height

$$\mathrm{MinimumDeckHeight} = \frac{\mathrm{Stroke}}{2} + \mathrm{Connectingrodlength} + \mathrm{Pistoncrownheight} \quad (8.1)$$

A compact design requires minimizing the deck height, which means reducing the stroke, connecting rod length, or piston crown height. Reducing the stroke to minimize deck height requires increasing the bore, which will affect the bore-to-stroke ratio, and also possibly the vee angle. The minimum piston height will be based on the need to fit the ring pack above the piston pin, the need to control ring temperature, and the need to ensure structural integrity of the piston. In the diesel engine the need to incorporate the combustion chamber in the piston bowl will further impact the minimum height. These considerations will be further detailed in Chap. 15 on piston and ring pack development. The final variable determining the static deck height is connecting rod length. Reducing the connecting rod length reduces both rotating and reciprocating forces, but increases side forces (and thus friction and wear) and the second order component of the reciprocating forces. As the connecting rod length is reduced, a further consideration is counterweight-to-piston skirt clearance. This is shown with the BDC view in Fig. 8.11. As the piston approaches BDC the counterweights located at some of the crankcase throws swing an arc that approaches the piston. It should be noted that the relationship between counterweight and piston locations is dependent on both the engine configuration (number and layout of cylinders) and on counterweight design. The reader is referred back to Chap. 6 for a discussion of counterweight design and position. The minimum connecting rod length for counterweight-to-piston skirt clearance is then determined by the necessary skirt length. In most automobile engines today the skirts are cut back to allow shorter connecting rods and the resulting deck height reduction. In heavy-duty engines such cutouts are not used

Fig. 8.12 Piston at TDC

in order to maximize skirt durability. Looking at the production engines in Chap. 1, those in Fig. 1.6a and c have skirts cut-outs and shorter connecting rods. The spark-ignition engine in Fig. 1.6b, and all of the diesel engines in Fig. 1.8 use full piston skirts and longer connecting rods.

The equation presented above is for all components in their static state. It is important to realize that while the engine is running, there are dynamic forces that will affect the piston-to-head clearance, and must be accounted for. There will be dynamic deflection of the crankshaft, connecting rod, and piston due to inertia forces. Also, the piston will rock in its bore around the piston pin due to secondary piston motion, and thus may cause the leading or trailing edge of the piston to be higher than the cylinder deck.

A related issue associated with deck height is the potential need to increase the stroke of the production engine at a later date. Referring again to Fig. 8.11, if the stroke is increased and the deck height is to remain unchanged, the connecting rod must be shortened or the piston pin must be raised in the piston. Since deck height is a key block layout dimension it must remain fixed in all but the lowest volume engines or the retooling costs will be huge. The least expensive approach is typically to raise the piston pin height, but this may not be possible. In summary, if there is any possibility that the engine's stroke will later be increased, design margin should be included in either the piston or the selection of connecting rod length.

8.3.2   Vee Angle

The decision pertaining to connecting rod length is further complicated in the vee engine configuration, based on the combination of desired angle between the banks and the chosen bore-to-stroke ratio. If both connecting rods in the vee share the same crankpin, near BDC the adjacent pistons come close to touching as shown in Fig. 8.9 previously. To address this, the vee angle may be increased or the deck height may be increased. Increasing the deck height moves the working portion of the cylinder bore away from the crank centerline, which increases connecting rod length for a given stroke. Finally, the bore-to-stroke ratio for a given displacement can be changed.

Since the engine block accounts for up to a third of engine weight, it is desirable to minimize the weight of this component. A more compact cylinder block with a short deck height, enables a lighter weight engine. Several combinations are shown in Fig. 8.13 and the critical clearances are identified. Larger bore, shorter stroke, wider vee angle designs generally package better in order to maintain acceptable deck height dimensions in vee engines. This can be seen in the figure, where each of the engines shown has the same displacement per cylinder. A larger vee angle drives a shorter deck height, for a constant bore and stroke, as shown in Fig. 8.14.

8.3.3   Cylinder Bore Spacing

In order to realize the most compact longitudinal cylinder block design, whether I-4 or V-8, it is important to minimize the cylinder bore spacing. In automobile engines the key limiting dimension is head gasket land width between cylinder bores. This can be influenced by cylinder liner choice, and will be discussed further in Sect. 8.5, Cylinder Design Decisions.

Another key decision is whether the cylinders will be separate or *siamesed*, which means they are cast as a unit and there are no cooling jacket between cylinders as shown in Fig. 8.15. Separate cylinders will allow a coolant passage between cylinders, which improves cooling uniformity, but increases block length. Siamesed cylinders will increase block strength and reduce block length, at the sacrifice of cooling between cylinders which may increase thermal bore distortion.

The combination of main and rod bearing width and crankshaft web thickness are also cylinder spacing considerations, and their roles in defining the engine layout are seen in

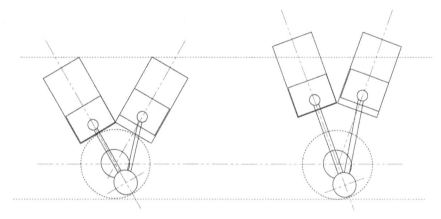

Effect of changing vee-angle, with same bore and stroke

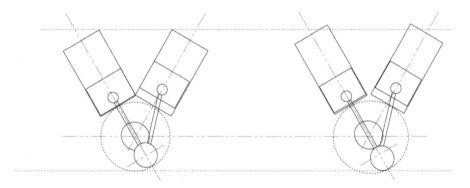

Effect of changing bore and stroke, with same vee-angle and same displacement

Fig. 8.13 The effect of vee angle and bore-to-stroke ratio on the deck height of a vee engine

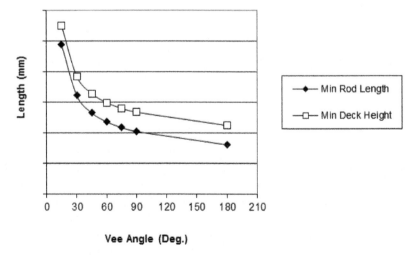

Fig. 8.14 Deck height as a function of vee angle for a constant bore/stroke

Fig. 8.15 Example of siamesed cylinders

Figs. 8.16 and 8.17. The bottom view of the cylinder block is shown for an in-line engine
in Fig. 8.18. Note that two connecting rod bearings share a single rod throw on the vee
engine crankshaft. One bank is therefore placed one rod bearing width behind the other
from front to back. This can also be seen in the production V-8 engines shown in Figs. 1.6c
and 1.8a. This benefits the deck height, as the vee can be narrower for a given connecting
rod length since the bores are not directly in line. An additional manufacturing consideration is the need for the cylinder hone to run out below the cylinder bottom for complete
machining, and clearance must be left from the tool to the crankshaft bearing bulkhead.

Another consideration in this layout of the engine is the need to include a thrust bearing
surface at one of the crankshaft main bearings. While the majority of the loads seen by
the crankshaft are perpendicular to the main bearing axis, some loading is seen along this

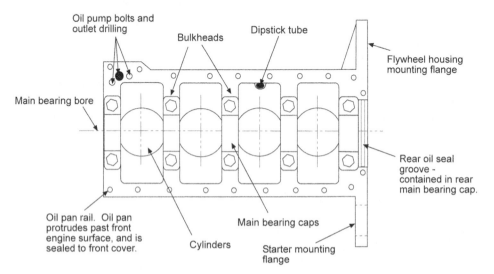

Fig. 8.16 Bottom view of the crankcase of an in-line engine

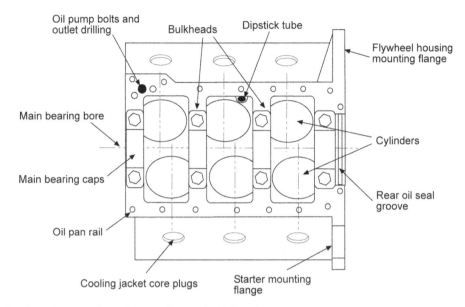

Fig. 8.17 Bottom view of the crankcase of a V-6 engine

Fig. 8.18 Bottom view of an inline engine

axis as well. The greatest load along the axis is seen when the clutch in a manual transmission application is disengaged. Other loads along the axis are contributed by the torque converter in automatic transmission applications, helical gears driven by the crankshaft (manual transmission gears, or gear-driven camshaft and accessory drives), bending of the crankshaft throw, block deflection, and dimensional misalignments. The thrust bearing is most often placed at the second main bearing from the rear of the engine. Loads along the crankshaft axis are highest at the rear of the engine, but the rear bulkhead must also accommodate the crankshaft seal. In engines having especially stiff crankshafts the thrust bearing may be placed at the center bulkhead. The thrust surface provides the fore and aft datum for crankshaft machining, and centrally locating this surface allows the crankshaft machining tolerance to be split evenly between the front and back cylinders, allowing closer dimensional control. The thrust bearing requires a wider main bearing, and if the journal bearing area is not to be compromised at this bearing engine length must be increased slightly. Finally, the front and rear of the crankshaft must accommodate oil seals. While looking at the bottom view of the engine layout it is important to consider oil pump and pick-up location as well. On any given engine the oil pump will be placed in a location convenient to be driven from the crankshaft. Although the pump location remains constant a given engine may be designed with the option of either a front or rear oil sump for different vehicle applications. If this is the case, a symmetric pan rail bolt pattern may allow a single oil pan to be used, with different pick-up tubes for the different sump locations.

8.3.4   Other Block Dimensions

Other critical clearances when laying out cylinder block are:

- Connecting rod cap and cap bolts to pan rails clearance throughout the stroke,
- Connecting rod to cylinder bore clearance throughout the stroke,
- Crankshaft counterweight to the cylinder bottom,
- Crankshaft counterweight to the piston bottom at BDC,
- Crankshaft counterweight to the operating oil level in a wet sump,
- Piston to piston clearance in a vee engine at BDC, and
- Piston to cylinder head clearance at TDC

When laying out the engine, it is important to consider the fact that under loading components will deflect, and this must be accounted for in the design. Additionally, components have machining tolerances and casting tolerances that may adversely stack up.

The minimum crankcase width at the pan rails is determined by the combination of the engine's stroke and the necessary counterweight size and connecting rod motion envelop. Counterweight diameter is invariably reduced by splitting the counterweight mass across a given connecting rod throw, as well as by wrapping the counterweight around a greater arc. As this is done a greater total crankshaft mass is required to achieve the required balancing. This was previously discussed in Chap. 6. When determining pan rail location, the designer must also consider whether there could be a need for increased displacement at some later time. If the stroke is ever to be increased this must be considered, and design margin must be allowed in the pan rail width dimension to ensure that the larger stroke crankshaft (and correspondingly larger counterweights and connecting rod orbit) will still clear the pan rails. The pan rails are often used for fixturing during block machining operations. With in-line engines the pan rails are generally the widest portion of the block, and the block machining line will be designed to clear the pan rails, but with little further margin. Both of these facts mean that increasing this dimension later would result in extremely expensive tooling modifications.

8.4   Crankcase Bottom End

Among the early crankcase design variations that may be considered is that of block skirt length, as depicted in Fig. 8.19. The extended skirt design adds weight to the block and may increase the overall engine height. However, it also allows the main bearing caps to be tied more rigidly to the remainder of the block, and it simplifies oil pan design and sealing as the entire oil pan sealing surface can be made flat. With the short skirts machining complexity is minimized if the oil pan mating surface is at the main bearing centerline so that the bearing caps are mounted on the same machined plane as the oil pan. This requires the oil pan to include semi-circular openings or angled surfaces (and resulting complex sealing surfaces) at its front and rear in order to clear the crankshaft. The long skirt design is typical in heavy-duty engines as shown in Fig. 8.20, while most passenger car engines utilize the shorter skirt design.

Another important consideration is that of main bearing cap constraint. Most automobile engines use two-bolt caps, an example of which is shown in Fig. 8.21. Also shown in the figure are various four-bolt cap designs. The parallel bolt design is commonly seen on high performance automobile engines with short block skirts. The cross-bolt design significantly increases cap rigidity and block stiffness, at the sacrifice of load transmission to outer surfaces potentially increasing NVH emissions. It is especially common in extended skirt vee engines, where the vee configuration results in firing forces having significant horizontal components. The angled bolt design allows similar cap rigidity with the shorter block skirts. However, with this bolt geometry, machining and dimensional control of the cap are especially difficult. Occasionally on inline engines, the cylinder head bolts go all the way through the block and anchor the main bearing cap. While making assembly and servicing more difficult this approach places the majority of the block in compression, improving management of firing forces. In Fig. 8.22, a ladder frame or bedplate bearing cap is shown. In this case the entire cylinder block splits along the crankshaft main bearing centerline, and a single piece replaces the individual main bearing caps. This design

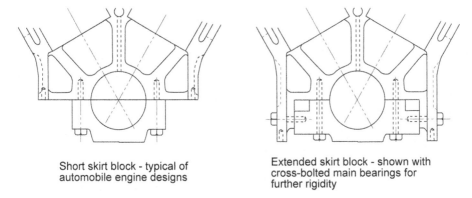

Short skirt block - typical of automobile engine designs

Extended skirt block - shown with cross-bolted main bearings for further rigidity

Fig. 8.19 Skirt design options with conventional main bearing caps

Fig. 8.20 Cross-bolted main bearing caps

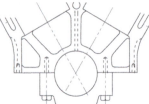

Two-Bolt Main Bearing Cap

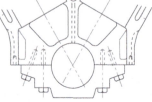

Four-bolt main bearing cap with angled bolts addressing horizontal loads in 'vee' engines

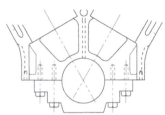

Four-bolt main bearing cap as is most commonly seen in high-performance engines

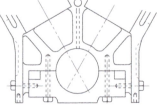

Four-bolt main bearing cap with extended skirts and cross-bolting for added rigidity in heavy-duty engines

Fig. 8.21 Main bearing cap design variations

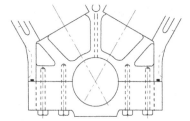

Ladder Frame
All main bearing caps and lower block skirt
are cast as a single piece, mating to the
block at the main bearing bore centerline.

Fig. 8.22 Ladder frame construction for the crankcase of an automotive engine

significantly increases cylinder block stiffness, and minimizes or eliminates cap separation. However it can be difficult to seal because it requires the use of sealant as the use of a gasket would impact bearing clearance. It is especially common in engines with aluminum blocks where design modifications to increase stiffness are required. Examples of this design are shown in Figs. 1.6b and 8.24.

In blocks made of aluminum, the main bearing cap is frequently made of cast iron for strength and rigidity. Gray iron limits bearing clearance growth. Additionally, an upper main bearing insert made of cast iron is occasionally cast into the main aluminum engine block to form a composite casting. No matter what material they are made from, main bearing caps and crankcase main bearing bores are always machined as a set, and identification is usually provided to make sure each cap mates to its original bore upon reassembly.

The horizontally opposed engine requires significantly different crankcase geometry, and the basic features of two designs are shown in Fig. 8.23. In both of these designs the cylinder sections are cast separately and bolted to the crankcase. This is not absolutely necessary, but is typical in order to reduce casting size and complexity. One challenge with this design is tightening the connecting rod fasteners with the block assembled. The design on the right shows the entire crankcase splitting at the main bearing centerline. In the design on the left main bearing caps similar to those used with in-line and vee engines are used.

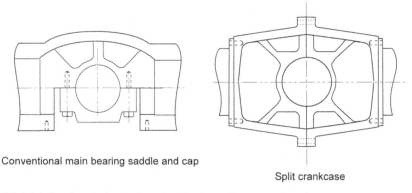

Conventional main bearing saddle and cap

Split crankcase

Fig. 8.23 Main bearing design options for a horizontally-opposed engine

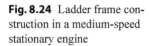

Fig. 8.24 Ladder frame construction in a medium-speed stationary engine

8.5   Cylinder Design Decisions

The discussion now returns to the cylinder section of the block. The goal of the cylinder liner is to provide a good wear surface for the piston rings, and to maintain cylindricity to aid piston ring sealing. If the liner deviates from round through machining, static head bolt loads, dynamic firing loads, or thermal stresses, the engine will experience higher oil consumption, higher blow-by, and higher engine friction. There are three general approaches to the cylinder liner: *integral liner, dry liner,* or *wet liner*. Depending on the cylinder architecture, the block may be *closed deck* or *open deck*. At the top of the cylinders the cooling jacket may be sealed with an upper block surface referred to as the *firedeck,* on closed deck engines. The cylinder head is then bolted to this surface. In open deck engines, this surface is eliminated—the cylinder head is sealed at the top surface of the cylinder walls and at the outer wall, and coolant makes direct contact with the cylinder head. Referring to the production engines introduced in Chap. 1, those shown in Figs. 1.6, 1.7, and 1.8b are all parent bore designs. The engines in Figs. 1.6a and 1.8b have open decks, while those of Figs. 1.6b, c, 1.7, 8.1, and 8.2 have closed decks.

8.5.1   Integral Cylinder Liner

For automobile engines the most common approach is the parent material bore or integral liner design, in which the cylinder walls are cast integrally with the cylinder block. This is most common in cast iron blocks and closed deck applications. The cylinders are surrounded by a cooling jacket, which in turn is surrounded by the outer wall of the block. This design minimizes the cylinder bore spacing, and shortens the total length of the block, as well as providing the shortest possible axial cylinder length. It has the further advantages of good heat transfer from the cylinder surface to the coolant, and reduced cylinder bore distortion compared to dissimilar materials used in other applications. An integral cylinder liner is the least expensive to manufacture, but offers the least flexibility for engine overhaul.

Alternately, a coating or thin film may be applied to the parent bore cylinder walls on an aluminum cylinder block to increase wear resistance. While running a parent bore liner may save as much as 0.5 kg per cylinder, the challenges include added cost and higher scrap rate due to adhesion problems. Specifically, coatings will not adhere to porosity exposed by machining of the parent bore.

8.5.2   Dry Cylinder Liner

A dry cylinder liner is one that is made from separate material than the block, and is inserted or cast into the crankcase. This technique can be used on both open and closed deck blocks. No contact with the coolant occurs, as this is completely encased in the parent block material. This reduces the opportunity for leaks, but reduces the heat transfer from the liner through the block to the coolant. A dry liner variation is the removable liner that cold be replaced during an engine rebuild. This approach was common in earlier truck and tractor engines but is seldom seen today. Controlling temperature is extremely difficult, leading to high piston ring temperatures and bore distortion.

If the block is cast from aluminum, cylinder liners made of a higher hardness material may form the running surface, with the block cast around these liners, as shown in Figs. 8.25 and 8.26. Material choices are typically iron, steel, or high silicon aluminum alloy, and it is critical to make sure the aluminum adheres to the liner so that proper thermal contact is maintained. This is achieved with outer surface features and by heating the liners to over 200 °C prior to casting the rest of the block around them.

Fig. 8.25 Cast in cylinder liner

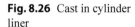

Fig. 8.26 Cast in cylinder liner

8.5.3   Wet Cylinder Liner

Most heavy-duty engines and some passenger car diesel engines in production today use wet liners, as shown by the engine in Figs. 1.8a, and those in Figs. 8.27 and 8.28. In such engines, the cylinder liners are pressed or slip fit into the surrounding structure such that they are in direct contact with the engine coolant, and sealed at the top and bottom. In order to provide sufficient structural rigidity, the block casting includes walls separating the cooling jackets for each cylinder. The wet liner engine holds the advantages of having the cylinders easily replaced when the engine is rebuilt, often with the engine in place in the vehicle, and direct contact of the cylinders with coolant in individual jackets for optimum temperature control. However the cylinder spacing and thus the overall length of the engine must be increased and the possibility of coolant leakage is increased. Care must be taken with dissimilar materials, and differential thermal growth will affect both the radial fit, and axial length of the cylinders. Due to manufacturing variation, the protrusion heights of the sleeves at the cylinder deck may be different and the head gasket must tolerate this. Alternatively, the top of the cylinders can be machined after installation, to guarantee the same protrusion height. Of course engine cost is also substantially higher than that of a parent bore design.

8.5.4   Cylinder Cooling Passages

The most significant thermal loading on the block is near the top of the cylinder, and between adjacent cylinders. Cooling jacket design will be covered in greater detail in Chap. 13, but a few observations should be noted here. In the case of the integral liner and dry liner engines the cooling jacket is an open passage surrounding all of the cylinders. Coolant supply and return locations are carefully optimized to achieve as even a flow distribution

Fig. 8.27 Wet liner and piston

around each cylinder as possible. The wet liner engine requires separate jackets surrounding each cylinder. These jackets must be fed from a separate header, increasing complexity but resulting in relatively even flow distribution. A simple design rule for gasoline engines is to size the coolant passages around the cylinder assuming that 10% of the fuel energy supplied the engine is rejected to the coolant at the cylinders as shown in Eq. 8.2.

$$Q_{coolant} = \frac{q_{heat}}{c_p \cdot \Delta t_{coolant} \cdot \rho_{coolant}} \tag{8.2}$$

Where:

$Q_{coolant}$	Volumetric flow rate of coolant (velocity * area)
q_{heat}	10% of fuel Lower Heating Value (LHV)
c_p	Specific heat of coolant
$\Delta t_{coolant}$	Desired temperature rise of coolant
$\rho_{coolant}$	Density of coolant

Fig. 8.28 Wet liner installed

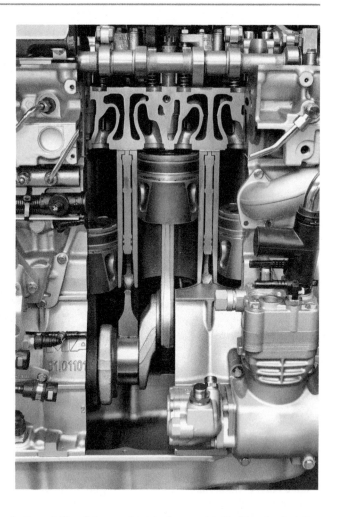

The cross sectional flow area is the width of the coolant jacket multiplied by the height. The height of the coolant jacket around the cylinder ranges from 30 to 60% of the piston stroke for an aluminum block and 50–100% of the stroke for a cast iron block due to the difference in thermal conductivity. Cylinder liner temperature is typically highest near the cylinder head, as this portion is exposed to combustion. The combustion pressure and temperature are highest when the piston is near its TDC position. As the piston approaches BDC, temperature and pressure have dropped; this portion of the cylinder is exposed for a shorter period of the operating cycle. The coolant velocity can be obtained from pump performance data. It is important to keep the coolant jacket small for engine packaging, and also to keep the velocity of the coolant high. It is also beneficial to minimize the total amount of coolant in the engine, as this reduces engine warm up time and improves catalyst light-off.

If the cooling jackets are to be sand cast, core prints must protrude through the outside surfaces of the block to locate the core. This will be necessary both to hold the cores in place and to remove the core sand after casting. A lost foam casting will have the same requirements. It is general practice to make the resulting openings in the outside surface of the block circular. They are then machined and stamped steel *freeze plugs* are pressed in during block assembly to seal the cooling jackets. The open deck design identified earlier in this section allows the cooling jackets to be cored from the top surface of the block, and core prints or freeze plugs are not required. In engines designed with wet liners the cooling jackets are cored as part of the cylinder openings, and separate cooling cores are not needed. In these engines a water header may be cored along an outside surface of the block to feed coolant to the individual jackets around each cylinder.

Simplified top- and side-view layouts of cylinder banks for parent bore and wet liner engines are provided in Fig. 8.29. The layout in Fig. 8.29a depicts the key design features that determine the required length of the cylinder bank for the parent bore engine. In the previous section main and rod bearing width and crankshaft web thickness were identified as important parameters defining the required cylinder spacing; this can clearly be seen in the figure. Another important parameter is cooling jacket design. The jacket shown in the figure must be of sufficient width that the casting dimensions can be maintained in production. If the block has been sand cast or cast with the lost foam process, sand clean out must also be readily achieved. The casting thickness at the cylinder wall is determined by

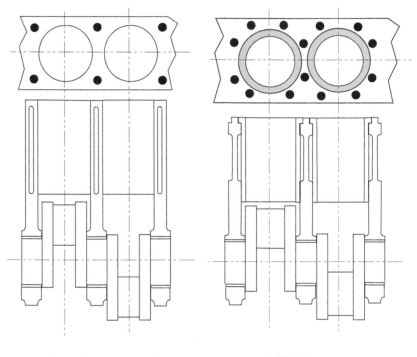

a. Parent Bore   b. Wet Liner

Fig. 8.29 Layout variables determining cylinder bore spacing in a parent bore and a wet liner engine

casting dimensional control, structural requirements, and by expectations that the cylinder may be re-bored during engine rebuild. In order to minimize cylinder spacing the cylinder walls may be cast siamese, assuming sufficient crankshaft bearing area can be maintained. The cooling jackets are placed only around the perimeter of the cylinder bank. In addition to reducing cylinder spacing the siamese casting increases structural stiffness, but the resulting temperature profiles must be carefully assessed. This may be improved by machining a slit or small cross drillings near the head gasket, to allow a minimal coolant flow between cylinders at the hot spot, as shown in Fig. 8.30. Finally, it should be noted that four head bolt bosses have surrounded each cylinder in the parent bore engine. This is typical practice in automobile engines (although a few examples can be found using five bolts per cylinder). The bolt bosses are placed at each corner, where they have little or no impact on either cylinder spacing or intake/exhaust port restriction. It is beneficial to move the threaded region for the head bolts deep in the block, below the firedeck, by using a long counterbore. This has the advantage of enabling longer head bolts, reducing the stress concentration of the threads, and reducing the amount that the thread distortion impacts cylinder roundness. It also moves the threads away from the firedeck, were thermal creep relaxation may be an issue.

Fig. 8.30 Slit machined between siamesed bores

The simplified top- and side-views are repeated for a wet liner engine in Fig. 8.29b. As with the parent bore engine the crankshaft webs and main and rod bearing widths are parameters that may limit the minimum cylinder spacing, but in this engine several additional factors must also be considered. It should be noted that even though these further factors lead to greater cylinder spacing the design may still be limited by the crankshaft parameters. This can be explained by the typically higher loading and greater durability expectations imposed on the wet liner engine. The outside diameter of the cylinder must include a step that positively seats the liner in the block. The step may be located either above or below the cooling jackets, but in either case adds to the spacing required between cylinders. As was previously described, the wet liners usually require an additional wall cast between each cylinder, with a cooling jacket on either side of the wall. The cooling jackets (or at least the portion between each cylinder) may have machined surfaces, allowing narrower passages and thus reducing the required cylinder spacing somewhat as compared to the cast jackets discussed in the case of the parent bore engine. The combination of bore diameter, peak cylinder pressure requirements, durability expectations (specifically head gasket life), and cylinder head stiffness determine the clamping requirements at the critical head gasket seal. With the larger bore diameters in combination with high load factors and long durability expectations it is typical to use between six and eight head bolts per cylinder. The head bolts must be relatively evenly spaced, so it can be expected that head bolt boss location may also impact cylinder spacing.

8.6 Camshaft Placement Decisions

While there is increasing interest in the "cam-less" engine, virtually all four-stroke engines in production today continue to use a camshaft and mechanical valve actuation. It thus continues to be important to consider where to place the camshaft in the engine layout. The trade-offs that go into determining camshaft placement will be detailed in Chap. 17, where the camshaft and valvetrain are discussed in detail. Briefly summarizing that discussion, high performance and high engine speed make overhead camshafts—placed above the valves, in the cylinder heads—attractive. In some heavy-duty diesel engines separate lobes on the camshaft are used to actuate the fuel injectors; in these engines high camshaft loading and high stiffness requirements also make overhead cams attractive. However, overhead cams result in complex drive systems, make the engine taller, and are typically more expensive than cam-in-block systems. The remaining discussion in this section will address the details of camshaft placement in a cylinder block, as is typical with overhead valve (OHV) engines.

The first consideration is that of where to place the camshaft within the block. If the camshaft is to be placed in the block it will be desirable to place it as close to the crankshaft as possible in order to simplify the drive system. The simplest drive system is a direct gear coupling, with a gear on the nose of the crankshaft meshing directly with one having twice as many teeth on the nose of the camshaft. As camshaft placement moves further from the crankshaft the gear diameters increase until their sizes become prohibitive and intermediate gears or a belt or chain-drive is substituted. With in-line engines the camshaft will be placed above the crankshaft along one side of the block. With vee engines it is most convenient to use a single camshaft placed in the vee between the two banks of cylinders. Whereas the camshaft placement in the in-line engine allows direct gear drive, placement in the vee necessitates greater distance between the camshaft and crankshaft, generally requiring one of the alternative drive systems. A few exceptions should be noted: First, as a means of gaining some of the increased stiffness and reduced inertia of the overhead cam engine while maintaining a simpler drive system, the camshaft may be placed near the firedeck to keep the pushrods short. Second, on some vee engines separate camshafts are used for each bank of cylinders. In these cases the cams are placed on the outside of the block as there will not be sufficient space for two camshafts in the vee.

The camshaft is designed to spin in a series of journal bearings under fully hydro-dynamic lubrication. The bearings are placed at the same positions relative to the front and back of the engine as the crankshaft main bearings. This allows the same bulkheads to carry the bearings, and in many cases allows the same oil cross-drillings to be drilled to reach both the main bearings and the cam bearings. Unlike the crankshaft bearing the camshaft bearing shells are continuous, and are pressed into the journals. This requires that the radius of each bearing be made larger than the radius of the cam lobe at maximum lift, such that the camshaft can be installed from the front or back of the engine. It is important to consider possible future valve lift increases and cam lobe radii for future engine performance enhancements.

The camshaft bearing bore is cut through the entire length of the engine. In most engines the camshaft is installed from the front of the engine, and the bore is sealed at the rear, typically with a pressed-in plug. In a few cases the camshaft is driven from the rear of the engine and the installation is reversed. The camshaft must be held in place against fore and aft motion. This is most often accomplished with a bearing plate placed behind the cam gear. The plate is then bolted to the block surface through access holes in the gear, and provides a thrust bearing surface against fore and aft motion.

The cam lobe acts on a solid tappet, a hydraulic lifter, or a roller follower. Each tappet or hydraulic lifter is clearance-fit into a bore in the block. Pressurized oil must be supplied to each tappet bore, both for lubrication of the part sliding in the bore and in the case of hydraulic lifters, supplying oil for valve lash adjustment. Lifter design will be further covered in Chap. 17. Pressurized lubricant is also supplied to the cam bearings. Through drainback from these locations, and other locations above the camshaft, enough lubricant must be supplied to provide a hydrodynamic film at the cam lobe interface to the lifter. However, it is also important not to "bathe" the camshaft in oil as this would significantly increase friction as well as foaming of the oil. It is therefore important to provide sufficient openings below the cam lobes for rapid drainback to the crankcase.

8.7 Positive Crankcase Ventilation

If the example of a single cylinder engine is considered, the movement of the piston up and down in its bore can be seen to significantly change the volume in the crankcase cavity below the piston from BDC to TDC. If the crankcase were completely sealed, this cyclical volume change would have little effect on total work over the 4-stroke cycle, as the crankcase volume would act as a spring absorbing work and then returning it. Once a second cylinder is added to the same crankpin, two different scenarios can play out depending on vee angle. If the vee angle is very small (0–45°) this change in crankcase volume is magnified, if the vee angle is large (between 90 and 180°) the effects start to cancel as air and oil mist are transferred back and forth between cylinders across the crankcase cavity. As additional pairs of cylinders are added to form a V-4 arrangement, the movement of air is driven from one pair of cylinders in a bay to the other. As further engine cylinders are added to the vee arrangement, the percent change in crankcase cavity volume from TDC to BDC is reduced. The transfer of air and oil mist across main bearing bulkheads is termed *bay-to-bay breathing.* As engine speed is increased the restriction to flow between the bays becomes increasingly important, and needs to be considered for its impact on parasitic losses.

Returning to the single cylinder engine, the crankcase is not actually perfectly sealed. The cyclical crankcase pressure will blow oil past the crankshaft seals near BDC, and will draw air past the piston rings near TDC. The flow of air past the rings depends on what the crankcase cavity pressure is with respect to pressure above the piston rings. As pressure builds in the combustion chamber, a percentage of combustion gases will leak past the piston rings into the crankcase cavity below the piston, which is known as piston ring *blow-by*. Over repeated cycles of combustion, positive pressure will eventually build up in the crankcase cavity. The solution to this issue is a one way check valve that allows ventilation of the gases that build up in the crankcase, known as the *positive crankcase ventilation* (PCV) valve.

In addition to blow-by gases in the crankcase, the mass movement of air within the crankcase entrains oil mist in the air, and adds hydrocarbons to the blow-by gasses. Early engines allowed this positive crankcase pressure to ventilate to the atmosphere. However, this combination of gasses is now considered an environmental pollutant. To limit the amount vented to atmosphere, these gasses are now typically vented back into the intake tract of the engine, so that they may be combusted prior to being released. This lessens the environmental emissions, but creates a new problem as the presence of oil in the intake air can reduce knock margin. However, this still releases some emissions to the atmosphere, hence efforts to make the piston rings as tight as possible to prevent blow-by, and devices known as air/oil separators are used to filter as much entrained oil from the air as possible.

Typically, these devices to separate the air and the oil have a torturous path that the air can navigate but the heavier oil cannot. Sometimes a coalescing media or sponge-like material is used to help separate the two by causing the oil entrained in the air to condense, and drain back to the sump. Usually air/oil separators are plumbed into a still cavity in the oil deck or valvetrain cavity of the cylinder head to isolate this flow from the more turbulent crankcase area, and to allow maximum time for the oil to separate from the air on its own. This drives the need for communication passages between the crankcase cavity and the valvetrain cavity of the cylinder head. For convenience, the oil drain backs from the head are typically used for this function also. It is important to consider the non-steady state air flow that will be moving back and forth in these passages, when sizing the diameter. As oil attempts to drain from the cylinder head, positive crankcase pressure may blow it back into the head if the passages are sized too small.

As the piston moves through its stroke, the piston and rings may move opposite and out of phase with each other due to piston motion or gas flow into and out of the piston rings. This relative movement of the rings is known as ring flutter, and significantly affects piston ring sealing. Drawing a slight vacuum in the crankcase cavity helps to stabilize the rings against the bottom of their grooves in the piston, and improves sealing. This improved sealing also improves in-cylinder emissions, as it is more difficult for combustion gases to flow into and out of the piston crevice volumes between the rings, which leads to unburned hydrocarbons.

8.8   Recommendations for Further Reading

The following paper is an excellent recent example of cylinder block design and analysis. The emphasis is on weight reduction through material removal, and is intended to be applicable regardless of the base material selected (see Osman 2012).

Originally published in German, the AAM-Applications series made available through Verlag Moderne Industrie includes sect. 1 on Powertrain. The focus is on aluminum parts, and includes excellent material regarding design approaches to the cylinder block, cylinder liners, pistons, and cylinder heads. English translations are available on the internet. At the time of this writing the articles could be found at the following link: http://www.pdfengineeringbooks.yolasite.com/resources/Aluminum_Applications_Power_train.pdf

References

Osman, A.: Design Concept and Manufacturing Method of a Lightweight Deep Skirt Cylinder Block. SAE 2012-01-0406 (2012)

Cylinder Head Layout Design

9

9.1 Initial Head Layout

Several of the decisions discussed in previous chapters determine the starting point for cylinder head layout. In Chap. 6 the trade-offs determining cylinder bore were presented, as were those that determined the number of cylinders making up a bank. In Chap. 8 the variables that go into determining cylinder spacing were discussed, and the number and approximate placement of the head bolts was introduced. The question of camshaft placement was also introduced in Chap. 8, and will be taken up again later in this chapter.

The primary functions of the cylinder head in a four-stroke engine are to:

- Contain combustion pressure
- Route air/fuel into and out of the cylinder in most efficient manner
- Support components (sparkplug, valvetrain, fuel injector, pumps)
- Route fluids (oil, coolant, and sometimes fuel to injectors)
- Support other engine accessories and brackets

Figure 9.1 reviews the initial layout decisions and provides a starting point for the further layout of the cylinder head. In Figs. 9.1a and 9.2 the layout typical in automobile engines is depicted—a slab cylinder head with four head bolts per cylinder. In the larger heavy-duty diesel engines the possibility of individual heads for each cylinder might be considered. The resulting alternative layout is depicted in Figs. 9.1b and 9.3. It should be immediately apparent from these figures that the first detriment resulting from individual cylinder heads is increased engine length. Each cylinder head must be individually sealed, requiring additional bolts to be placed between each cylinder, and the cast walls of each head require further spacing as well. A further disadvantage is increased block movement since the individual heads contribute far less to block rigidity. Individual heads also add

© Springer Vienna 2016
K. Hoag, B. Dondlinger, *Vehicular Engine Design*, Powertrain,
DOI 10.1007/978-3-7091-1859-7_9

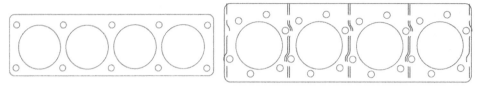

Fig. 9.1 Cylinder and head bolt layout for slab and individual head designs

Fig. 9.2 Slab cylinder head

Fig. 9.3 Individual cylinder
heads

complexity to the cooling circuit. Finally it would be especially difficult to incorporate
an overhead camshaft if that were desired. These disadvantages are sometimes worth the
price in larger heavy-duty applications as they are traded off against several benefits. The
first is that associated with modular design—using many of the same components in various in-line and vee configurations of the same engine. Casting complexity is also reduced
with the smaller individual heads. Another advantage in larger, heavier engines is that of
serviceability. Finally the individual heads may hold durability advantages as it is easier

to manage the firedeck thermal loads over an individual cylinder than over an entire bank of cylinders. This topic will be further explained in the discussion of thermal loading in Chap. 10.

While the subject of material choice was discussed earlier, it requires further discussion specific to cylinder heads. In today's automobile applications cylinder heads are predominantly cast from aluminum. A small percentage of automobile engines continue to be cast gray iron, and iron becomes almost universal in heavy-duty truck, agricultural and construction engines. Aluminum is a desirable material to manufacture cylinder heads from because of its high thermal conductivity and two to three times higher specific strength than gray cast iron. The higher specific strength of aluminum can help lower the center of gravity of a vee type engine. However, aluminum has twice the thermal expansion of a steel fastener, so care must be used when designing cylinder head bolts, especially near the exhaust ports. Cast iron remains mandatory in long life, highly loaded applications.

9.2   Combustion Chamber Design Decisions

Having made the initial layout decisions the cylinder head design work continues with the details of the combustion chamber. A discussion of combustion system optimization is beyond the scope of this book, and the reader is referred to texts on engine thermodynamics, combustion and exhaust emissions. However, because the cylinder head design so significantly influences the combustion system, a brief overview will be presented here. The design object of the cylinder head is to enable efficient cylinder filling and create efficient combustion in conjunction with the ports, valves, and piston. As with the piston design, to be covered in Chap. 15, cylinder head design will be markedly different between diesel and spark ignition engines, and each will be discussed separately in the paragraphs to follow.

9.2.1   Spark-Ignition (SI) Combustion Chambers

In spark-ignition engines, whether fueled with high octane liquid fuels, natural gas, propane, or hydrogen, a relatively uniform mixture of air and fuel (homogenous charge) is ignited with a high energy electrical spark to initiate the combustion process. A spherical flame front then progresses in every direction from the spark plug until all of the mixture has burned. The rate at which the flame front travels, and hence the rate of energy release is critical to engine efficiency, power output, and emission control. Key parameters in determining the energy release rate are the location of the spark plug and the distance from that location, to each location on the combustion chamber walls (in other words, the chamber geometry). Along with the intake port, the combustion chamber geometry also strongly impacts turbulence within the cylinder—another critical parameter governing flame speed. The clearance volume in spark-ignition engines is typically recessed into the cylinder head. The piston is usually either flat, or for higher compression, domed.

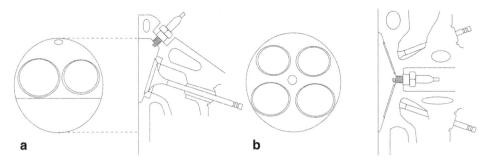

Fig. 9.4 a Two-valve wedge combustion chamber. **b** Four-valve pent roof combustion chamber

Two general combustion chamber geometries are especially prominent in automotive spark-ignition engines—the two-valve "wedge" chamber as shown in Fig. 9.4a, and the four-valve "pent roof" chamber of Fig. 9.4b. For maximum engine performance and efficiency it is desirable to release the energy as rapidly as possible, and a hemispherical combustion chamber with a centrally located spark plug provides the geometry that allows the flame front to most rapidly progress through the reactant mixture. A truly hemispherical chamber, or radial valve arrangement, requires each valve to be placed on a different complex angle (two directions) relative to the cylinder centerline, creating dimensional control challenges and increasing machining costs. A radial valve arrangement also allows a larger valve head diameter for the cylinder bore. The "pent roof" chamber closely approximates the hemispherical chamber while allowing the valve centerlines to be angled in single directions from the cylinder centerline. Each of the engines shown in Figs. 1.6 and 1.7, and Fig. 9.6, uses a four-valve pent roof combustion chamber. The wedge chamber provides rapid energy release and a lower cost, two-valve configuration as shown in Fig. 9.5. The flat portion of the cylinder head firedeck results in rapid "squish" of the mixture into the wedge-shaped region of the chamber as the piston approaches TDC at the end of the compression stroke. This squish of the fuel air mixture into the remaining combustion chamber volume increases mixture motion and increases the speed of combustion. Adjustments from this general geometry can be effectively utilized to enhance turbulence and flame speed during combustion.

The combustion chamber recess is typically contained in the casting mold, and the final surface is that obtained as cast. In some cases the designer may choose to machine the combustion chamber surface to its final dimensions. This allows closer control of the geometry and volume (and hence the resulting compression ratio), but increases manufacturing cost.

Important clearances must be maintained when laying out the valve positions. Valves must clear pistons, and the minimum clearance does not always occur at piston TDC or maximum valve lift. The valve head and valve seat may be recessed further away from the combustion chamber to add piston clearance, but will increase combustion chamber volume and reduce the compression ratio. In a pent roof combustion chamber, valves must clear each other during exhaust-to-intake valve overlap.

Fig. 9.5 Cross-section of
wedge head

Another important design parameter is the size and configuration of the spark plug. Tight packaging constraints have resulted in a design trend toward smaller spark plugs. A small number of standard thread specifications and diameters used throughout the industry allow cost control and reduce the number of spark plug specifications and part numbers. New designs are necessarily constrained in almost every case to the existing available spark plug specifications. It is good design practice to make the spark plug end flush with the combustion chamber, and to prevent threads from protruding into the combustion chamber. Exposed threads may be sources for pre-ignition, and carbon buildup over time will make plugs difficult to remove. Because the spark plug is threaded into the cylinder head and must be easily removable the design space must allow for the hex head of the spark plug and socket wrench clearance. As the spark plug extends upward through the cylinder head from the firedeck surface, space must also be provided for the spark plug wire or plug-mounted ignition coil. These concerns must be kept in mind as the upper portions of the cylinder head are designed.

Fig. 9.6 Cross-section of
pentroof head

9.2.2   Direct-Injection Spark-Ignited (DISI) Combustion Chambers

An increasingly frequent alternative combustion system seen in vehicular engines is the
Direct-Injection Spark-Ignition (DISI), or Gasoline Direct-Injection (GDI) engine. The
concept was first used to enable ultra lean burn due to stratified combustion. The lean
burn approach is precluded from automotive applications today because exhaust emission
standards cannot be met through in-cylinder controls alone, and the excess oxygen in the
exhaust requires complex aftertreatment systems for NO_x control. Using direct-injection
with a stoichiometric combustion system holds several attractions and is now widely used.
The attractions include improved volumetric efficiency, resulting in increased specific
power output, and improved knock margin, allowing a compression ratio increase and
efficiency improvement. The fuel is injected directly into the cylinder but early enough to
approach homogeneity of a stoichiometric mixture prior to initiating combustion. There
are two dominant forms of combustion chamber design, ***wall-guided*** and ***spray-guided***. In

wall-guided or piston-guided systems the injector is typically placed immediately below the intake valves; fuel is sprayed toward the piston crown and its geometry is used to induce mixing. In spray-guided systems the injector is placed adjacent to the spark plug. With either approach significant combustion system optimization is required to achieve the required homogeneity.

From the standpoint of cylinder head design, the key requirement is the placement of both a fuel injector and a spark plug directly in the combustion chamber. Additionally, some designs use air-assisted fuel injection in which case a compressed air passage must supply the base of the fuel injector.

9.2.3   Diesel or Compression-Ignition (CI) Combustion Chambers

In the diesel engine fuel is injected directly into the cylinder near the end of the compression process, and ignited as it vaporizes and mixes with the hot pressurized air in the cylinder. The fundamental criteria for diesel combustion chamber design are those of maximizing the contact area and mixing rate between the fuel and air. Dividing the injected fuel into a series of spray columns, and centrally locating the injector in the combustion chamber volume best distributes the fuel to the available air. With modern electronic control of fuel injectors, multiple injections per cycle are possible to aid in noise and emissions control. A flat cylinder head, with the fuel injector centered in a piston bowl is universally used in the *direct injection* engine. Production examples of the four-valve configuration are seen in Figs. 1.8b and 9.7; this configuration lends itself especially well to the diesel combustion chamber design goals. A narrow valve angle is used to reduce combustion chamber volume in the cylinder head to increase the static compression ratio for diesel combustion. The four-valve cylinder head makes it possible to place the injector on the cylinder bore centerline. The fuel spray plumes angle into the piston bowl which is also centered on the cylinder bore. The piston bowl protrudes upwards at the center, which for a given compression ratio maximizes the volume in the outer region of the bowl. The angle of the fuel spray plumes is slightly shallower or slightly deeper than the angle of the base of the piston bowl. Air is "squished" toward the spray plume by the piston rim as the piston approaches TDC. The combination of this air motion and the spray velocity and direction create a vortex containing the mixing fuel and air near the outer edge of the piston bowl.

In order to maximize valve area in a two-valve engine it becomes necessary to offset the fuel injector from the bore centerline as shown in the production engine of Fig. 1.8a. The piston bowl is similarly offset, and the rationale for combustion chamber design remains similar to that of the four-valve engine just described.

In both two- and four-valve engines the valve faces are not angled, or angled very little relative to the firedeck plane. This too follows directly from the combustion chamber design objectives. The high compression ratio requirements results in a small TDC volume, and symmetry at each spray plume is required for performance and emissions optimization.

Fig. 9.7 Direct-injection diesel combustion chamber

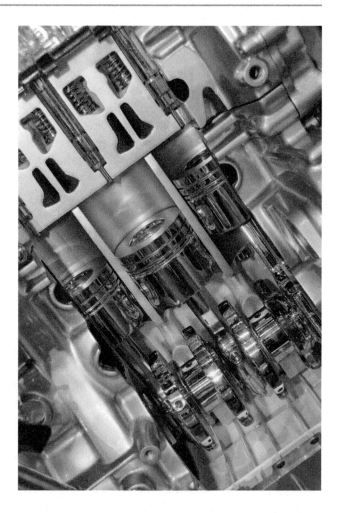

Many older diesel engines were referred to as ***indirect injection*** engines and used a
pre-chamber in the cylinder head such as that shown in Fig. 9.8. As the flat-topped piston
approaches TDC the majority of the air in the combustion chamber is forced through a
narrow passage into the pre-chamber. The air in this chamber is highly turbulent, and
readily mixes with the fuel, allowing considerably lower fuel injection pressures. Combustion initiates in the pre-chamber and a high velocity combusting jet returns to the main
chamber during expansion. The pre-chamber engine holds the advantages of lower cost
fuel injection systems (due to the lower required injection pressure), lower combustion
noise, and inherently lower soot emissions. However, the gas transferring in and out of
the pre-chamber generate high pumping losses, and the intense fluid motion increases
heat rejection to the cylinder head. The net result is a severe fuel efficiency penalty. With
the advances in fuel injection systems resulting in higher injection pressure and electronic
control of injection timing and rate new vehicular engines are exclusively direct injection.

Fig. 9.8 Indirect injection
diesel combustion chamber

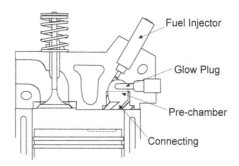

The indirect injection engine is now limited to small industrial diesel engines, and older automobile engines—those designed prior to the mid-1980s.

As with the spark plugs required in spark-ignition engines the diesel engine faces the packaging requirements of fuel injector placement. There are historically a wide variety of fuel injection systems and injector designs. They range in diameter from a few millimeters to 25 mm or more (although always narrower at the tip). Sealing the fuel injector against the high cylinder pressure at the firedeck surface is an important design requirement. High contact pressure against a metallic gasket or the head surface is typical. The injector must be supplied with fuel, either through a steel line entering the injector near its top, or in some cases through drilled passages in the cylinder head, sealed with o-rings in grooves on the injector, above and below the fuel supply drillings. Many injectors require both supply and return fuel lines, creating further packaging challenges.

Most automotive diesel engines also require glow plugs to aid in starting the engine on cold days. The glow plugs are small electrically heated elements protruding through the cylinder head firedeck into each cylinder. Glow plugs can be seen adjacent to the injectors in the engines in Fig. 1.8a and b.

Due to the high combustion pressure of diesel combustion, and the many components that protrude through the fire deck, the combustion chamber flexes like a membrane and high cycle fatigue is a key design consideration. Cylinder head fatigue development will be further addressed in Chap. 10.

9.3   Valve, Port and Manifold Design

Poppet valves placed in the cylinder head—referred to as *overhead valves*—are now universally used in vehicular engines. *Flathead* engines with the valves in the cylinder block were commonly seen on automotive engines through the 1950s, and are still seen in small industrial engines. Various alternatives to the poppet valve—most notably sleeve valves—held some interest in the early development of engines but are no longer seen in production. The discussion here will be limited to overhead valve designs using poppet valves.

The basic features of the poppet valve assembly are shown in Fig. 9.9. The valve itself consists of two sections. The valve head is the portion that seals the combustion chamber.

Fig. 9.9 Valve assembly details

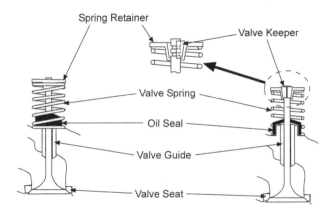

It is typically constructed of a steel alloyed with nickel, chromium, and other materials selected for high temperature strength and hardness. The valve stem may be constructed in one piece with the head or of a separate material welded to the head. A separate material is sometimes selected to reduce cost since the alloy properties of the head are not required at the cooler operating temperatures of the stem. Some of the newest valve designs use a thin wall tube for the valve stem, reducing the valve weight and resulting inertia. In most engines a high hardness seat is pressed into the cylinder head beneath each valve. The seat is made of a similar alloy to the valve head, and is used to minimize wear resulting from the impact loads seen at valve closure and sliding between the valve and seat. At least 75 % of the heat transfer out of the valve is to the valve seat at the narrow band of contact between the two. Because this is a quite restrictive heat transfer path, and because the valve alloys have relatively low thermal conductivity, valve temperatures are high and must be carefully monitored in engine development.

The valve stem may ride directly in the cylinder head, or in a valve guide pressed into the head as shown in Fig. 9.9, and typically uses the ratios listed in Eqs. 9.1–9.3 as a starting point for design. Exhaust stems are typically thicker proportionally to handle the higher temperature. The goal is to have as small a stem diameter and length as possible to reduce valve weight and reduce port flow restriction. The valve assembly is completed with a valve spring, retainer and keepers as shown in the figure. Valve durability and the design of the various valve train components will be further discussed in Chap. 17.

$$\frac{\text{Valve Guide Length}}{\text{Valve Guide } \varphi} \approx 6.7 \tag{9.1}$$

$$\frac{\text{Int. Valve Head } \varphi}{\text{Int. Valve Stem } \varphi} \approx 5.8 - 7.0 \tag{9.2}$$

$$\frac{\text{Exh. Valve Head } \varphi}{\text{Exh. Valve Stem } \varphi} \approx 5.3 - 5.6 \qquad\qquad (9.3)$$

The first valve and port design questions faced by the cylinder head designer are those of the number and size of intake and exhaust valves. Especially in the spark-ignition engine the output of the engine at any given speed is a direct function of volumetric efficiency, and these design questions are answered based on cost versus efficiency trade-offs. Volumetric efficiency is determined by a number of parameters including: valve event timing, valve lift versus crank angle, port and in-cylinder heat transfer, intake port design and runner length, and the various parameters that determine exhaust flow and residual. A discussion of these parameters is beyond the scope of this book; it is sufficient to observe that for any combination of the remaining parameters volumetric efficiency is maximized by maximizing the available intake and exhaust flow area. Two further observations each lead to a question that the designer must address:

1. While each valve contributes to the needed intake or exhaust flow area each also requires material (restriction) for the valve seat. Recognizing that each valve requires a circular opening within the circular cylinder bore what will be the optimum number of valves beyond which flow area is reduced? This question is further impacted by whether the domed combustion chamber shape allows any increase in surface area, and by whether the valves and seats can be extended slightly beyond the bore diameter (with angled stem axes and cutouts along the upper edge of the cylinder bore.
2. Engine breathing requires both that new charge can be introduced and that spent products can be exhausted. What ratio of intake flow area to exhaust flow area will maximize engine breathing (maximize volumetric efficiency)?

The second question will be addressed first as the optimum ratio of intake to exhaust area impacts the choice of the number of valves. While at first glance one may say that the intake and exhaust area should be equal since the same mass must go out as that coming in; however two factors modify this answer. First, the mass flow through each valve is driven by the pressure ratio across the valve at any instant in time. These pressure ratios are continuously changing and must be considered for the individual engine. One especially important factor is the large pressure differential seen across the exhaust valves when they are first opened. The pressure in the cylinder was elevated by combustion, and while it drops during expansion it remains well above exhaust port pressure when the exhaust valve opens. The **blowdown** resulting from this initial pressure differential is effective in exhausting on the order of half of the combustion products at the beginning of the exhaust stroke. The other important factor is that of the thermodynamic state of the exhaust products versus that of the intake charge. The specific volume (volume per unit mass) of the exhaust gas is significantly higher than that of the intake charge thus requiring greater flow area for an equivalent mass flow rate. The net impact of these opposing factors is dependent on the specific engine, and for maximum performance should be determined for the particular case being considered. Initial sizing ratios for a naturally aspirated engine are shown in Eq. 9.4.

$$\frac{\text{Total Intake Valve Area}}{\text{Total Exhaust Valve Area}} \approx 1.2 - 1.3 \tag{9.4}$$

The discussion now returns to the question of the number of valves per cylinder. Two valves will offer the lowest cost and least design complexity, but clearly will not well utilize the available cylinder head flow area. The four-valve configuration is commonly used in high performance and premium engines, and very effectively utilizes the possible flow area. Three- and five-valve engines are sometimes seen, with one more intake than exhaust valve, keeping in mind the optimum ratio of areas discussed earlier. It is generally found that while there is a rapidly diminishing return as the number of valves is increased the five-valve configuration holds a slight performance advantage over the four-valve engine. Further increases (to six or more valves) actually detract from the breathing area of a circular bore engine, as the space needed to package the valve seat starts to reduce the diameter of the intake valve. Generally, the limit on valve head diameter is valve seat insert diameter. The valve seat insert diameter is typically limited by the cylinder wall on the outside of the bore, and the minimum material permissible between seats towards the inside of the bore. This material limit is usually the bridge of material between exhaust ports on a four-valve head. Typical valve head diameters are summarized in Table 9.1, with fewer data points available for three-valve and five-valve heads:

This discussion began with specific reference to the spark-ignition engine. Diesel engine performance is much less sensitive to volumetric efficiency since power output is determined by the amount of fuel injected into an environment always containing excess air. However, engine efficiency is impacted by the work required to pump air into the cylinder and exhaust products out, so available flow area remains important as a means of maximizing in-cylinder pressure during the intake stroke and minimizing it during exhaust. Two- and four-valve designs are both common, with the four-valve engine offering improved efficiency and combustion chamber design advantages at the expense of increased cost and complexity. Most new diesel engines use the four-valve configuration—not necessarily for improved engine breathing, but because it allows the injector to

Table 9.1 Valve head diameter as a function of piston diameter

Valve Configuration	D_{intake}/D_{piston}	$D_{exhaust}/D_{piston}$
Two-valve	0.43–0.53	0.35–0.45
Three-valve (2I, 1E)	0.37–0.39	0.41–0.43
Four-valve	0.35–0.42	0.28–0.37
Five-valve (3I, 2E)	0.30–0.32	0.32–0.34

be centrally located, improving in-cylinder air utilization. The ratio of intake to exhaust flow area is often one-to-one.

Once the number and flow area of the valves has been determined the designer must consider the details of valve and seat geometry. The performance of the engine is sensitive to the effective flow area available at any crank angle position during the intake and exhaust processes. The effective flow area is in turn dependent on the combination of valve lift versus crank angle and flow area versus lift as depicted in Fig. 9.10. Valve lift versus crank angle is determined by the camshaft lobe design and valve train capabilities, or the capabilities of an electronic, hydraulic, or pneumatic valve operating system. These systems will be discussed further in Chap. 17.

Effective flow area versus valve lift is determined by the cylinder head designer through a combination of port, valve, and seat design. The geometric flow area is limited

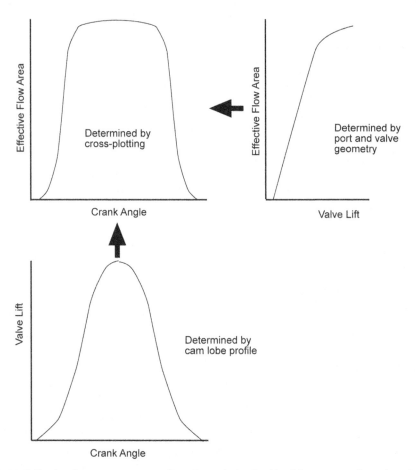

Fig. 9.10 Effective flow area versus crank angle, as determined by lift versus crank angle and effective flow area versus lift

by three different regions as valve lift increases—first by the annulus between the valve and the seat (valve curtain area, Eq. 9.5), then by the geometry of the valve seat and the port-side of the valve head, and finally by the port itself (geometric port area, Eq. 9.6). Typically the intake valve is not lifted more than one quarter of its diameter as shown in Eqs. 9.7 and 9.8, as there are decreasing gains in flow rate above this ratio. The greater the valve is lifted, the higher the stress on the valvetrain, and the greater the likelihood of valve-to-piston or valve-to-valve contact. However, in racing applications this ratio may increase above 30 %.

$$\text{Valve Curtain Area} = (\pi \cdot D_V) \cdot L_V \qquad (9.5)$$

$$\text{Geometric Port Area} = \frac{\pi}{4}(D_P^2 - D_S^2) \qquad (9.6)$$

$$\text{Max Int. Valve Lift} \approx \frac{L_{V,\text{max}}}{D_V} \approx 0.25 - 0.31 \qquad (9.7)$$

$$\text{Max Exh. Valve Lift} \approx \frac{L_{V,\text{max}}}{D_V} \approx 0.30 - 0.36 \qquad (9.8)$$

Where:

D_V = Valve head diameter (max)
L_V = Valve lift from seated
D_P = Minimum diameter of the valve seat
D_S = Diameter of the valve stem

The actual flow area is less than the geometric area due to flow separation across the valve and seat. This is typically the critical restriction in the air intake system, and its performance will govern the volumetric efficiency of the engine, and ultimately engine performance. There are many different ways to describe the efficiency of the valve-port arrangement, and there is not an industry standard. Different methods use different references: valve curtain area versus valve lift, port area versus valve lift, and local speed of sound (Mach number) versus engine speed. It is important when reviewing published data to understand the reference used as it may not be possible to compare one source to another. For illustrative purposes, one definition of Discharge Coefficient is shown in Eq. 9.9. Design modifications can be made experimentally with port and valve models made from plastic resins, rapid prototypes, wood, or with actual castings. Modeling clay is often used to rapidly consider design changes. Computational fluid models are increasingly being used. Once they are correlated to experimental measurements they can be used to rapidly assess the effects of design changes.

$$\text{Discharge Coefficient } (C_D) = \frac{\text{Effective Flow Area (measured)}}{\text{Geometric Flow Area (calculated)}}$$

(9.9)

$$C_D = \frac{A_e}{(\pi \cdot D_V) \cdot L_V}$$

The discussion of flow optimization begins at the interface between the valve and seat. The valve seating angle is an important flow parameter with which to begin, as depicted in Fig. 9.11. Basic seat angles range from 15 to 45°. The shallower seat angle allows effective flow area to increase more rapidly as the valve initially lifts from the seat, and it reduces wear due to impact loads. However, the shallower seat angle typically increases flow separation as lift is increased, and makes sealing more difficult. Increased seat angles are generally more advantageous unless valve and seat wear rates become too high. Angles above 45° are not used as these result in valve sticking in the closed position and increase the required opening force. A 0.5–2.0° difference between the valve angle and the seat angle is typically applied to aid in sealing, and the sealing face width is typically 1.25–2.0 mm for intake and 1.5–2.5 mm for exhaust seats. Flow separation can often be reduced and the flow coefficient increased by grinding the seat at three or more angles—a base 45° angle at which the valve seats, and shallower 30° and steeper 60° approach angles from the cylinder and port respectively. This is less frequently used in production due to the difficulty of controlling the resulting valve seat width, both on the production line and in the field. In high performance applications which are less cost sensitive, a radius and up to a seven-angle valve seat geometry are sometimes cut.

While considering the valve to seat interface an important design parameter is the port-side surface of the valve head. Minimizing material in this region directly improves flow area but is dependent on the structural capabilities of the chosen valve material. Intake and exhaust valves have different requirements in this area since the flow is reversed.

Fig. 9.11 Valve and seat geometry details

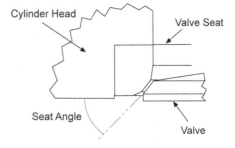

A final important area to consider is the transition from port to valve seat. The intake and exhaust ports are typically sand cast into the cylinder head, and the port cores can shift during casting. The seats inserts are typically machined from stock or formed from powdered metal, and have much better dimensional control. If the cast port is misaligned with the assembled seat insert, a step change in diameter is formed and a significant disruption to airflow can occur. The production solution is typically to make the intake port slightly smaller in diameter than the valve seat, to insure this diameter step change is in a favorable direction. Since the flow direction is reversed on the exhaust port, the opposite is true, and the port is typically larger than the exhaust valve seat insert. At added cost, the transition between the cast port and valve seat insert can be machined after seat installation to reduce or eliminate this step.

As the discussion moves to the design of the intake and exhaust ports a number of engine performance considerations drive the design choices. While thermodynamics and engine performance development are not central emphases of this book it will be necessary to make brief mention of several engine performance concepts in order to explain the rationales of port and manifold design. The reader is encouraged to look carefully at the port designs of the example engines presented in Figs. 1.6, 1.7, and 1.8 while reviewing the following paragraphs where each of the port design considerations are taken up in turn.

9.3.1 Intake Port Swirl

The intake port is responsible for contributing to the bulk in-cylinder flow characteristics needed for optimal combustion. One such flow parameter is swirl—a large scale rotation of the air about the axis of the cylinder bore centerline. Swirl is generated through the design of the intake port using either a helical or tangential port. With the helical intake port the swirl is initially generated about the axis of the intake valve stem and expands to the cylinder axis as it enters the cylinder. The tangential port shown on the left in Fig. 9.12

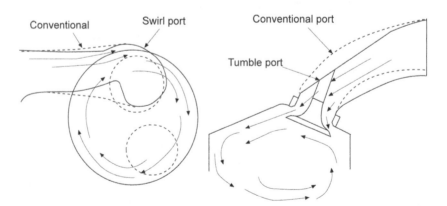

Fig. 9.12 Controlling swirl and tumble flows through intake port design

relies on the combination of port angle and the cylinder wall to create the swirling flow as the charge enters the cylinder. Swirl generated during the intake process does not decay but is maintained throughout compression, even increasing in velocity as the charge density increases late in compression. Swirl is used in many diesel engines to stretch the fuel sprays and increase mixing. It is sometimes used in spark-ignition engines to stabilize the flame front when high turbulence might tend to extinguish the flame.

9.3.2   Intake Port Tumble

Another bulk flow parameter governed by intake port design is tumble. Depicted on the right in Fig. 9.12 tumble is a large scale rotational flow perpendicular to the axis of the cylinder. Unlike swirl, tumble decays as the piston approaches BDC. The flow breaks up into small scale turbulence with the inertia of the bulk flow providing the needed turbulent energy. Tumble is commonly utilized in spark-ignition engines, especially four-valve pent roof arrangements, where the small scale turbulence is an important parameter controlling flame speed. In particular combustion systems reversing the tumble direction may be desirable. Reverse tumble is seen in some direct-injection spark-ignition engines where the intake charge interacts with the fuel injector spray. This requires drawing the intake charge across the top of the cylinder head.

As can be seen from Fig. 9.12, the angle at which the intake port enters the combustion chamber heavily influences the degree of tumble flow. Besides impacting combustion, there are many other packing constraints on the port and valve angle, such as location and type of valvetrain, valve spring pocket, cylinder head bolts, coolant passages, spark plug or fuel injector location, and intersection with the intake manifold.

Port geometries are typically developed using a prototype single cylinder representation of the cylinder head attached to a flow bench—a device used to simulate the pressure differential caused by piston motion, but under steady state flow conditions. Bulk swirl, and to some extent tumble, can be quantified on the flow bench by installing laminar flow elements and measuring the torque about the flow axis of interest.

9.3.3   Intake Port and Intake Manifold Length

Another important parameter in spark-ignition engines is intake runner length. The drop in pressure across the intake valve as it is first opened, and as the piston begins its downward travel, creates an acoustic expansion wave that travels up the length of the intake runner at the local speed of sound. This is the 'signal' for the bulk air to begin moving into the cylinder. As the wave encounters a sharp increase in area, such as a plenum or branch in the intake manifold, a compression wave is reflected back towards the valve. Depending on engine speed and runner length the pressure waves may travel back and forth in the intake manifold from one to several times per intake cycle, and are superimposed on the bulk air motion. At combinations of engine speed and intake runner length that result in a positive pressure wave arriving at a cylinder near the middle of an intake stroke, the resulting increased pressure aids in filling the cylinder. The volumetric efficiency is improved over a small range of engine speeds as the pulse arrives a bit before of after mid stroke. For a given runner length there will be "trough speeds" in between those at which pressure pulses are well timed. At these trough speeds the intake supply pressure is reduced, and volumetric efficiency is penalized. Many automobile engines today implement variable runner length. The simplest and most commonly used approach incorporates an upstream valve that when opened creates a shorter path to the plenum, improving high-speed volumetric efficiency while the longer runner provides low speed tuning. The engine shown in Fig. 1.6c uses a rotating drum in the intake plenum, allowing continuously variable intake runner length, optimizing volumetric efficiency over an engine speed range.

Intake tuning can be modeled using Helmholtz resonance equations for a single tube as described in Eq. 9.10. It can be seen from the equation that reducing the runner length improves high-speed volumetric efficiency while increasing runner length improves the volumetric efficiency at low speeds.

$$N = \frac{a}{2\pi}\sqrt{\frac{A}{L \cdot V_{eff}}} \tag{9.10}$$

Where:

N = Engine speed where beneficial tuning peak occurs (Hz)
a = Local speed of sound (m/s)
A = Cross-sectional area of intake port or runner
L = Length of intake port and manifold runner
V_{eff} = Helmholtz resonator volume (usually 1/2 cylinder volume + combustion chamber volume)

Today most of this work is done using engine cycle simulation programs that allow geometry details to be described and allow the effects of interaction between cylinders, surface roughness, and port and manifold heat transfer to be included.

This design parameter is generally less important in diesel engines as the performance is not directly dependent on volumetric efficiency, and the engine is typically turbocharged to provide the needed intake supply pressure.

9.3.4   Intake Port Surface Roughness and Flow Area

It is intuitive that minimizing surface roughness, maximizing flow area at any cross-section, and ensuring smooth transitions are all necessary to minimize flow losses and pressure drop. The only modification to such thinking is to recognize that as cross-sectional area increases the magnitudes of the pressure pulses discussed in the previous section are reduced. Increasing the flow area without considering the effect on the pressure pulses and intake tuning can actually have detrimental results.

9.3.5   Intake Port Heat Transfer

In a naturally aspirated engine the incoming charge temperature is below that of the port walls, resulting in heat transfer from the walls to the charge. Heat transfer to the charge increases its temperature, reducing knock margin, and reduces its density, hurting volumetric efficiency. Especially in spark-ignition engines it is thus important to minimize the heat transfer from the intake port walls to the charge. Maintaining the port walls as cool as possible, keeping the port length as short as possible, and minimizing surface roughness and turbulence all contribute to this goal. At first glance the goal of reducing port length and that of controlling runner length for intake tuning will seem contradictory. Two further points will clarify the design goals. First, one must consider the relative contribution of the two parameters to engine performance. It is often the case that the impact of intake tuning is far greater than that of heat transfer (especially at higher engine speeds) and thus takes precedence. Second, intake tuning is governed by total runner length to the plenum chamber. The heat transfer effects can often be minimized by placing the majority of the runner in an intake manifold bolted to the cylinder head. The manifold can be designed to be thermally isolated from the cylinder head, keeping its temperature below that of the cylinder head. In each of the engines shown in Figs. 1.6 and 1.7 the length of the intake port is relatively short, and the manifold contacts the head only at the gasketed joint, with the manifold runner surrounded by ambient air (see especially Fig. 1.7d).

9.3.6   Fuel Injector Placement, and Intake Manifold Design

Many automotive spark-ignition engines produced today have port fuel injection. Injector placement is an important design consideration. Because of the objective of rapid fuel vaporization the injector is placed to spray the fuel against the back surface of the intake valve head. This too can be clearly seen in all of the engines of Figs. 1.6 and 1.7. The valve head temperature is significantly higher than that of the port walls due to its exposure to the combustion chamber and its restrictive heat transfer path. One of the challenges of the direct-injection engine results from giving up this hot surface that aids in fuel vaporization.

Flow uniformity between cylinders is an important parameter in manifold design. Interactions between the cylinders of a multi-cylinder engine present a complicating factor that can be an advantage or disadvantage, as pulses from neighboring cylinders can interact. Uniformity is also impacted by the air flow path through the intake manifold. In general, non-uniformities can be minimized by using equal-length runners feeding into

Fig. 9.13 Intake manifold with port injectors

Fig. 9.14 Reinforced plastic intake manifold

a large plenum chamber as shown in Fig. 9.13 on an aluminum manifold. The larger the plenum volume, the better the pressure recovery downstream from a throttle. As plenum chamber volume is increased cylinder-to-cylinder interactions are reduced, however so is engine responsiveness. Further discussion of this possibility will not be provided here, but requires the engine designer to work closely with those conducting analysis of the intake system flow dynamics.

Intake manifolds have historically been cast from gray iron or aluminum. More recently reinforced plastic is increasingly used to reduce cost, reduce heat transfer, and attenuate sound as shown in Fig. 9.14. In order to maximize performance over a wider speed range four-valve and five-valve engines have sometimes used different runner lengths and separate ports. Intake manifolds with variable length runners that can be changed during operation are sometimes seen, as shown in the engine of Fig. 1.6c. Diesel engines in today's automotive and heavy-duty applications are almost universally turbocharged, and most incorporate charge air coolers as well.

9.3.7   Exhaust Port Heat Transfer

As with the intake port it is important to minimize exhaust port heat transfer. In this case the gas temperature is significantly higher than that of the port walls, and heat transfer is in the other direction, from the gas to the walls. Exhaust port heat rejection contributes significantly to the overall cooling system load, and should be minimized for this reason alone. The metal temperature around the exhaust port is higher than anywhere else in the engine, and is typically a limiting factor in overall engine performance. With increased metal temperature comes creep, and loss of bolt and head gasket preload. Because catalyst efficiency is strongly temperature-dependent it is important to minimize heat transfer from the exhaust prior to the catalytic converter to reduce the time for the catalytic converter to come up to temperature. Since the majority of engine emissions are generated in the first 30 s of engine operation, the time it takes for the catalyst to light-off is critical. Finally, in turbocharged engines it is important to minimize heat transfer in order to maximize the energy available at the turbine inlet.

Cast-in ceramic port liners are sometimes considered as a means of reducing heat rejection, but have a poor cost-to-benefit trade-off. A more attractive approach is to insert a stainless steel sleeve into the port, leaving an insulating air gap between the port wall and the cylinder head. This approach is seen in Fig. 1.7d. Finally, in some cases the exhaust manifold protrudes into the port to form this barrier to heat transfer into the head. It is worth stating again, it is recommended to minimize the length of the exhaust port within the cylinder head.

9.3.8   Exhaust Port and Exhaust Manifold Length

As with the intake runner the exhaust runner length can be tuned for optimum cylinder scavenging at a particular engine speed. In this case the pressure pulse entering the port when the exhaust valve is opened travels down the runner at the local speed of sound. The local speed of sound is significantly higher in the exhaust tract due to the much higher than ambient temperature of the gas, as shown in Eq. 9.11. When it encounters a sudden volume increase a reflected expansion wave travels back up the runner. This negative pressure pulse can be timed to arrive at the valve before completion of the exhaust event, thus aiding in scavenging or removing exhaust gas from the cylinder. This technique is considerably important in racing applications but cannot easily be applied to exhaust systems for on-highway application due to packaging and cost constraints. The many small flow channels of the catalytic converter effectively dissipate the pressure pulses, with little reflection. In spark-ignition engines it is important to minimize exhaust volume upstream of the catalytic converter in order to increase the catalyst temperature as rapidly as possible. In turbocharged engines the overriding concern is to get the exhaust pulses to the turbine inlet with as little dissipation as possible. This will be discussed further in the paragraphs on exhaust manifold design.

$$a = \sqrt{kRT} \quad or \quad \frac{a_1}{a_2} = \sqrt{\frac{T_1}{T_2}} \qquad\qquad (9.11)$$

Where:

a = Speed of sound in a gas
k = Ratio of specific heats
R = Specific gas constant
T = Absolute temperature

9.3.9 Exhaust Port and Manifold Surface Roughness and Flow Area

In order to minimize flow restriction and heat transfer smooth transitions and minimal surface roughness are again important. Again, the cost versus performance trade-off points to as-cast surfaces in all but racing applications. Dimensional stack-up at the port-to-manifold transition is worthy of careful consideration in most applications. Flow area should be large enough to minimize restriction, but increasing the flow area beyond what is necessary dissipates the pressure pulses (important in turbocharged and racing applications) and increases thermal mass (important in systems using catalysts).

9.3.10 Exhaust Port and Exhaust Manifold Design

One of the most critical elements of exhaust manifold design is that of thermal load management. In order to minimize heat rejection to the cooling system and thermal stresses within the cylinder head the exhaust manifold is typically constructed as a separate component bolted to the head. The bolts clamping the manifold to the head are often made relatively long and placed in towers to reduce sensitivity to thermal growth, as shown in Figs. 9.3 and 9.15. This is done to minimize clamping load changes with temperature by increasing bolt stretch. The bolt hole diameters in the manifold are made large relative to the bolts and the gaskets are selected to maintain their seal while allowing sliding as the manifold grows and contracts with temperature. Typically a graphite gasket is used to seal the exhaust manifold because it can tolerate this movement. With long cylinder banks it may be necessary to use either bellows sections or slip joints to accommodate growth and shrinkage of the manifold.

One approach to exhaust manifold construction is casting from alloyed gray or ductile iron. The ductile iron is more expensive but provides the higher tensile strength that may be necessary to accommodate the high tensile loads of the thermal load cycles. Cast manifolds are used in some automobile applications and virtually all heavy-duty engines. Another approach increasingly seen in automobile engines is to construct a welded manifold

Fig. 9.15 Cast exhaust
manifold

Fig. 9.16 Tubular exhaust
manifold with close-coupled
catalyst

from either tubular or stamped stainless steel, as shown in Fig. 9.16. The primary driver
toward this approach is the resulting lower mass and reduced thermal inertia enabling
rapid catalyst warm-up. As emission standards become increasingly stringent further design modifications are made to close-couple the catalyst. This moves the catalyst up to
the exhaust manifold exit, and occasionally incorporates the catalyst inside the exhaust
manifold itself.

Fig. 9.17 Cylinder head bolt
clearance to valvetrain

A further exhaust manifold design consideration for turbocharged engines is referred
to as ***pulse conservation.*** Each time an exhaust valve opens the initial pressure difference
across the valve results in a high-pressure pulse. The objective of pulse conservation is to
transfer the additional energy of the pressure pulse to the turbine inlet with as little dissipation as possible, and without interfering with the exhaust flow from adjacent cylinders.
To accomplishing this objective, cylinders are grouped such that no two cylinders share
an exhaust manifold section with $<240°$ crank angle between firing. In order to minimize
pulse dissipation sudden volume changes within the exhaust manifold passages must be
avoided, and restriction and surface roughness should be minimized.

9.4   Head Casting Layout

The discussions of combustion chamber design and valve and port design in preceding
sections provide the information necessary to begin detailing the cylinder head layout.
The number and placement of the head bolts, introduced in Chap. 8, must be considered in
conjunction with port layout. Access to tighten head bolts must also be considered during
valvetrain layout, as shown in Fig. 9.17. Several examples of firedeck, port, and head bolt
column layouts are shown in Figs. 9.18, 9.19, and 9.20. A two-valve spark-ignition automobile engine design with a wedge combustion chamber is shown in Fig. 9.18. On this
in-line engine the intake ports enter on the same side of the head as the exhaust ports exit.
This is referred to as counter-flow, and careful attention must be paid to thermal stresses
in this region. This engine uses four head bolts per cylinder. The layout shown in Fig. 9.19
is that of a four-valve spark-ignition engine with a pent roof combustion chamber. This
double overhead camshaft design has cross-flow ports that minimize flow restriction. A
four-valve heavy-duty diesel engine layout is shown in Fig. 9.20. This engine also uses a
cam-in-block design, with greater packaging space required for the push rod columns and
the greater number of head bolts per cylinder.

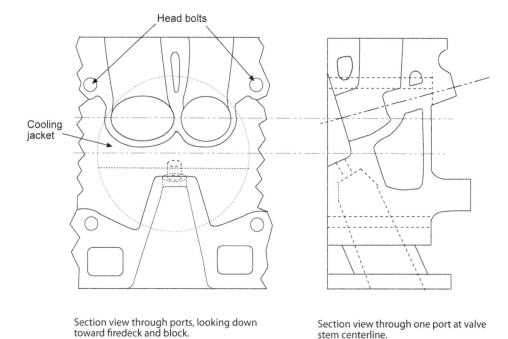

Section view through ports, looking down toward firedeck and block.

Section view through one port at valve stem centerline.

Fig. 9.18 Cylinder head cross-sections for two-valve wedge combustion chamber design

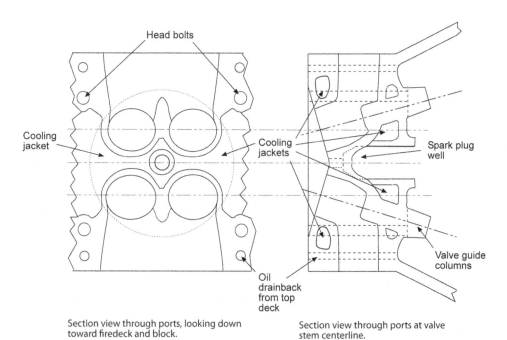

Section view through ports, looking down toward firedeck and block.

Section view through ports at valve stem centerline.

Fig. 9.19 Cylinder head cross-sections for cross-flow head with four-valve pent roof combustion chamber design

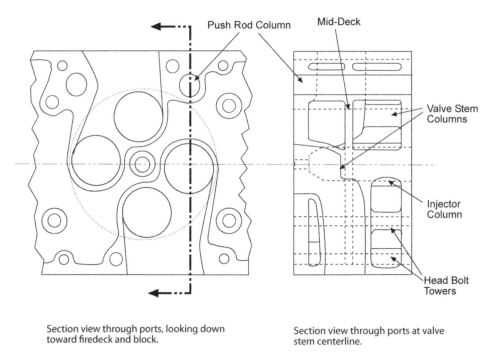

Push Rod Column   Mid-Deck

Valve Stem Columns

Injector Column

Head Bolt Towers

Section view through ports, looking down toward firedeck and block.

Section view through ports at valve stem centerline.

Fig. 9.20 Cylinder head cross-sections for four-valve direct-injection diesel cylinder head design

Cylinder head design details will be impacted by the casting process to be used in production. At this writing the vast majority of production cylinder heads, whether aluminum or gray iron, are sand cast. An increasing number of new cylinder heads are being produced using the lost foam casting technique. The need for close dimensional control of often complex port shapes and the ability to cast cooling jackets precludes most other casting technologies. Die casting is sometimes seen with air cooled aluminum heads and relatively straight port passages. If the combustion chamber is to be recessed into the cylinder head as with most spark-ignition engines its negative will be included in the casting mold. In most cases the combustion chamber surface will remain as cast. If close dimensional control is required for consistent combustion chamber volume, the entire surface may be machined. This is quite typical in high performance and racing engines that operate close to their knock limit. In sand cast cylinder heads cores will be used to create the intake and exhaust ports. A conceptual look at port coring was shown in Fig. 7.4. The valve seat recess will then be machined into the fire deck surface. Some further machining of the port may be done, primarily to optimize flow in the vicinity of the valve seat and to control dimensions at the transitions to the intake and exhaust manifolds. The cooling jackets will also be cored, and the remaining surfaces of the cylinder head will be defined by the casting molds. If the head is cast using the lost foam process both the ports and the cooling jackets will be packed with loose sand that is then removed from the finished casting.

As with the cylinder block, the design of the cylinder head is very iterative. Once head configuration and material choice are made, the cylinder head layout may proceed approximately as follows:

1. Target combustion chamber volume to achieve required compression ratio
2. Cylinder head bolts (quantity and location)
3. Number of valves per cylinder and valve layout (wedge, pent roof, radial; valve angle)
4. Valvetrain configuration and location (DOHC, SOHC, OHV, etc.)
5. Valve specific geometry (valve diameter, length, head height from the deck)
6. Squish pads to aid combustion
7. Spark plug and fuel injector position
8. Port layout, intake and exhaust
9. Manifold surfaces, intake and exhaust
10. Coolant deck, core prints, transfer ports
11. Oil deck and drainbacks
12. Valve cover and breather arrangement

As this portion of the cylinder head is detailed the designer is faced with many questions concerning wall thicknesses and radii. Unfortunately it is impossible to provide general guidelines for these dimensions as they are dependent on many factors requiring the designer to work closely with the foundry engineer, the structural analyst, and in many cases specialists in noise control. From the standpoint of casting processes critical considerations include distortion, core shift and core strength (sand casting), sand cleanout, and required draft (outer surfaces of sand castings). Distortion of the casting impacts the amount of machining that will be required later to meet the final dimensional requirements. Casting designs having greater distortion require additional material in the as-cast dimensions that increase the material and machining costs of the part. If the port or cooling jacket cores shift during the casting process some wall thicknesses are increased and others are reduced from the nominal dimensions. The amount of core shift must be carefully determined by measuring cut prototype dimensions as the casting is developed. The molds and cores must be optimized to minimize shift, and the wall dimensions adjusted as required to account for any remaining shift. Core strength and sand cleanout requirements both work to define minimum cross-sections of cored passages. This along with core print placement will become especially important in designing the cooling jackets. Casting draft is mentioned here for completeness; it must be considered in determining the outside dimensions of sand and die cast parts.

Cylinder head dimensioning is also impacted by the complex combination of thermal and structural loads to which the head will be exposed. The section on fatigue discussed in Chap. 3 and depicted in Fig. 3.4 provides a look at a few of the dimensions that must be determined, and allows identification of some of the loads and stresses to which the head is exposed. The thickness of the firedeck, from the combustion surface to the base of the water jacket, serves as an important example. This section of the head is exposed to a tremendous thermal gradient from the combustion gas temperature on one side to the coolant on the opposite. It is simultaneously exposed to the rapid cycling of pressure within the cylinder. The surface is constrained by the clamping load sealing the head to the cylinder block, and by the rest of the cylinder head structure. Finally, there are the loads associated with press fitting the valve seats into the cylinder head. This combination of loads clearly makes it impossible to identify general guidelines for cylinder head dimensioning. These loads will be further discussed along with cylinder head durability validation in Chap. 10. The design process typically begins with dimensions selected from experience, and similar to those used in previous, successful cylinder head designs. Experimental and computational tools to be discussed in Chap. 10 are then used to optimize the design.

9.5   Cylinder Head Cooling

One of the greatest challenges of cylinder head design is that of thermal load management, yet cooling jacket placement and optimization often receives relatively little consideration. "Lay out everything else, and then place the cooling jackets wherever they fit" is a common design instruction. More recently many designers are discovering that overcooling can actually reduce the fatigue life of the head due to severe temperature gradients. Other important heat transfer considerations include the role of peak temperature in the onset of knock in spark-ignition engines, and the role of total coolant volume in engine warm-up time. As a general summary, it is important to keep the coolant velocity sufficiently high to avoid film boiling in the highest heat flux regions near the exhaust ports, and to not place cooling jackets where they are not needed. Cooling jacket flow optimization will be further discussed in Chap. 13. In addition to heat transfer considerations the designer must plan for proper filling and venting of the cooling jackets, and ensure complete sand clean-out during the casting process.

In the sand casting process the cooling jackets are created with cores that must then be broken out after the casting solidifies. The dimensions discussed in the previous section are met through the placement and geometry of the coolant cores. The cores must be designed to create the outer surfaces of the head bolt columns, the valve stem columns, and any material surrounding the spark plug or fuel injector if these cavities are placed such that they traverse upward through the head.

The cores are held in place with core prints through which the core sand is then removed. The resulting openings in the outer surfaces of the head are then machined to be circular, and stamped steel plugs are pressed into them to seal the cooling jackets. In some larger engines these openings may be sealed with gasketed cover plates. In the case of lost foam castings similar openings are required to remove the loose sand that filled the cooling jackets during casting. It is important to design the cooling jackets such that all of the sand can be easily removed. During the prototype casting stage this must be validated by sectioning head castings and ensuring that no sand remains in the casting. In some cases one factor in determining the core print locations is that of inserting cleaning rods or test instruments to ensure that sand is completely removed from critical locations. It is best to avoid thin or small diameter core sections, as these can break and be difficult to clean out.

In most engine designs the coolant enters the cylinder block first and then is transferred into the head through various drilled or cast passages surrounding the cylinders. The coolant then passes through the cylinder head jackets in one of two ways. In most automobile engines the coolant passes by each cylinder from one end of the head to the other. The coolant may be fed into the head entirely from one end, or through progressively sized passages surrounding each cylinder. In larger heavy-duty engines the coolant enters the cylinder head through passages surrounding each cylinder and is then transferred up through the head and collected again above each cylinder. The most common cooling circuits are each discussed in Chap. 13. In large heavy-duty engines it is also common to place an intermediate deck in the cylinder head, using upper and lower cooling jacket cores.

Filling the engine with coolant and proper venting will be discussed further in Chap. 13. From the standpoint of cooling jacket design it is important that air can always escape from the top of the jackets as the engine is filled, and that there are no locations where pockets of air or steam can be trapped. Such pockets may result in hot spots and overheating. If a large pocket of air flows as a unit to the water pump it may cause the pump to loose its prime and the coolant flow to stop, overheating the engine.

9.6 Oil Deck Design

The upper portion of the cylinder head, above the cooling deck, is referred to as the *oil deck*. Unlike the closed system of the coolant deck, the oil deck is an open system consisting of liquid oil and air/oil vapor. Pressurized lubricant for the valvetrain is discharged into the oil deck, where it collects and drains back to the crankcase. It is important to make sure all oil drains back to the crankcase, and that oil does not form stagnant pools which lead to aging, sludge, and varnish buildup. The oil deck includes provisions for the valve spring assemblies and valve train, and is also part of the crankcase breathing system. The oil deck will be sealed with a valve cover, so provisions are made for a gasketed joint around the perimeter of the valve train. Many of the design features of a complex oil deck can be seen in the cut-away in Fig. 1.7d and in the top view in Fig. 9.21.

Fig. 9.21 Oil deck, showing oil drain back on lower *right* corner

9.7   Recommendations for Further Reading

Few recent technical publications are available regarding the layout design of cylinder heads. The reader is referred generally to papers summarizing new engine releases, and also to the references following Chap. 10. The latter includes several papers covering cylinder head analysis.

The following paper provides a design overview, including both concept layout and structural analysis of a cylinder head with an integrated exhaust manifold (see Kuhlbach et al. 2009).

As a relatively new design approach, several publications can be found regarding reinforced plastic intake manifold design. The follow paper is recommended as a helpful overview (Tanaka et al. 2005).

References

Kuhlbach, K., Mehring, J., Bormann, D., Friedfeldt, R.: Cylinder head with integrated exhaust manifold for downsizing concepts. MTZ. **70**, 12–17 (2009)
Tanaka, H., Isomura, R., Katsuragi, S.: Use of CAE for Development of Plastic Intake Manifold. SAE 2005-01-1517 (2005)

Block and Head Development

10

10.1 Durability Validation

The concepts of reliability and durability were previously introduced in Chap. 3. That chapter was introduced with the example of the cylinder head to emphasize the importance of ensuring the durability of such major structural components. In this chapter discussion returns to the cylinder block and head, applying the concepts of Chap. 3 to the durability validation of these components. Both the cylinder block and cylinder head are expected to perform without failure for the engine's life to overhaul. In automobiles the expectation is extended to include reuse in at least one engine overhaul and in heavy-duty applications to several overhauls. This expectation translates into one of an extremely low failure rate (less than one in many thousands) after many miles or hours of operation. If this expectation is not met the engine quickly gains a bad reputation from which it is extremely difficult to recover. This is especially devastating when one considers the investment in tooling for a new engine. A battery of tests and analysis must be devised to absolutely ensure that these durability expectations are met—an extremely challenging endeavor, and a critical path in the timeline of engine development.

The challenge just described is made especially difficult by the multiplicity of loads seen by the block and head, and the complexity of the paths over which these loads are distributed. The remaining sections are intended to break down the types of loads, and provide examples of the tools and tests that can be applied in the durability validation process.

10.2 High-Cycle Loading and the Cylinder Block

The reciprocating engine is designed to provide a sealed chamber in which the combustion reaction can be harnessed to produce work. As the structural components sealing the combustion chamber the cylinder block and head must absorb the reaction forces resulting

© Springer Vienna 2016
K. Hoag, B. Dondlinger, *Vehicular Engine Design*, Powertrain,
DOI 10.1007/978-3-7091-1859-7_10

Fig. 10.1 Firing pressure load
paths through the cylinder
block and head

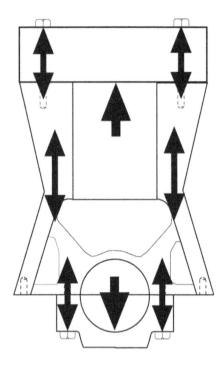

from the combustion process and work transfer. The primary load paths through which
the cylinder pressure reaction forces are carried are identified in Fig. 10.1. The cylinder
pressure acts simultaneously upward on the cylinder head and downward through the
piston, connecting rod, and crankshaft to the main bearing caps as indicated in the figure.
The cylinder head is clamped to the block with the head bolts that see a tensile force and
transmit this force into the block. The main bearing bolts see tensile forces that are again
transmitted back into the cylinder block. An especially important observation to be made
in Fig. 10.1 is that the firing load path seen by the cylinder block between the head bolts
and main bearing bolts is tensile. The cast aluminum or cast gray iron of the cylinder block
are best suited to compressive loads. Managing this tensile load path requires distribution
over a large cross-section, and as specific output increases may require special design
considerations—through-bolting that directly link the head bolts and main bearing bolts,
or alloy changes are examples. The entire assembly is rapidly loaded and unloaded as each
cylinder experiences cyclic pressure loading. The effects of these pressure forces and various other forces on the block and head will be further addressed in turn.

The discussion begins with the forces transmitted into the block at the main bearing
bores. In addition to cylinder pressure the resultant force transmitted to each main bearing bore includes contributions due to reciprocating and rotating forces generated by the
piston assemblies, connecting rods, and crankshaft. Each of these forces was previously
identified in Chap. 6. The resultant force at each main bearing bore varies continuously—
both throughout the engine operating cycle and as engine speed and load are changed. An

Fig. 10.2 Bulkhead and main bearing bore deflection due to firing pressure

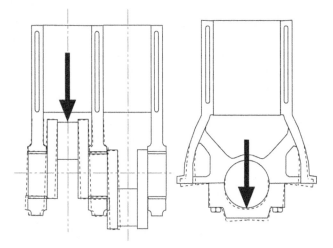

exaggerated picture of the reaction force at the moment of peak firing pressure is depicted in front and side views in Fig. 10.2. At this point in the operating cycle, when the peak firing pressure is reached, the main bearing bores adjacent to the particular cylinder are loaded near the center of the bearing caps as shown. The bearing caps and main bearing bolts experience tensile loading, as does the portion of the cylinder block in the load path between the threaded main bearing bolt bosses and those of the cylinder head bolts. The main bearing bulkhead experiences compressive loading in the region between the main bearing bolts. Looking at these loads in the side view, the main bearing bulkheads on either side of the firing cylinder experience bending away from the cylinder. This loads the portions of the bulkhead surface toward the cylinder in tension, and portions on the opposite surfaces in compression.

Assuming the engine operates in a four-stroke cycle the next revolution of the crankshaft reverses the direction of loading on each of the surfaces just discussed. At this time in the cycle the crankshaft is again approaching TDC, but during valve overlap when cylinder pressure is low. The resultant force is dominated by the decelerating reciprocating mass, and the crankshaft load is upward into the center of the block portion of the main bearing bore. The continuous operation of the engine over the four-stroke cycle results in a repeated cycling of the loads shown in Fig. 10.2 with their opposites. The main bearing bulkheads experience opening and closing as depicted in the front view and fore and aft bending depicted in the side view. Over the life of the cylinder block this load cycle will be experienced many millions of times, and the fatigue life of the block must be sufficient to ensure that cracking does not occur.

The forces applying load to the crankshaft and from there to the cylinder block can be readily calculated based on the equations introduced in Chap. 6. The distribution of crankshaft loads into the various main bearing journals is more difficult to quantify since in order to do so one must take into account the crankshaft stiffness. This too can be accomplished through structural analysis or careful measurement. High cycle mechanical loads whose magnitude and direction can be readily determined lend themselves well to rig testing. Such tests are commonly used for crankcase (skirt and bulkhead) durability validation. For example, a section of the crankshaft might be placed across two main bearing journals in the position near TDC at which the peak loads are seen, and fixed against rotation. Alternating load is applied to the crankshaft at the rod bearing location using a servo-hydraulic actuator. The cyclic frequency of the alternating load is significantly higher than that seen in engine operation, and the load magnitude can be increased a small percentage to further accelerate the test. The most difficult aspect of such testing is that of fixturing the block to maintain approximately the same stiffness that would be seen in the operating engine. For this effort it is important to match the measurements recorded with strain gauges at various locations in an operating engine. In new engine designs an initial assessment of crankcase loading can be made using structural analysis techniques (finite element or boundary element) before prototype cylinder blocks are available for rig testing.

Another important rig testing technique that can be used to simulate the load paths depicted in Fig. 10.1 is hydraulic pulsator testing. Depending on the needs of the particular project all or some of the loads shown in Fig. 10.1 can be simulated. The cylinder is filled completely with grease or hydraulic oil and sealed using a dummy piston and cylinder head or cover plate. The piston position can be fixed using the connecting rod and crankshaft. The actual cylinder head and gasket can be used, or if interest is only in the crankcase loads a cover plate can be used. The oil is then cyclically pressurized to represent cylinder pressure. This test is commonly used to evaluate head gasket durability as well as the effects of pressure-induced loading throughout the cylinder block. In some cases the test has been used to look at interactions between cylinders and the effects of a misfiring cylinder.

10.3   Modal Analysis and Noise

Closely connected to the subject of high-cycle crankcase loading is that of the block's role in noise attenuation and control. While a detailed discussion of noise control is beyond the scope of the book the basic concepts will be introduced here with an eye to the role of the cylinder block.

Noise is a cyclic pressure fluctuation carried through the atmosphere. A given noise signature—how it is perceived by the listener—is dependent on both its amplitude and its frequency. Noise sources in the engine include the combustion process, gear and chain drives, the valve train, secondary piston motion, and the intake and exhaust processes. The paths by which the noise is transmitted through the engine and radiated to the atmosphere are summarized in Fig. 10.3. Cyclical movements of large flat surfaces of the engine create and modify the pressure pulses sent into the atmosphere thus impacting the noise signature. As shown in Fig. 10.3 the block acts as both a transmitter to other parts of the engine and directly as an emitter. The block transmits high-cycle forces to the oil pan, front

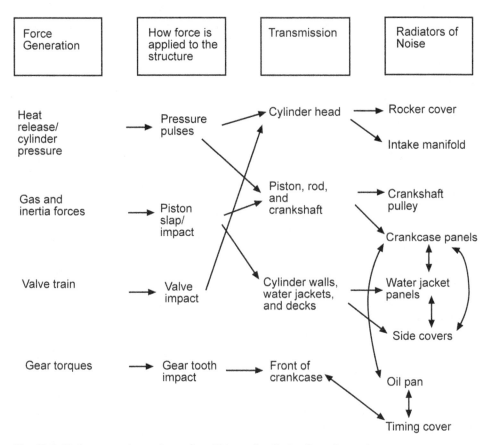

Fig. 10.3 Noise generation and transfer within, and radiation from the engine

cover, and through the cylinder head to the valve covers. At certain frequencies dependent on their structure each of these components will resonate, causing large increases in the amplitude of their movement, and thus large increases in the resulting pressure pulses and noise.

Several example block mode shapes are depicted in Fig. 10.4. These modes and the frequencies at which they occur are important both from the standpoint of noise control and block fatigue. Clearly at the frequencies at which resonance occurs the increased block motion also results in increased strain, leading in turn to fatigue damage. The critical frequencies and mode shapes of a given engine can be identified by mounting accelerometers at various locations on the cylinder block and sweeping through the operating speed range. The movement of the block under resonance can be visually observed using a strobe light.

Modifying the design of the component to change its stiffness will change the frequencies at which resonance occurs. Cast iron blocks possess relatively high stiffness, suggesting that resonant frequencies will be fairly high. This may limit, but in most cases not eliminate the resonant modes that occur over the operating speed range of the engine.

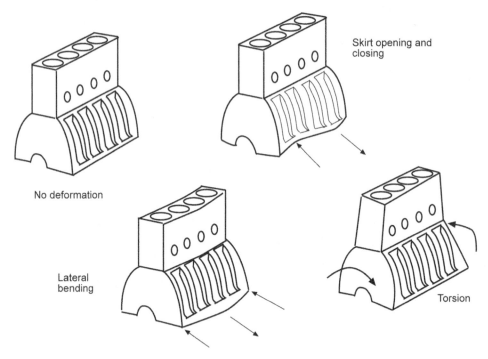

Fig. 10.4 Block mode shapes resulting from high cycle loading at various frequencies

A similar design constructed from aluminum will resonate at lower frequencies and as a result may require a greater amount of ribbing to increase its stiffness. Oil pans and gear covers made from stamped steel are far less stiff than the cylinder block, and will resonate at relatively low frequencies. Ribbing of these components must be carefully placed in order to minimize noise and address the problem of fatigue cracking.

In addressing resonance the block cannot be considered in isolation, but must be looked at as an assembly with any other components that significantly impact its assembly stiffness. Block movement is greatly impacted by cylinder head stiffness. In larger engines the use of individual heads for each cylinder versus a single slab head for a bank of cylinders has an important effect on required block stiffness. In engines having a cast oil pan or a cast front cover these too will impact the assembly stiffness and resulting block movement.

Another important issue pertaining to crankcase movement is that of main bearing journal movement and its effects on the crankshaft and bearing loads. Especially in aluminum engines specific steps may need to be taken to minimize journal movement. Design variations such as the ladder frame block were previously introduced in Chap. 8 when discussing the crankcase design. In some cases an additional frame is added, tying the main bearing caps together, or a perimeter frame may be added immediately inboard of the pan rails; these approaches are shown in Fig. 10.5.

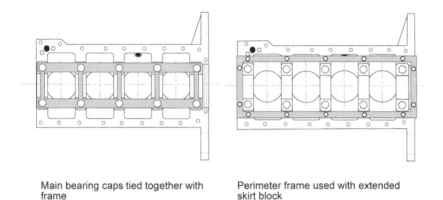

Main bearing caps tied together with
frame

Perimeter frame used with extended
skirt block

Fig. 10.5 Approaches for increasing stiffness and reducing movement of the crankcase

10.4   Low-Cycle Mechanical Loads

A more difficult category of loads to identify in cylinder block development is that of
low-cycle mechanical loads. Examples include application-dependent loads that occur in
reaction to the torque output, and vehicle shock loads transmitted through the suspension.
The forces generated in reaction to the engine's torque output are depicted in Fig. 10.6.
While the steady-state loads and even the loads experienced under any given transient can
be readily calculated, the challenge lies in the variety of duty cycles to which the engine
might be exposed by different operators. Fatigue life estimation under low-cycle loading,
through use of cumulative damage theory, was presented in Chap. 3. However, such life
estimates are entirely dependent on the particular duty cycle to which the engine has been
exposed. How many rapid and how many slow accelerations is the engine anticipated to
experience over its life? How many accelerations occur on up- or down-hill gradients?
Will the vehicle pull a trailer during any of these accelerations? How often will the vehicle
be bogged down in snow or mud, and how aggressively and rapidly will the driver shift the
transmission from first gear to reverse and back? These questions give just a few examples
of the wide array of variables that impact fatigue life under the low-cycle drivetrain loads.
The required approach must be to design very conservatively to ensure that acceptable

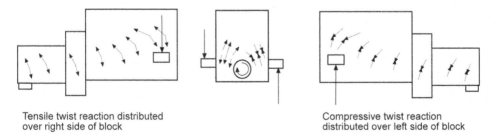

Tensile twist reaction distributed
over right side of block

Compressive twist reaction
distributed over left side of block

Fig. 10.6 Block loading in response to driveline torque

fatigue life is achieved under an aggressive array of loads. A combination of strain gauge measurements in vehicles exposed to aggressive loads, and either computational or bench test techniques can be used to validate the durability of the engine under such loading.

The next category of vehicle-induced loads that must be considered in block validation is that of forces arising from bumpy terrain, and transmitted through the suspension. Components mounted in such a way that their mass is cantilevered from the mounting pad are especially susceptible. Examples include oil filter heads, oil coolers, turbochargers, starter motors, dipstick tubes, and intake and exhaust piping. Experimental development is often done using a vibration table. The component is instrumented with an accelerometer, and measurements are taken versus time in a vehicle operated on a bumpy road. The component and its mounting structure are then placed in a fixture on a shaker table and a rig test is developed to match the loads seen in the field. This method can be very effective in determining the fatigue life of the mounting flange, and may be used by itself or in conjunction with structural analysis tools to rapidly improve the design as needed. For this technique to be effective it is imperative that not only the component but the mounting fixture stiffness duplicate that seen in the engine.

10.5   Block and Head Mating and the Head Gasket

The cylinder head gasket joint is among the most critical and challenging aspects of any engine design. Due to its proximity to the combustion chamber thermal loading becomes important as well as mechanical loading. The clamping forces at the head bolts vie with the unloading forces of cylinder pressure, and the combustion seal must be maintained in the face of this high-cycle alternating force over the life of the engine. Any sliding that occurs between the gasket and either the block or head surface results in fretting, thus aggravating wear and the potential for failure of the seal. In addition to sealing the combustion chamber the same gasket is called upon to maintain the seal at various oil and coolant transfers between the block and the head.

Gasket design will be discussed further in Chap. 14, but the specific case of the combustion seal is introduced here. Figure 10.7 provides an example of the clamping load as it varies around the perimeter of a cylinder. The clamping load, actually acting in the vertical direction is laid out horizontally around the perimeter to better visualize its variation. The length of each vector from the cylinder center line to the arrowhead represents the vertical load seen at that location of the seal ring. The clamping load is seen to be at its maximum adjacent to each of the head bolts, and to drop off between the bolts. The amount by which it drops off is determined by cylinder head and block stiffness along the mating surface, and the distance between the head bolts. The clamping load is further affected by temperature changes—as the components heat up their thermal expansion results in a clamping load increase. Increasing the head bolt lengths reduces the clamping load change with

Fig. 10.7 Sealing pressure distribution around the cylinder perimeter

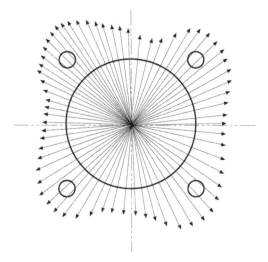

temperature. Finally, as cylinder pressure increases its effect is to push the head away from the block and reduce clamping load. This aspect will be further discussed in Chap. 14. The increased head bolt length required to minimize temperature effects results in a greater clamping load reduction due to cylinder pressure. Relatively long head bolts are generally used because the engine must maintain its combustion seal over a range of operating temperatures. The design is then optimized to maintain the combustion seal with these long bolts. The design objective can be summarized as maintaining the clamping load below that which would crush the gasket beyond limits required for maintaining its fatigue life, and maintaining the clamping load above that required for sealing. Around the entire perimeter of each cylinder the clamping load must be maintained between these two limits under all operating conditions.

In engines using removable cylinder liners the additional variables of liner protrusion, liner stiffness, and block stiffness at the liner seat impact the clamping load along with those discussed previously. In Fig. 10.8 two engine designs are shown—one having the liner seat above the cooling jacket, and the other placing it below the jacket. The various paths impacting stiffness are shown. The lower seating position reduces the alternating load seen by the head gasket because of the increased length of liner compliance between the head gasket and liner seat. However, when the clamping forces at the head bolts are applied a significant portion of the ring travel area of the liner is placed in compression, increasing the potential for distortion. This approach also places additional regions of the cylinder block between the head bolt bosses and the liner seat in tension.

Hydraulic pressure testing was previously described, and provides an effective rig test for fatigue life validation. Structural analysis techniques are increasingly used to aid in developing the head gasket joint. The modeling techniques are complex due to the need

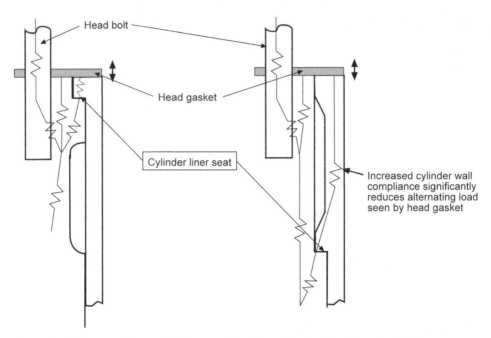

Fig. 10.8 The impact of removable liner stop location on head gasket alternating load

to consider both the elastic and plastic properties of the gasket material as well as the complexities associated with modeling the bolted joint. Load sensitive paper is commonly used to experimentally quantify the clamping load variation, and gasket manufacturers have developed strain-gauged gaskets for engineering tests that provide further insights. Strain gauges and ultrasonic techniques are available to determine the vertical load at each head bolt, since the correlation between applied torque and clamping load is quite poor.

Thermal loading at the joint between the cylinder head and the block is another important source of cylinder block loading. The cylinder head is exposed directly to the combustion chamber, resulting in thermal growth of the firedeck surface as engine load is increased. The cylinder block has little direct contact with the combustion chamber in parent bore engines, and none in engines with removable liners. In addition the cylinder walls are surrounded by cooling jackets. The cylinder block thus remains much cooler than the cylinder head firedeck, and constrains the growth of the head. This places an important compressive load on the cylinder head firedeck surface, and places the block fire deck in tension. In engines with a single cylinder head covering an entire bank of cylinders the compressive stresses progressively increase from the end cylinders to those at the center of the bank. The tensile stresses in the block tend to be especially high at the outside corner head bolts. Individual heads significantly reduce these stresses. In engines with long banks of cylinders having a single head the head is sometimes saw-cut between cylinders to provide stress relaxation.

10.6 Cylinder Head Loading

The various loads seen by the cylinder head were identified in Chap. 3, with reference to Fig. 3.4, repeated here as Fig. 10.9. Beginning with assembly loads the firedeck experiences a significant tensile offset due to the press-fit of the valve seat into the cylinder head. The press-fit dimensions are selected to ensure that the seats will not lose their press over the wide range of temperatures seen at this joint from cold ambient start-up to full load operation. The next important assembly load is that resulting from head bolt clamping. In many cases the resulting compressive loads seen by the head bolt towers approaches or slightly exceeds the compressive strength of the material, causing some yielding of the bolt towers. This region of the head sees very little alternating load, so one would not expect fatigue cracking at the towers themselves. However, the high compressive stress within the towers creates high shear stress at the regions where the towers blend into port or deck surfaces. This is often a source of fatigue cracking.

Turning next to the alternating loads seen by the cylinder head the most important are high-cycle cylinder pressure loads and low-cycle thermal loads. As the cylinder pressure increases during each operating cycle the bridge regions of the firedeck surface see a tensile load resulting from the constrained bending of the bridge. While the gas temperatures in the combustion chamber swing over an extremely wide range during each operating cycle this swing occurs so rapidly that wall temperatures remain virtually constant. This will be examined further in Chap. 13 where engine cooling is discussed. It is therefore the low-cycle temperature changes that occur with changing engine load and engine warm-up that determine the thermal loads seen by the cylinder head and other combustion chamber components. These loads result in compressive stresses in the thin bridges of the firedeck that approach and often exceed the compressive yield strength of the head casting. Much

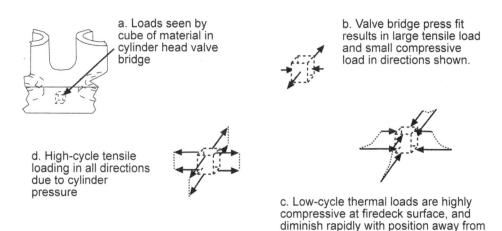

a. Loads seen by cube of material in cylinder head valve bridge

b. Valve bridge press fit results in large tensile load and small compressive load in directions shown.

d. High-cycle tensile loading in all directions due to cylinder pressure

c. Low-cycle thermal loads are highly compressive at firedeck surface, and diminish rapidly with position away from firedeck.

Fig. 10.9 Summary of cylinder head firedeck loading at the valve bridges

of the cylinder head development effort must focus on these thermal loads, and they will be discussed in further detail in the next section. The other load sources identified here must be considered in interaction with the thermal loads. In addition to those already identified, valve train loads, and more significantly fuel injection loads with some diesel fuel systems, may also make high-cycle load contributions of sufficient magnitude that they cannot be ignored.

10.7   Thermal Loads and Analysis

In order to best appreciate the importance of thermal loading in the cylinder head it is instructive to look at the temperature gradient through the firedeck—from combustion chamber surface to cooling jacket surface. Peak temperatures seen in the combustion chamber are well over 2500 K. While the combustion chamber walls are fortunately unable to respond to the rapid excursions to such temperatures the cycle averaged gas temperature typically ranges from 900 to 1100 K. A firedeck on the order of 8–10 mm thick separates the hot combustion gas from coolant that will be maintained at temperatures between 360 (heavy-duty diesels) and 385 K (automobile engines). Convective resistance between the combustion gas and the head surface, and again between the cooling jacket surface and coolant reduce the gradient actually seen by the head but the remaining gradient seen through the firedeck is quite typically 260–290 K with grey iron alloys and 130–150 K with aluminum alloys. Such steep temperature gradients over relatively narrow sections give rise to high compressive loads because the combustion chamber surface attempts to expand with increasing temperature, but is constrained by the much cooler material beneath the surface. While not as severe the radial temperature gradient on the combustion chamber surface, in combination with constraints imposed by the cylinder block, further restrict the thermal growth of the firedeck's hottest regions. The maximum temperature is typically seen in thin bridges surrounding the exhaust valves, or between exhaust valves and the spark plug or fuel injector. Maximizing the cylinder head fatigue life requires controlling the temperature in these maximum temperature regions, and controlling the temperature gradients to minimize severity in any one region of the head.

The thermal fatigue cycle of interest for cylinder head development is that which occurs from shutdown or cold idle to full-load operation. It is with each occurrence of this cycle that the cylinder head undergoes fatigue damage. Generally the most useful tool for thermal fatigue validation is structural modeling. Once the model is constructed only two cases need to be considered—full load and either idle or shutdown. Model construction and careful validation of the full-load temperature profile require the greatest resource expenditure. It is typical to use a relatively coarse model of the entire cylinder head assembly for determining the boundary conditions, and then a much finer model of the specific region of interest. The edge nodes of the sub-model are deflected based on the distortion predicted for the complete assembly model. Artificially high stresses are noted at the edges of the sub-model, but these rapidly decay, providing an accurate stress profile at the location of interest. The result is a relatively small but highly detailed model that can now be used to assess changes in fatigue life with various design, load, or material property changes. When assessing a new engine design it is recommended that the modeling results be compared with a similar analysis of a production engine for which the fatigue life can be quantified. In other words, no matter how carefully the model has been developed it cannot be used alone to determine absolute fatigue life.

It is extremely difficult to accurately simulate fatigue loading using a component rig test, but engine testing can sometimes be used effectively. A cyclic test can be developed in which the engine load and coolant temperature are both varied over wide ranges, and the cycle rate is selected to best optimize the combination of thermal load swing and re-quired test time. Such tests are generally referred to as *deep thermal cycles*, and a plot of temperature versus time for such a test is presented in Fig. 10.10. Caution in the use of such tests must be emphasized. The tests can identify severe problems, but the resulting design modifications may or may not be sufficient. Because the cylinder head passes such a test can it be concluded that it will not fail in the field? Such conclusions can only be drawn in cases where the test has been well correlated to field history. Do cylinder heads that have cracking problems in the field crack on this test? Do cylinder heads that perform well in the field pass this test? Only when both of these questions can be answered affir-matively will such a test be dependable for fatigue life validation.

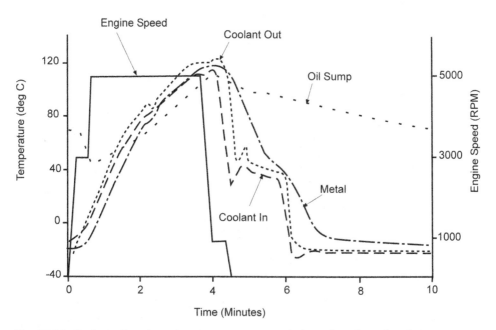

Fig. 10.10 Coolant, oil, and metal temperature response during a deep thermal cycle test

10.8   Recommendations for Further Reading

The majority of published studies of high cycle loading and vibration in cylinder blocks focuses on NVH development. The following papers begin with two general studies of cylinder block movement and vibration, and two intended to address main bearing journal loads. These are then followed by papers addressing NVH (see Ishihama et al. 1981; Honda et al. 2000; Knoll et al. 1997; Vorwerk et al. 1993; Ebrat et al. 2003; Okamura and Arai 2001; Stout 2001).

A general presentation on NVH development in engines can be found in the following paper (see Beidl et al. 1999).

A recent overview of cylinder block vibration analysis and testing is given in the following paper (see Novotny and Pistek 2010).

A critical aspect of cylinder block and head structural development pertains to the head gasket joint. The following papers provide recent analyses and tests focusing on this joint (see Popielas et al. 2000, 2003a, b).

Cylinder head thermal loading receives considerable attention in the published literature, and the following recent studies are recommended (see Gocmez and Pischinger 2011; Gocmez and Lauer 2010; Su et al. 2002; Takahashi et al. 2002; Lee et al. 1999; Koch et al. 1999).

The following paper discusses a particular aspect of structural development—that of cam bore distortion and machining (see Liu et al. 2004).

References

Beidl, C.V., Rust, A., Rasser, M.: Key Steps and Methods in the Design and Development of Low Noise Engines. SAE 1999-01-1745 (1999)

Ebrat, O., Mourelatos, Z., Hu, K., Vlahopoulos, N., Vaidyanathan, K.: Structural Vibration of an Engine Block and a Rotating Crankshaft Coupled Through Elastohydrodynamic Bearings. SAE 2003-01-1724 (2003)

Gocmez, T., Lauer, S.: Fatigue Design and Optimization of Diesel Engine Cylinder Heads, CIMAC Congress, Bergen, Norway, Paper No. 90 (2010)

Gocmez, T., Pischinger, S.: A contribution to the understanding of thermomechanical fatigue sensitivities in combustion engine cylinder heads. Proc. Inst. Mech. Eng., Part D: J. Automob. Eng. **225**, 461–477 (2011)

Honda, Y., Wakabayashi, K., Kodama, T., Kihara, R.: A Basic Study on Reduction of Cylinder Block Vibrations for Small Diesel Cars. SAE 2000-01-0527 (2000)

Ishihama, M., Hayashi, Y., Kubozuka, T.: An Analysis of the Movement of the Crankshaft Journals during Engine Firing. SAE 810772 (1981)

Knoll, G., Schönen, R., Wilhelm, K.: Full Dynamic Analysis of Crankshaft and Engine Block with Special Respect to Elastohydrodynamic Bearing Coupling. ASME 97-ICE-23 (1997)

Koch, F., Maassen, F., Deuster, U., Loeprecht, M., Marckwardt, H.: Low Cycle Fatigue of Aluminum Cylinder Heads—Calculation and Measurement of Strain under Fired Operation. SAE 1999-01-0645 (1999)

Lee, K.S., Assanis, D.N., Lee, J., Chun, K.M.: Measurements and Predictions of Steady-State and Transient Stress Distributions in a Diesel Engine Cylinder Head. SAE 1999-01-0973 (1999)

Liu, E.A., Winship, M., Ho, S., Hsia, K.-T., Wehrly, M., Resh, W.F.: Engine Cambore Distortion Analysis from Design to Manufacturing. SAE 2004-01-1449 (2004)

Novotny, P., Pistek, V.: New efficient methods for powertrain vibration analysis. Proc. Inst. Mech. Eng., Part D: J. Automob. Eng. **224**, 611–629 (2010)

Okamura, H., Arai, S.: Experimental Modal Analysis for Cylinder Block-Crankshaft Substructure Systems of Six-cylinder In-line Diesel Engines. SAE 2001-01-1421 (2001)

Popielas, F., Chen, C., Obermaier, S.: CAE Approach for Multi-Layer-Steel Cylinder Head Gaskets. SAE 2000-01-1348 (2000)

Popielas, F., Chen, C., Ramkumar, R., Rebien, H., Waldvogel, H.: CAE Approach for Multi-Layer-Steel Cylinder Head Gaskets—Part 2. SAE 2003-01-0483 (2003a)

Popielas, F., Chen, C., Mockenhaupt, M., Pietraski, J.: MLS Influence on Engine Structure and Sealing Function. SAE 2003-01-0484 (2003b)

Stout, J.: Engine Excitation Decomposition Methods and V Engine Results. SAE 2001-01-1595 (2001)

Su, X., Lasecki, J., Engler-Pinto, C.C., Tang, C., Sehitoglu, H., Allison, J.: Thermal Fatigue Analysis of Cast Aluminum Cylinder Heads. SAE 2002-01-0657 (2002)

Takahashi, T., Nagayoshi, T., Kumano, M., Sasaki, K.: Thermal Plastic-Elastic Creep Analysis of Engine Cylinder Head. SAE 2002-01-0585 (2002)

Vorwerk, C., Busch, G., Kaiser, H-J., Wilhelm, M.: Influence of Bottom End Design on Noise and Vibration Behavior of 4-Cylinder In-Line Gasoline Engines. SAE 931315 (1993)

Engine Bearing Design

11

Within any given engine a large number of bearings incorporating several design, material, and operational variations are seen. While needle, ball, or roller bearings are sometimes seen in automotive and heavy-duty engines the majority of engines for these applications use primarily plain bearings. Roller bearings are receiving increasing attention due to their lower oil supply requirements and potential for reduced friction and parasitic losses in automotive engines. Increased space requirements, and in the case of connecting rod bearings increased rotating mass, must be weighed against potential attractions. Automotive engine examples of ball and roller bearing are seen in valvetrain and balancer shaft bearings. The discussion presented in the following sections will be limited only to plain bearings.

11.1 Hydrodynamic Bearing Operation

Crankshaft rod and main bearings, camshaft bearings and (if used) roller follower bearings operate under fully hydrodynamic lubrication under all conditions except engine start-up. Bearing areas at each of these locations are determined such that hydrodynamic operation is ensured under maximum load conditions. The camshaft and roller follower bearings are continuous shell, single-piece bearings. In most multi-cylinder engines assembly requires the use of split bearing shells at the crankshaft main and connecting rod bearings.

The bushings found at the piston pin, rocker levers, and (if used) cam followers experience stop-and-start motion even when the engine is operating at a steady speed. The stop-and-start motion precludes hydrodynamic operation, so these bearing surfaces are designed to operate under mixed film lubrication. Bearing area and material selection decisions are based on controlling the wear rates within limits defined by expected engine life.

© Springer Vienna 2016
K. Hoag, B. Dondlinger, *Vehicular Engine Design*, Powertrain,
DOI 10.1007/978-3-7091-1859-7_11

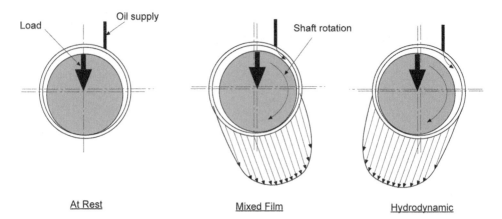

Fig. 11.1 Plain bearing operation, showing start-up and fully hydrodynamic conditions

Hydrodynamic bearing operation is summarized in Fig. 11.1. In each of the cases shown in the figure the shaft is loaded in the vertical direction. The bearing clearance is exaggerated to aid in viewing the operating characteristics. In the panel at the left the shaft is at rest and thus its contact with the bearing is centered about the vertical. The shaft is offset from the bearing centerline by the radial clearance between the shaft and journal (bearing surface).

In the center panel of the figure the shaft begins rotating. Pressurized oil is now being fed to the bearing as shown, and the spinning shaft carries oil into the gap between the shaft and bearing. While the oil film "lifts" the shaft away from the bearing the clearance (oil film thickness) is not yet sufficient to completely separate the surfaces. Because of the metal-to-metal contact between the shaft and bearing the shaft tends to "walk up" the bearing, and the ***contact patch*** is centered upstream from the direction in which the load is applied. The contact patch is the region between the shaft and the bearing over which load is distributed. The minimum oil film thickness occurs at some location within the contact patch. During this time of mixed film lubrication the shaft offset within the journal is the difference between the radial clearance and the minimum oil film thickness, and the direction of the offset is upstream of the direction from which the load acts.

As shaft speed increases and additional oil is supplied the film thickness increases sufficiently to completely separate the shaft from the bearing. The film thickness required for complete separation is dependent on the shaft and bearing surface finishes. The transition to fully hydrodynamic operation is shown in the panel on the right side of Fig. 11.1. Once hydrodynamic operation is achieved the contact patch shifts downstream of the applied load direction as shown. The shape and location of the oil film pressure distribution can be theoretically derived from fluid mechanics for ideal hydrodynamic operation.

From this discussion the variables impacting hydrodynamic operation can be identified. The combination of the magnitude of applied load and the bearing diameter and width define the resulting oil film pressure. The oil viscosity, the radial clearance between the shaft and journal, and the shaft speed work to define both the leakage rate and the film thickness with given oil film pressure; the oil supply rate must be maintained at or above the leakage rate. Finally, the shaft surface finish determines the minimum oil film thickness required for hydrodynamic operation.

It should be noted that this description of hydrodynamic bearing operation assumes zero deflection of either the shaft or the journal. Engines often experience significant journal deformation, resulting in what is termed *elasto-hydrodynamic* operation. The elasticity of the journal may be sufficient to result in two or three distinct minimums in the oil film thickness, and two or three distinct film pressure peaks. Block movement and shaft deflection may further impact the instantaneous relationship between the shaft and journal centerlines. Machining tolerances impact the initial alignment between shaft and journal and must also be controlled. Each of these effects will be further discussed later in this chapter.

As shown in the middle and right panels in Fig. 11.1 the load distribution causes a distribution of film pressures. The film pressures are well above the supplied oil pressure; this is an important point, necessitating that whenever possible the oil supply drilling be placed upstream of the contact patch and not within it for two reasons. First, because pressures within the contact patch are above the supply pressure oil flow from a drilling directly into the contact patch would be effectively shut off when the bearing was loaded. Second, the diameter of the oil drilling would be lost from the bearing contact area.

Before leaving the general discussion of hydrodynamic bearing operation the various operating limits for plain bearings must be considered. The general limits are identified in Fig. 11.2 as a function of shaft speed and bearing load. The axes in this figure are shown without numbers since the specific limits are strongly dependent on the engine and the lubricant properties. As has already been discussed as load is increased the bearing is limited by asperity contact. Since this is the limit determined by the transition to and from hydrodynamic operation it should not be surprising that higher loads can be accommodated

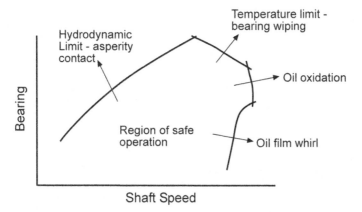

Fig. 11.2 Limiting regimes for plain bearing operation

as the shaft speed is increased. At high shaft speeds the shear forces within the lubricant result in significant temperature increases. The speed and load are limited by temperature limits of the bearing itself—the bearing surface may reach the temperature at which the material locally melts and streaks or wipes along the shaft. At slightly lower loads the temperature may not be sufficient to melt the bearing surface, but a longer term phenomenon is lubricant oxidation breakdown due to the elevated temperature. Finally, at high shaft speeds and light loads shear within the oil film may result in turbulent vortices causing the oil to spin or whirl within the film. This increases friction, oil foaming and cavitation, and may then result in damage to the bearing surface.

11.2   Split Bearing Design and Lubrication

As was stated at the outset of this chapter the need to assemble the bearing journal around a shaft necessitates the use of split bearings for the crankshaft main and rod bearings in multi-cylinder engines. The basic features of split bearing design are summarized in Fig. 11.3. The journal and the bearing shell are split along the shaft centerline as shown. If the joint shown represents a main bearing, the bearing cap is bolted to the block after installation of the crankshaft; if the joint shown is a connecting rod the rod is clamped around the shaft, and is held in place with the rod bearing cap. In either case the bearing shell is clamped in position as the cap is installed. The two halves of the bearing shell are each made slightly larger than their respective halves of the journal, resulting in a bearing *crush load* when the cap is clamped in place. The crush load is intended to create a uniform outward force around the perimeter of the bearing shell, holding it firmly in place. This is a critical design variable, as too little crush results in bearing shell movement against its journal, and fretting wear on the back surface of the shell. With time the

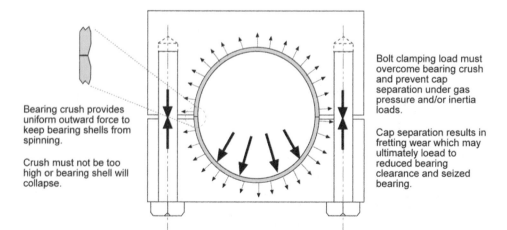

Bearing crush provides uniform outward force to keep bearing shells from spinning.

Crush must not be too high or bearing shell will collapse.

Bolt clamping load must overcome bearing crush and prevent cap separation under gas pressure and/or inertia loads.

Cap separation results in fretting wear which may ultimately loead to reduced bearing clearance and seized bearing.

Fig. 11.3 Split bearing shell design principles

bearing loses load carrying capability due to loss of supporting material, and this may in turn lead to a bearing fatigue failure due to overload of the remaining material. If the crush load is too great it may damage the bearing shell during assembly, and may force bearing material against the shaft. High crush load also contributes to cap separation. Firing forces on the main bearings and inertia forces on the rod bearings both act in the direction of the cap, tending to pull the cap away from its seat. If these forces (along with the reaction to bearing crush) are high enough to momentarily separate the cap from its mating surface fretting wear results at this joint. Over time the resulting material loss reduces the bearing clearance, and may ultimately lead to bearing scuffing or seizure.

The peak loads resulting from firing forces act on the cap portion of the main bearing. Main bearing bolt design options for preventing cap separation and minimizing unwanted journal movement were discussed in Chap. 8. A further main bearing design challenge is that of supplying lubricant where it is most needed. As was discussed in the first section of this chapter ideal placement of the oil supply drilling is immediately upstream of the contact patch. In the case of the main bearing this would require placing the oil drilling in the main bearing cap, adding considerably to cost even if space were available. The practical requirement is to supply the main bearings with oil from cross-drillings from the oil rifle. The cross-drillings supply oil into the block portion of the main bearing journal, from where it must traverse along the bearing surface to where it is most needed. A design guideline is thus to place the location of the cross-drilling as close as is practical to the cap mating surface on the side of the engine that will result in crankshaft rotation carrying the oil directly toward the cap. Since this design guideline impacts placement not only of the cross-drillings but the oil rifle drilling, the oil pump, and the camshaft (any component or system requiring pressurized oil) it may not be possible to achieve in every engine design. Another design option is to place a groove in the center of the main bearing shell, carrying oil from the supply drilling around the entire bearing perimeter. This holds the advantages of supplying oil directly to the needed regions of the bearing, and better supplying the crankshaft drilling to the connecting rod bearings, but it also has the disadvantage of reducing bearing area by the width of the oil groove. Since bearing area is most needed (loads are highest) at the mid-position of the bearing cap another option is to place the oil groove only in the upper bearing shell, with a smooth transition eliminating the groove as the mid-position in the cap is approached. This option is very attractive from a design standpoint, but is chosen with caution because of the risk of reversing the bearing shells in the field.

The need for a thrust bearing surface at one of the crankshaft main bearings was previously discussed in Chap. 8. In most automotive engines the thrust surfaces are incorporated in one of the main bearing shells. In larger engines separate thrust bearings will be mounted adjacent to the main bearing shells on one bulkhead.

As with the main bearings, the direction and magnitude of loading seen by the rod bearings changes continually throughout the engine operating cycle. This is depicted in Fig. 11.4. The peak firing loads are seen just after TDC in the expansion stroke, and act near the center of the rod. A second peak results from the reciprocating inertia forces and

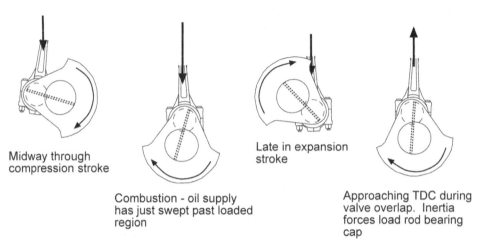

Midway through
compression stroke

Combustion - oil supply
has just swept past loaded
region

Late in expansion
stroke

Approaching TDC during
valve overlap. Inertia
forces load rod bearing
cap

Fig. 11.4 Rod bearing loading at various points in the four-stroke operating cycle

is seen near TDC during the valve overlap period. It acts near the center of the rod cap. These forces will be further discussed in the next section.

The connecting rod bearing faces the unique situation of receiving its lubrication from a drilling in the spinning crankshaft. Pressurized lubricant is supplied to the main bearings, and from there passes through angled drillings in the crankshaft to reach the connecting rod bearings. This holds the advantage of sweeping oil around the entire bearing perimeter as the crankshaft spins. Unfortunately it also requires that the oil drilling break out in or very near the loaded region associated with the peak firing forces. The design objective is to angle the drilling such that it breaks out ahead of the TDC position. The oil drilling should sweep across the central region of the rod bearing, supplying a film of oil immediately before the peak loads are seen. The placement of the drilling is optimized within crankshaft geometry constraints.

11.3   Bearing Loads

The magnitude and direction of the forces acting on each connecting rod bearing can be calculated as a function of crank angle using the equations presented in Chap. 6. The forces include the resultant of the cylinder pressure and reciprocating forces transmitted through the connecting rod, and the rotational force contributed by the mass of the lower end of the connecting rod. Each of these forces and their calculation is identified in Fig. 11.5. The resulting bearing load is often depicted using polar load diagrams, two examples of which are given in Fig. 11.6. The trace shown in either figure is that of the tip of a vector whose origin is at the rod bearing centerline. At any crank angle position the length of the vector is proportional to the magnitude of the force applied to the bearing, and the direction indicates the direction in which the force acts. Using the polar load diagram one can clearly see the peak bearing force resulting from combustion and nearly

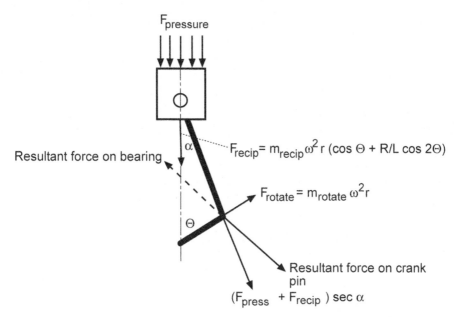

Fig. 11.5 Mechanisms leading to the resultant load seen by connecting rod bearings

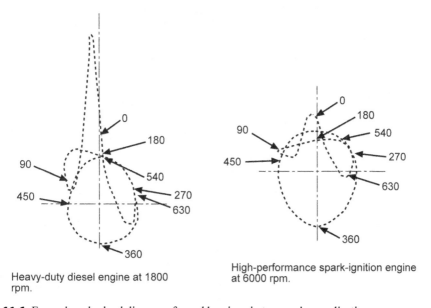

Heavy-duty diesel engine at 1800 rpm.

High-performance spark-ignition engine at 6000 rpm.

Fig. 11.6 Example polar load diagrams for rod bearings in two engine applications

centered along the rod axis (just after 0° crank angle in the figures, where zero has been chosen as TDC between compression and expansion). One crankshaft revolution later (360° crank angle) the piston is again at TDC during valve overlap. The cylinder pressure is low and the inertia force resulting from deceleration of the piston assembly results in a load peak centered in the connecting rod cap. The relative magnitudes of firing pressure and inertia forces are quite different when comparing the two polar load diagrams in Fig. 11.6. The diagram shown on the left in Fig. 11.6 is that for a heavy-duty diesel engine, where cylinder pressures are quite high, and because the engine speed is relatively low, the inertia force is considerably lower. The case shown on the right in Fig. 11.6 is that of a high-performance spark-ignition engine. The engine speed is quite high, and the cylinder pressure is well below half that in the diesel engine. Several items of information important to the engineer can be gained from these polar load diagrams. The first is of course the peak load magnitude as this will be required in order to size the bearing and ensure that fully hydrodynamic operation is achieved. Second, the peak inertia load will be important in developing the connecting rod cap and the joint between the cap and rod. Finally, while the loads are much lower along the parting plane between the rod and cap they are not zero and very little bearing load can be handled along this plane. This becomes especially important in engines where the parting plane is placed on an angle. The polar load diagram provides guidance in ensuring that the parting plane is selected at a location where loading is as low as possible.

Figure 11.7 progressively identifies the loads seen by the main bearings. At least in theory the same calculation methods can be carried over to the main bearings, however the situation quickly becomes much more complex. The loads generated at each connect-

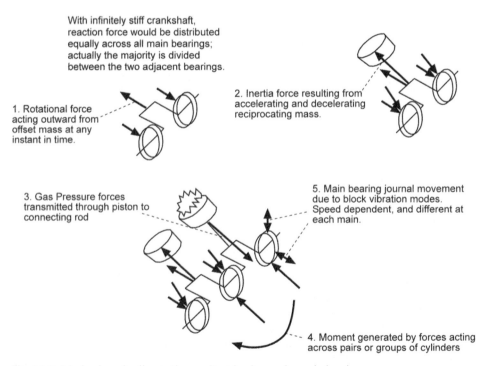

Fig. 11.7 Mechanisms leading to the resultant loads seen by main bearings

ing rod journal are the same as those seen at the rod bearings with the exception that the rotating mass now includes that of the crankshaft as well as the bottom end of the rod. If the crankshaft and cylinder block were infinitely stiff the load generated at each cylinder would be evenly distributed across all of the main bearings. In reality the majority of the load is divided between the two adjacent main bearings, with the remaining bearings seeing progressively smaller loads. In Chap. 6 moments resulting from dynamic couples between cylinders or groups of cylinders were identified. These too can be calculated, and their effects minimized or eliminated through crankshaft counterweight design as was discussed in Chap. 6. Up to this point the calculation of main bearing loads remains quite straight-forward though a bit challenging if the effect of crankshaft stiffness on load distribution across the main bearings is to be included. Unfortunately the situation is far more complicated. As a result of the block dynamics discussed in Chap. 10 the main bearing bulkheads shift with block mode shape—a function of block stiffness and engine speed. While these shifts are small they are often sufficient to significantly further load or unload a given main bearing.

Two examples of main bearing polar load diagrams are presented in Fig. 11.8. Because of the effect of block bending and twisting modes the polar load diagrams can vary significantly not only from one main bearing to another within a given engine, but at any given bearing as the engine speed changes. The diagram shown on the left in Fig. 11.8 is that for a vee engine, and at least two firing and two inertia peaks can be identified, but each is further skewed by block motion, and loads generated at throws further away on the crankshaft. The diagram on the right in Fig. 11.8 shows the simpler diagram of an in-line engine, but significant movement of the main bearing centerline has shifted the majority of the loading to one bearing quadrant. It should be clear from these examples that accurate characterization of the main bearing loads requires a coupled analysis that includes the effects of block motion and crankshaft bending.

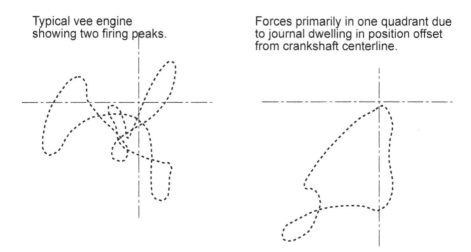

Typical vee engine showing two firing peaks.

Forces primarily in one quadrant due to journal dwelling in position offset from crankshaft centerline.

Fig. 11.8 Example main bearing polar load diagrams

While further examples are not shown here, similar bearing loading calculations can be made for the cam bearings, rocker levers, piston pins, and each of the other bearing locations within the engine. The cam bearings are designed for fully hydrodynamic lubrication. Many of the other bearing surfaces are subjected to continual start-and-stop motion, and thus metal-to-metal contact. Bearing design for both scenarios will be further discussed in the sections to follow.

11.4 Classical Bearing Sizing

Whenever possible, the foremost goal in developing a bearing is to ensure fully hydrodynamic operation under all conditions. This means that the oil film thickness throughout the loaded region must be sufficient to completely separate the "peaks" in the machined surface of the shaft from those on the bearing surface. The variety of parameters that must be considered in ensuring hydrodynamic operation were introduced in the opening section of this chapter. Most of these parameters will be further discussed as the classical computations of bearing film pressure and thickness are presented in this section, and advanced techniques designed to address deviations from the theoretical ideal are covered in the next section.

If one can assume that sufficient oil is present, that the bearing is under steady load, and that neither the shaft nor the journal experience any deflection a nearly exact solution for oil film thickness and pressure distribution can be derived from Reynolds' equation. The derivation was initially done for infinitely long bearings (no edge leakage, and uniform pressure across the bearing length), and was extended by Ocvirk (see the Recommendations for Further Reading) to include "short" bearings. The derivation is not repeated here, but the equation for film pressure as a function of position along the bearing width and around the perimeter is as follows. Pressure distributions as shown in Fig. 11.1 result.

$$P = \frac{3\mu U}{r c_r^2}\left(\frac{w^2}{4} - z^2\right) \frac{\left(\dfrac{e}{c_r}\right)\sin\theta}{\left(1+\left(\dfrac{e}{c_r}\right)\cos\theta\right)^3} \qquad (11.1)$$

where:

P = film pressure
μ = absolute viscosity of the oil
U = surface velocity
R = bearing radius
c_r = radial clearance
w = bearing width

z = positive or negative distance from the center of the bearing
e = eccentricity of shaft centerline relative to journal centerline
θ = angular position from the location of maximum film thickness

A graphical presentation, shown in Fig. 11.9, and based on the dimensionless Sommerfeld number was developed by Ocvirk. The following example demonstrates its use in bearing sizing.

Example
A peak load of 222,400 N is applied to a bearing of 100 mm diameter and 50 mm width. The radial clearance is 0.05 mm, and the oil viscosity is 0.03 Nsec/m². The shaft is spinning at 1800 rpm. The Sommerfeld Number as applied in Fig. 11.9 is calculated as:

$$S = \left(\frac{0.03 \, \text{N sec}}{\text{m}^2} \right) \left(\frac{(0.1\text{m})(0.05\text{m})}{(222,400\text{N})} \right) \left(\frac{1800\text{rpm}}{60 \, \text{sec/ min}} \right) \left(\frac{0.05}{0.00005} \right)^2 = 0.02$$

Note that 'N' as used in the Sommerfeld number calculation is in units of revolutions per minute rather than a surface velocity. From the graph the eccentricity ratio is estimated as 0.89. The shaft offset is thus 0.89 times the radial clearance, or 0.0445 mm. The predicted oil film thickness is thus $0.05 - 0.0445 = 0.0055$, or 5.5 micron.
This bearing sizing technique significantly over-predicts the oil film thickness in reciprocating engine bearings (typically by a factor of two or more). Nevertheless, it remains an

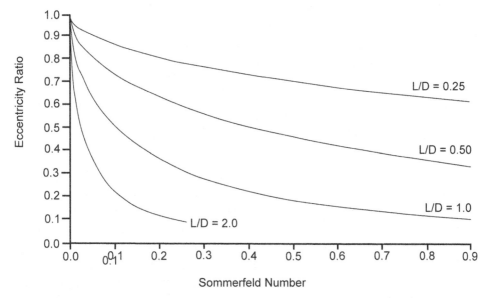

Fig. 11.9 Eccentricity ratio versus Sommerfeld Number as calculated by Reynolds equation for various bearing lengths

Rigid Journal   Elastic Journal Deflection

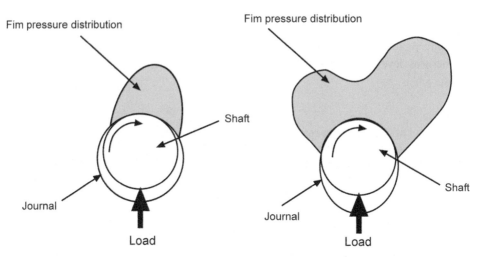

Fig. 11.10 Film pressure distributions for ideal rigid shaft and journal and with significant journal deflection

important tool due to its ease of use, and its effectiveness for relative comparisons. While it may not accurately calculate the actual minimum oil film thickness it can be used to estimate how the film thickness in a new engine will compare to that known to be sufficient in an existing engine.

There are two important reasons for this over-prediction. One is the unsteady loading seen in reciprocating engines. Each rapid load increase forces oil from the bearing, and the dynamic load cycle may not allow the film to fully recover before another rapid load change is experienced. The various deflections of the shaft and journal result in further oil film thickness changes. An important example is shown in Fig. 11.10 where applying a load to a rigid shaft results in journal deflection (exaggerated in the figure). The shaft itself generally deflects much less than the journal. As the shaft is loaded the journal deflects creating pinch points of minimum clearance as shown in the figure. Although the applied load in this case is in the vertical direction two zones of minimum clearance are seen—one in the upstream, and one in the downstream direction. Because the maximum film pressure and minimum film thickness both occur downstream of the direction of loading there is an additive effect on the resulting pressure and film thickness as shown in the figure.

11.5   Dynamic Bearing Sizing

The advent of modern computer capability has allowed automation of the Reynolds solution described in the previous section. Various commercial computer codes are available for predicting oil flow and leakage, film pressure distribution, and oil film thickness versus both time and position. Most of the available codes are designed to conduct the hydrodynamic calculations, and have options for elasto-hydrodynamic calculations—with significant run-time increases. Several examples of the typical output from such codes are provided in Fig. 11.11 for a connecting rod bearing in an in-line diesel engine. The coordinates chosen for these figures are shown in Fig. 11.11a. The 'x' direction is along

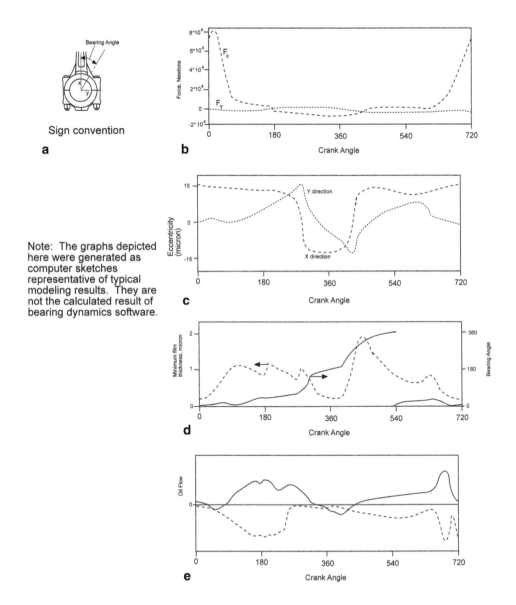

Note: The graphs depicted here were generated as computer sketches representative of typical modeling results. They are not the calculated result of bearing dynamics software.

Fig. 11.11 Depiction of example calculations produced with journal bearing design software. Magnitudes and trends representative of diesel engine rod bearing, and rigid journal, hydrodynamic solution

the cylinder axis, and the 'y' direction is through the split between the connecting rod and cap. The 'bearing angle' is chosen relative to the centerline of the rod, with 0° at the center of the rod. Engine crank angle is given relative to TDC between compression and expansion (0°). The applied load is shown in Fig. 11.11b. For this in-line engine it is almost entirely in the 'x' direction, with a slight 'y' direction component in reaction to the piston side forces. This load would be seen to increase with connecting rod length reduction. The next plot (Fig. 11.11c) shows eccentricity in both the 'x' and 'y' directions. Eccentricity is the distance between the shaft and journal centerlines, and is given in microns. The combination of eccentricity in 'x' and 'y' directions results in the minimum oil film thickness, plotted in Fig. 11.11d along with the bearing angle at which it occurs. Notice that one minimum is seen shortly after TDC on the expansion stroke, and that it is located near the center of the bearing shell in the connecting rod. A second minimum occurs 360° crank angle later and is located near the center of the bearing shell located in the rod cap. The final plot (Fig. 11.11e) shows the rates of oil supply to and leakage from the bearing. When carefully validated to experimental results such analysis tools can be used to rapidly assess the effects of design modifications for connecting rod bearings. They can also be used effectively for camshaft bearings.

For the reasons discussed earlier far more effort is needed to couple such tools with block dynamics and crankshaft deflection if one is to predict main bearing performance. An example of such a coupled solution technique is given in Fig. 11.12. Of necessity the complete block model is rather coarse, so accurate predictions at the main bearing journals requires finer sub-models. The coarse and fine models of one bulkhead and main bearing

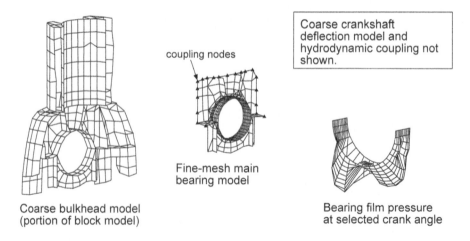

Fig. 11.12 Example of coarse block deflection model, and fine main bearing journal sub-model and resulting film pressure calculation. Based on the models presented by Knoll, G., (see Recommendations for Further Reading)

journal are shown in the figure along with the coupling nodes. Examples of the predicted journal deformation and film pressures are shown in Fig. 11.12c.

Careful engineering judgment must always be applied in bearing computational analysis, recognizing the assumptions that have been made and the effects that may have been ignored. For example, the effect of journal elasticity described previously and often important was ignored in the computations presented in Fig. 11.11. Any misalignment or dimensional variations in crankshaft machining are generally ignored as well. Although bearing analysis tools may provide design guidance engine testing and careful visual assessment remain critical aspects of bearing design validation, and will be further discussed in Sect. 11.7.

11.6   Bearing Material Selection

Ensuring fully hydrodynamic operation involves selecting the combination of contact area and surface finish that will completely separate the bearing from the shaft when the anticipated loads are applied. As power densities increase in many engine applications much attention is now being placed on micro-honing techniques to achieve hydrodynamic operation with thinner oil films. Several other factors must be considered in hydrodynamic bearing design. The factors driving bearing material selection will be discussed in the following paragraphs.

Fatigue Strength Each of the major bearings in a reciprocating engine is exposed to high cycle loads as defined in Chap. 3. Thus the fatigue strength of the bearing material is of crucial importance. The surface of an overloaded bearing will crack with time under fatigue loading. As the cracks spread portions of bearing material will break away, leaving a reduced bearing area and thus further overloading the bearing. Once fatigue cracking initiates the failure mechanism rapidly accelerates due to this material removal.

Scuff Resistance Although the rod, main, and cam bearings are designed to operate under hydrodynamic conditions they experience significant metal-to-metal contact during engine start-up. Under cold conditions, when the oil viscosity may be increased by an order of magnitude or more, the time delay before pressurized oil reaches the bearing may be several seconds. Local friction heating may result in micro-welding and tearing (scuff), or in severe cases seizure and a spun bearing. The bearing has literally welded to the spinning shaft, and the backing tears away from the journal (spins) as the shaft rotates. The propensity of a bearing to bond with the shaft is strongly impacted by material compatibilities, and bearing materials are chosen to minimize bond compatibility.

Wear Rate At the locations where hydrodynamic operation cannot be achieved (piston pins, rocker levers) an overriding material selection criterion is that of low wear rates under high loading with metal-to-metal contact. Wear rate is an important criterion for hydrodynamic bearings as well. With these bearings the wear occurs only during startup, but the criterion remains important because the engine experiences many cold starts over its lifetime (and in some applications many more hot starts). The primary problem resulting from high wear is loss of oil pressure as the increased clearances allow the oil to flow too quickly from the bearing. The increased clearance also results in impact loading between the shaft and journal, thus accelerating further wear. Some bearing materials have a softer overlay coating, and when the bearing wears they lose this coating.

Conformability Even with careful attention to dimensional tolerances machining variation can result in misalignments that consume the small design clearances required for bearing operation. The bearing material must be much softer than that of the shaft, allowing a new bearing to quickly conform and restore (or approach) design tolerances. This can be done to accommodate small machining variations but does not eliminate the need for careful dimensional control.

Embedability Another problem requiring a soft bearing surface is that of hard particles becoming trapped between the bearing and shaft. It is desired that any dirt and metal particles would be suspended in the lubricant and carried to the filter. However this is not always possible from the confined region of the bearing, and trapped particles can quickly score the bearing and shaft. The problem can be lessened if particles that cannot be removed are instead imbedded in the bearing.

Corrosion Resistance Fuel fragments, sulfur compounds, and oxides of nitrogen are acidic products of combustion, and build up in the oil due to combustion blow-by. Water vapor from combustion and accumulated from the air also build up in the oil. The bearing materials must therefore be resistant to corrosion from both water and acid.

Bearing materials are selected based on the criteria just described, always with the further objective of minimizing cost. Most bearings in use today are referred to as thin-wall bearings, and consist of relatively thin layers of the bearing material bonded to a steel shell. While there are many bearing alloys optimized to meet the most challenging requirements of particular applications they generally fall into three families—white metal bearings alloyed primarily with tin, copper, and antimony; alloys based on aluminum and silicon (often referred to as bi-metal); and the tri-metal bearings consisting of a bronze base and a soft overlay.

The white metal bearings hold the advantages of good scuff resistance and low cost but are limited to low load applications due to their poor fatigue strength. They have good resistance to acid corrosion, but are susceptible to corrosion due to water vapor. Because of the increasing demands placed on new engines this category of bearings is becoming less common.

Bearings alloyed from aluminum and silicon offer good fatigue strength, scuff resistance, and corrosion resistance at reasonable cost. The aluminum-silicon alloy is typically roll plated onto a low carbon steel shell. As silicon content is increased for fatigue strength the bearing becomes harder and loses conformability and embedability. Historically, aluminum-silicon bearings have not required an overlay, but the frequent stop-start cycles now being applied to many automobile engines make resin overlays attractive. These overlays are further discussed in the following paragraph.

The tri-metal bearing begins with a bronze base, cast or sintered onto the steel shell. The bronze provides excellent fatigue strength, but has high hardness and poor conformability. A soft overlay, until recently almost invariably alloyed from lead, tin, and copper provides scuff resistance and the needed conformability. A thin barrier of nickel is applied between the bronze and the overlay; this is necessary to keep the tin in the overlay from migrating into the bronze and reducing bond strength. There have been several recent advances in overlay materials. Ceramic overlays consisting of aluminum oxide along with copper and lead had become common, but are now being replaced by lead-free aluminum-copper or aluminum-copper-tin alloys, thermally sprayed or sputtered onto the bronze bearing. Another family of coatings is based on molybdenum disulfide resin, sometimes mixed with nylon. These coatings are alloyed with various modifications and applied by physical vapor deposition. The term, *sputtered bearings,* is often used among engine designers, and can refer to either the resin or lead-free metal overlays, both of which are applied by sputtering processes. As discussed earlier various resin alloys are now being applied as overlays to aluminum-silicon bearings as well, improving their capability in start-stop applications.

The most common materials for non-hydrodynamic applications are bronze alloys. The contact areas are designed large enough to achieve acceptably low wear rates for the desired engine life. These joints are especially sensitive to oil change interval due to particle build-up in the oil.

11.7   Bearing System Validation

Because of the limitations of analytical approaches discussed earlier experimental validation of each engine bearing system remains a crucial aspect of ensuring the needed durability. Rig testing is attractive due to its simplicity and relatively low cost, and several standard tests have been developed and are widely used. Two examples—the Sapphire test, and the Underwood test are depicted in Fig. 11.13. The Underwood test is especially widely used in the bearing industry. The test mechanically loads the bearing through the use of offset weights on spinning shafts as shown in the figure. The magnitude of the loads can be varied by changing rotational speed and the offset mass. The Sapphire test uses regulated pressure in a piston actuated by the shaft offset. The load magnitude again varies with shaft speed as well as with the regulated pressure. These tests are most effective for bearing material development, and are best used for back-to-back comparisons of a new bearing to an existing bearing. The rig test results do not correlate well to engine tests. One reason for this poor correlation is the difficulty simulating not only the loads but the actual deflections seen in the engine. Another very important reason is that bearing wear

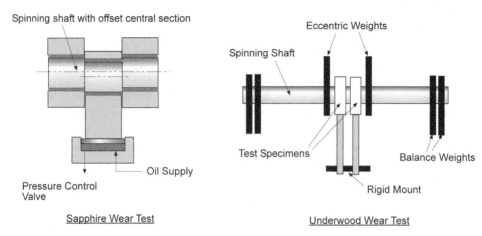

Fig. 11.13 Sapphire and Underwood journal bearing rig tests

is strongly dependent on shaft machining, and especially on whether the shaft is new or is being reused. This is explained by the fact that in a hydrodynamic bearing all of the wear is occurring during start-up, by asperity contact between the shaft and bearing. The surface profile of both parts changes greatly after initial operation. This is an important fact to keep in mind during durability testing conducted in an engine as well. Bearing wear with a newly machined crankshaft will not correlate well to that of replacement bearing shells in the same engine.

In addition to the standard rig tests there are many possibilities for custom tests to address specific design questions. An example test fixture providing various options for piston pin loading is shown in Fig. 11.14. The objectives of any such rig test development must always be carefully thought out. The tests are especially useful in determining the

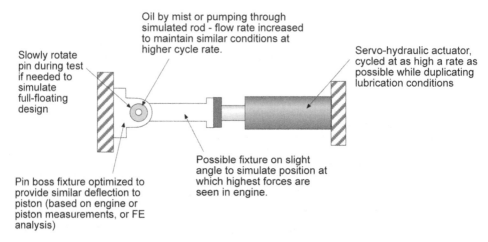

Fig. 11.14 Conceptual layout of a piston pin bushing rig test

limits of a particular bearing or lubricated joint, and in identifying sensitivities to particular variables. Correlation to the actual conditions seen in the engine is far more difficult.

Several categories of engine testing are also important for bearing development. The first is that of further defining the limits of a bearing system. For example, a bearing designed for hydrodynamic operation can be progressively loaded and with careful measurements the maximum load might be identified. In some cases the onset of metal-to-metal contact can be detected through temperature measurement using thermocouples in drillings approaching the running surface from the back of the bearing. A sharper rise in the rate of temperature increase indicates the onset of metal-to-metal contact due to overload.

The second category of engine testing is that of repeated shutdown and restart to generate wear of the hydrodynamic bearing surfaces. The key elements of this cycle are starting the engine and rapidly increasing the load and speed, and then shutting the engine down and allowing the oil to drain back. Various approaches as follows are used to increase the wear rates and reduce total test time:

- Machining back the surface of the bearings to reduce the contact area.
- Externally heating the oil to reduce its viscosity, or cooling it to delay its flow to the bearings.
- Injecting compressed air into the oil rifle to rapidly drain it during each shutdown.

Again, it must be kept in mind that if several tests are to be run with new bearing shells the crankshaft should either be replaced on each test (for most rapid wear), or the same "used" crankshaft should be used for each test.

One might question the purpose of a test designed to generate wear. In the case of tests that repeat the startup and shutdown sequence the typical reason for the test is to ensure that the bearings will not wear out over the startups anticipated over the life of the engine. If the test is accelerated the engineer is then faced with correlating the results from the aggravated test conditions back to "real world" conditions. The best approach for doing this is to compare the results of the new engine or bearing to results obtained under the same aggravated conditions on an existing engine known to perform well in the field.

The third category of engine testing is that done to assess wear rates at non-hydrodynamic bearing locations. Each bearing surface of interest is carefully characterized prior to the test and again at the test's conclusion. Wear rates are maximized by running the tests at high engine speeds and high load (often overloaded) conditions. Wear at these locations is especially sensitive to oil quality, and the wear rates can be further accelerated by running the tests with "dirty" oil, containing carbon and metal particles. Repeatability demands that the particle content of the oil is carefully monitored.

Especially in engine testing visual inspection is an important diagnostic tool and a valuable part of the design validation process. Wear rates, fatigue cracking, corrosion damage, and fretting are all identified through visual inspection. The wear pattern also provides the most direct indication of machining tolerance control problems.

As has been described earlier there are two locations where fretting impacts bearing life. Rod cap separation results in fretting between the rod and cap. This results in loss of running clearance and may ultimately cause bearing scuff or a spun bearing. Movement of the bearing shell in the journal, either because of cap separation or inadequate crush load, results in backside fretting of the bearing shell. This may lead to fatigue cracking of the bearing surface due to loss of support for the bearing shell.

Dimensional control problems can be identified through abnormal bearing wear patterns. Misalignment between the shaft and journal results in concentration of loads along one edge of the bearing. Misalignment between the connecting rod and cap results in loss of clearance and resulting bearing wear at the location of the split in the bearing shells. If the shaft diameter is not maintained constant the bearing wear will be concentrated at the locations of largest shaft diameter. A concave shaft (larger diameter at the edges of the shaft contact region) results in an "hourglass" wear pattern. A convex shaft (larger diameter in the middle of the bearing contact region) results in a "barrel" wear pattern.

Another potential challenge to bearing design is that of cavitation. The general mechanism of cavitation damage was discussed in Chap. 3. It has not been introduced previously in this chapter due to the difficulty of detecting the problem through either measurement or experimentation. It is identified through visual inspection as eroded pinholes. The damage is progressive—as the number and size of the pinholes increases the surrounding material is weakened, resulting in fatigue cracking and material loss. Cavitation damage on lubricated bearings results from a rapid pressure drop followed immediately by a rapid pressure increase. This sequence makes connecting rod bearings most susceptible. Adjusting the bearing clearance slightly can significantly impact cavitation. Oil flow rate and the oil drilling breakout location also have an impact. Bearing computational analysis can sometimes be used to predict the propensity for cavitation damage and to assess the impact of design changes by predicting the rate of film pressure change.

11.8 Recommendations for Further Reading

The references must begin with the classic paper in which the Reynolds formulation for short plain bearings was originally presented (see Ocvirk 1952).

The following text, written in German, provides a detailed look at bearing and lubrication system design (see Affenzeller and Gläser 1996).

There are now many published papers on EHD bearing operation—analytical and experimental. The following are recommended examples for readers interested in further study (see Peixoto and Zottin 2004; Mian and Parker 2004; Sato et al. 2002; Hanahashi et al. 2001; Thomas and Maassen 2001; Okamoto et al. 1999; Ozasa et al. 1995; Priebsch et al. 1995; Torii et al. 1992):

Backside bearing temperature measurements can be an important tool in bearing design. The following paper provides an example of its use (see Suzuki et al. 1995).

Another important question is where to place the oil supply drilling. The following paper presents a study on a rig test with fixed loading. While this is quite different than the situation in a rod bearing (where the oil is supplied through the spinning shaft) it presents the theory (it is also close to the situations of main bearings and cam bearings—they aren't statically loaded, but the oil supply drilling doesn't move). The next paper addresses optimization of the breakout location in crankshaft cross-drillings (see Okamoto et al. 1995; Goenka and Stumbo 1986).

When addressing main bearing analysis one must consider both elasticity of the crankshaft and journal as well as motion of the block bulkheads themselves. The following studies present such computationally intensive coupled solutions (see Knoll et al. 1997; Ebrat et al. 2003).

For a comparative look at plain and roller bearings the following paper addresses their application at the camshaft (see Mackay 2012):

Recent bearing material advances have focused to a great extent on removing lead from the overlays, and simultaneously improving unit load capability (see Nirasawa et al. 2009; Aufischer 2010):

References

Affenzeller, J., Gläser, H.: Lagerung und Schmierung von Verbrennungsmotoren. Springer-Verlag, Wien (1996)

Aufischer, R.: Diesel Engine Bearings for a Lead-Free Future. SAE 2010-32-0060 (2010)

Ebrat, O., Mourelatos, Z., Hu, K., Vlahopoulos, N., Vaidyanathan, K.: Structural Vibration of an Engine Block and a Rotating Crankshaft Coupled Through Elastohydrodynamic Bearings. SAE 2003-01-1724 (2003)

Goenka, P.K., Stumbo, R.F.: A Method for Determining Optimum Crankshaft Oil-Hole Location. SAE 860357 (1986)

Hanahashi, M., Katagiri, T., Okamoto, Y.: Theoretical Analysis of Engine Bearing Considering Both Elastic Deformation and Oil Film Temperature Distribution. SAE 2001-01-1076 (2001)

Knoll, G., Schönen, R., Wilhelm, K.: Full Dynamic Analysis of Crankshaft and Engine Block with Special Respect to Elastohydrodynamic Bearing Coupling. ASME 97-ICE-23 (1997)

Mackay, S.: Comparison Between Journal and Rolling Element Bearings in a Camshaft Application. SAE 2012-01-1324 (2012)

Mian, O., Parker, D.: Influence of Design Parameters on the Lubrication of a High Speed Connecting Rod Bearing. SAE 2004-01-1599 (2004)

Nirasawa, T., Yasui, M., Ishigo, O., Kagohara, Y., Fujita, M.: A Study of Lead-free Al-Zn-Si Alloy Bearing with Overlay for Recent Automotive Engines. SAE 2009-01-1054 (2009)

Ocvirk, F.W.: Short-Bearing Approximation for Full Journal Bearings. NACA Technical Note 2808 (1952)

Okamoto, Y., Mochizuki, M., Mizuno, Y., Tanaka, T.: Experimental Study for the Oil Flow Supplied from Oil Hole on Statically Loaded Bearings. SAE 950947 (1995)

Okamoto, Y., Kitahara, K., Ushijima, K., Aoyama, S., Xu, H., Jones, G.: A Study for Wear and Fatigue of Engine Bearings on Rig Test by Using Elastohydrodynamic Lubrication Analysis. SAE 1999-01-0287 (1999)

Ozasa, T., Yamamoto, M., Suzuki, S., Nozawa, Y., Kononi, T.: Elastohydrodynamic Lubrication Model of Connecting Rod Big End Bearings; Comparison with Experiments by Diesel Engine. SAE 952549 (1995)

Peixoto, V.J.M., Zottin, W.: Numerical Simulation of the Profile Influence on the Conrod Bearings Performance. SAE 2004-01-0600 (2004)

Priebsch, H., Loibnegger, B., Tzivanopoulos, G.: Application of an Elastohydrodynamic Calculation Method for the Analysis of Crank Train Bearings. ASME 94-ICE-1 (1995)

Sato, K., Makino, K., Machida, K.: A Study of Oil Film Pressure Distribution on Connecting Rod Big Ends. SAE 2002-01-0296 (2002)

Suzuki, S., Ozasa, T., Yamamoto, M., Nozawa, Y., Noda, T., O-hori, M.: Temperature Distribution and Lubrication Characteristics of Connecting Rod Big End Bearings. SAE 952550 (1995)

Thomas, S., Maassen, F.: A New Transient Elastohydrodynamic (EHD) Bearing Model Linkable to ADAMS®. SAE 2001-01-1075 (2001)

Torii, H., Nakakubo, T., Nakada, M.: Elastohydrodynamic Lubrication of a Connecting Rod Journal Bearing in Consideration of Shapes of the Bearing. SAE 920485 (1992)

Engine Lubrication

12

12.1 Engine Lubricants

The functions that come immediately to mind when one thinks of the lubricant and lubrication system are wear reduction and friction reduction. This is of course a correct list but an incomplete one. Several further important lubricant functions are described in the paragraphs that follow. It is important to keep all of these functions in mind in the design and development of an engine's lubrication system.

Temperature Control Regardless of whether an engine is air- or water-cooled there are important aspects of temperature control that rely on the lubricant as a heat transfer medium. The most important example is that of controlling piston crown and ring temperatures. The majority of heat transfer from the combustion chamber occurs early in the combustion process, when flame temperatures are at their peak, and the piston is still very near TDC. Little of the cylinder wall is exposed, and the walls seen by the combustion gas are the piston crown and the cylinder head. Much of the cylinder head surface is made up of the valves, and the valves are a high-resistance heat transfer path. Thus, the piston crown receives the greatest share of combustion heat rejection. A portion of this heat rejection is transferred through the ring pack to the cylinder wall, and to the engine coolant. The rest is transferred through the piston undercrown to the lubricant. Lubricant temperature and heat transfer mechanisms thus play an important role in controlling piston crown and ring temperatures. As specific power output increases the role of the lubricating oil in controlling critical piston temperatures becomes more and more important, and steps are taken to enhance this heat transfer mechanism.

Another important temperature control function follows from the fact that the vast majority of energy lost to friction is transferred to the lubricant. Oil flow rates through main and rod bearings and turbocharger bearings have an important impact on bearing operating

© Springer Vienna 2016
K. Hoag, B. Dondlinger, *Vehicular Engine Design*, Powertrain,
DOI 10.1007/978-3-7091-1859-7_12

temperature. Bearing temperature control is important both for the durability of the bearings (fatigue life, and protection against wiping), and its impact on local oil viscosity and oil film thickness.

Particle Removal Sources of particles in the engine's crankcase include metal particles resulting from engine wear, carbon particles resulting from combustion, and ingested dirt particles from the outside environment. Each of these particles are abrasive, and aggravate wear and damage at sliding interfaces throughout the engine. Lubricant additives include detergents and dispersants, intended to pull particles from component surfaces and hold them in suspension in the lubricant. As the lubricant flows through the engine it is pumped through a filter intended to trap the suspended particles, and allow their removal from the system.

Sealing The lubricant participates directly in completing several important seals throughout the engine. Its surface tension and viscosity are utilized to complete a seal while it serves its hydrodynamic lubrication function. The most obvious example is at the compression rings, where the lubricant film between the ring and cylinder wall serves along with the ring to completely close the gas exchange path between the combustion chamber and the crankcase. If the lubrication oil film is incomplete due to high cylinder wall distortion or lack of oil, high blowby of combustion gas into the crankcase will result.

Another important place where the lubricant completes the seal is on rotating shaft seals. Wherever a spinning shaft protrudes from the crankcase the seal between the engine structure and the shaft relies on a hydrodynamic film. The shape of the seal creates a film pressure distribution that pushes oil back toward the crankcase and not out of the engine.

Hydraulic Capabilities Lubricating oil is an excellent hydraulic fluid, and may be called upon to provide a hydraulic function in support of engine sub-system operation. The most common example is the hydraulic lifters seen in many engine valve trains. The lubricant is supplied under pressure to a hydraulic link, taking up lash in the valve train. Another example is the use of a hydraulic chamber, again supplied with pressurized engine lubricant, to control pressure on a chain tensioner for the cam drive system. A number of cam phasing and variable valve timing (VVT) devices rely on hydraulic controllers using engine lubricant.

Corrosion Control Finally, the lubricant additive package includes compounds that are intended to adsorb as coatings onto component surfaces. These adsorbed coatings have lubrication and wear reduction properties but are also formulated to protect the components against acid- and water-based corrosion. Ensuring that the lubricant is well distributed to provide the needed, complete coatings is another aspect of lubrication system design.

The lubricant base stock, making up between 75 and 85 % of the oil by volume consists of a blend of hydrocarbons selected to provide the starting point for viscosity and lubrication performance. The molecules making up this base stock may be refined directly from crude oil or created through chemical processing (synthetic lubricants). In either case the hydrocarbon formulations may be quite similar, with the constituent makeup more closely controlled in the synthetic oils. Depending on the hydrocarbon sources used in creating the synthetic oil it may also be free of the unwanted sulfur and ash which are found in crude oils and which prove expensive to eliminate.

The remainder of the lubricating oil is an additive package consisting of a variety of chemical compounds selected to provide the expected lubricant performance. While the specific additive package varies from one oil to another several types of constituents are commonly included and are described in the following paragraphs.

Dispersants and Detergents Detergents are additives designed to lift dirt and deposits from the engine component surfaces. Dispersants are designed to hold particles in suspension and transport them to the filter for removal. Their chemical make-up and the mechanisms by which they operate are similar. This family of additives typically consists of one or more alkaline-earth soaps, with the detergents based on calcium or a sulfonic acid. The dispersants are non-metallic polybutenes. Both the dispersants and the detergents are long-chain hydrocarbon molecules having a polar end that attracts and holds the particle.

Overbase Additives Over time the acidity of lubricating oil increases due to leakage of combustion products into the crankcase and continued reaction. Sulfur compounds, various hydrocarbons, and oxides of nitrogen are acidic and unless counteracted they will damage bearing surfaces. In order to minimize the effect of acid build-up the new oil contains additives with very high base numbers. Acid build-up will reduce the base number of the oil with time, and will eventually render the over-base additives ineffective. It is thus one of the parameters that determines oil change interval. The typical measures of acid build-up are the Total Acid Number (TAN) and Total Base Number (TBN). The over-base additives are created by modifying the alkaline-earth soap molecules of the detergent additives. The modification of these additive molecules increases their base number by ten to twenty times.

Anti-Wear and Anti-Corrosion Additives Wear inhibitors and rust inhibitors are both additives that are designed to adhere to the engine component surfaces as adsorbed films. The primary purpose of the wear inhibitors is to provide a lubricating surface during engine start-up, before pressurized lubricant reaches the particular surface. The additives most commonly used as wear inhibitors are zinc dialkyl-dithio-phosphates, or ZDDPs. Most aftermarket oil additives contain ZDDPs. The rust inhibitors are over-based sulfonate soaps.

It should be noted that the additives discussed so far reveal use of the same or similar additives to achieve more than one objective. This allows improved additive capability while reducing the total additive volume, in turn controlling cost and maintaining a sufficient hydrocarbon lubricant base.

Viscosity Index Improvers The viscosity of hydrocarbon lubricants has a tremendous temperature dependency—about two orders of magnitude over the range from cold ambient to full load oil operating temperature for an unmodified blend. This dependency makes it extremely difficult to ensure adequate oil flow under all operating conditions, and to properly balance the supply and leakage rates for acceptable lubrication of critical surfaces. Viscosity index improvers are long-chain polymers that coil up at low temperature and uncoil as the temperature is increased. When coiled up they have little impact on the viscosity of the oil, but as they uncoil they make an increased contribution to viscosity, thus reducing the rate at which the oil's viscosity drops with temperature. Common viscosity index improvers include polymethacrylate esters and ethylene-propylene copolymers.

Anti-Oxidants Over time the hydrocarbons in lubricating oil have a propensity to react with oxygen. The rate at which oxidation occurs increases exponentially with increasing temperature. These oxidation reactions result in the lightest molecules "boiling off," leaving heavier sludge and deposits behind. Oxidation control is achieved primarily through selective blending of the hydrocarbon base stock. It is for this reason that oxidation resistance is an inherent advantage of synthetic lubricants, where the base stock formulation can be most closely controlled.

Given enough time lubricating oil exposed to oxygen will break down even at ambient temperature. At oil sump temperatures the oxidation rate is significantly accelerated. The viscosity of the oil increases with time as the lightest fractions boil off during engine operation, and the remaining oil becomes heavier. Sludge, consisting of heavy hydrocarbons and in many cases water, may build up throughout the engine as the heavier fractions agglomerate. Fuel fragments entering the crankcase exacerbate the problem as they too are prone to react with oxygen and with other hydrocarbon fragments.

Oil Breakdown and Oil Change Interval The time and temperature history in a given engine determine the rates of oil breakdown and sludge buildup. Under cold operation incomplete combustion feeds reactive fuel fragments into the crankcase. At high engine operating and sump temperatures the reaction rates are increased. Sequences of extended idle and high load operation make an engine especially prone to sludge buildup.

The highest temperature regions such as the piston ring lands and grooves result in the most rapid reaction rates and are especially susceptible to deposit buildup, leading to the common problem of piston ring sticking. As will be discussed in Chap. 15 an important parameter in engine development is control of the ring pack temperatures.

In the preceding paragraphs two of the important criteria that determine oil change interval were identified. These criteria are acid buildup and lubricant breakdown due to oxidation. A third criterion especially important in diesel engines is that of soot buildup in the oil. Among the combustion products that may enter the crankcase are the solid soot particles generated in diesel combustion. These particles are abrasive and thus accelerate wear in regions of mixed film and boundary lubrication such as the valve train. Soot buildup will also lead to oil filter plugging if the oil change interval is insufficient.

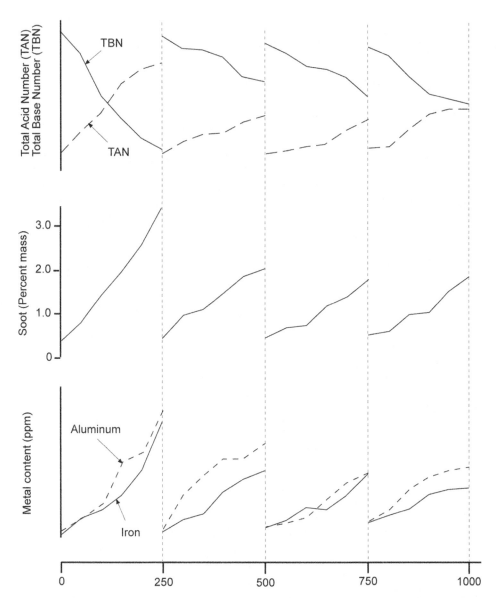

Fig. 12.1 Lubricant analysis conducted during engine durability test; lubricant tested every 50 h and changed every 250 h

Lubricant Analysis Lubricant analysis is often conducted at regular intervals throughout a durability test. In the diesel engine example shown in Fig. 12.1 oil samples were taken every 50 h of operation, and the oil changed every 250 h. The example shows the changes in TAN and TBN over time, soot build-up, and the build-up of two metals in the lubricant. Each time the oil is changed the new oil has a high base number and low acid number. Over time the base number drops and the acid number increases. In an aggressive durability test the base number is depleted far more than would be desired in the field. The acid number typically climbs much more rapidly during the first few hours than later in the test. This occurs because the rings have not yet seated in the new engine, and combustion gases initially enter the crankcase at a higher rate.

The lubricant is also sampled for each metal contained in the engine. Two metals, iron and aluminum, are shown in the example. Here too the evidence of initially higher wear rates is seen as the new components seat. Viscosity change, and soot buildup in a diesel engine, will also be monitored. It should be noted that new engine oil initially registers a few tenths percent soot; this is due to ash in the crude oil base stock.

Viscosity increase provides an indication of the rate at which the base stock is breaking down through oxidation, and along with the acid and base numbers and the soot buildup provides guidance for the oil change interval. Recommended oil change intervals provided to the customer are based on such measurements in conjunction with estimates of typical driving conditions. In large fleets such measurements are often made to maximize the oil change interval for the specific application. Finally it is important to note that of the various parameters that determine oil change interval only oxidation breakdown resistance is improved with synthetic lubricants. While synthetics are often seen as allowing an increased oil change interval this is not necessarily the case. Even though the oil may oxidize more slowly acid buildup may result in bearing damage, or in the case of the diesel engine soot buildup may cause increased wear and possibly filter plugging.

Uniform Oil Classification and Grading Both consumers and engine manufacturers need safeguards to ensure that the lubricant performs its functions as expected, and does not become the cause of engine failures. There are two important sets of measures to ensure that a lubricant's performance meets expectations. These are the viscosity classification system, and the performance classification system for diesel and spark ignition engines.

The viscosity classification system, administered by the Society of Automotive Engineers, categorizes lubricant performance versus temperature using two scales. The "winter" scale assigns a number between 0 and 25 W, in increments of five, based on low temperature cranking viscosity and a minimum high temperature viscosity. The second scale assigns a number between 20 and 60, in increments of ten, based on high temperature viscosity. Oils fall into particular "buckets" based on their performance relative to these scales, and can only be assigned one number from each scale. If they simultaneously meet the criteria of both scales they are referred to as "multi-viscosity," and are assigned a number from each scale (SAE 5W-20 or SAE 10W-40 for example). If they do not meet the criteria of any category on one of the scales they are assigned a single number from the scales whose criteria they meet, and are referred to as "straight weight" oils (SAE 20 or SAE 50 for example). The tests are quite straight-forward, and commercial PC-driven test cabinets are available from several suppliers.

Various lubricant performance standards have been developed by professional societies and in some cases engine manufacturers. The American Petroleum Institute (API) standards are taken as an important example. The API standards use two scales—the Service Classification for spark-ignition engines, and the Commercial Classification for diesel engines. Each classification system defines particular tests performed on production engines, and pass/fail criteria for each test. The tests are incremented over time, based on changing demands (often, but not exclusively emission-driven). With each test change the second letter designation changes. For example, SM, introduced in 2004 defines the current performance specification in the Service classification system. CJ, introduced in 2006 is the current Commercial classification. Examples of the tests included in these performance classification systems might include viscosity increase over time at elevated temperature, piston deposits and ring sticking, valve train or bearing wear, sludge and varnish levels, and oil consumption rates.

12.2 Crankcase Deposits

Before proceeding further into the engine lubrication system it is important to introduce the problem of deposits. Deposits can take various forms throughout the lubricated portions of the engine—everything from hard carbon deposits on the piston ring belt, to varnish on various surfaces, to sludge in the oil pan and under the valve covers. Lubricant oxidation was introduced in the previous section, where anti-oxidation additives were discussed. A more detailed description is given here, as oxidation reactions play important roles in deposit formation. Lubricants are hydrocarbons, and as such they have a propensity to react over time with oxygen. The reaction rates are strongly driven by temperature, and in this case described mathematically as an Arrhenius function—exponential in temperature. As the lubricant molecules react with oxygen lighter fractions may burn off, leaving heavier fractions behind. In the engine those heavier fractions result in sludge and deposits. At ambient temperature the oxidation reactions are very slow, and an open can of lubricant may only experience measurable deterioration over a timescale of years, but as the temperature increases the reaction rates increase exponentially—approximately doubling with every $17\,°C$ temperature increase. At oil sump temperatures this has an important impact on oil change interval. At piston ring land temperatures it has an important impact on deposit formation rates and ring sticking.

Deposits can be categorized in a number of ways including their impacts on engine operation, and the temperature regimes of their formation. In terms of impact on engine operation, four categories can be identified:

1. Those that restrict oil circulation. Deposits can build up at the oil pick-up in the sump, or around flow or drain-back passages, restricting flow and potentially endangering engine life.
2. Those that take the engine out of service prematurely. Deposits can cause ring or valve sticking, the former resulting in increased oil consumption, and the latter causing performance loss and potential sudden failure.

3. Those causing progressive engine performance loss. A wide variety of examples including increased blow-by, reduced compression due to valve leakage, PCV valve sticking, hydraulic lifter malfunction and others fall into this category.
4. Those not harmful to engine performance. One must be careful with this category, as visual deposits may hurt customer perception, or cause service technicians to draw unfavorable impressions of the product, even if no other detriments exist.

Another way to characterize deposits is based on temperature and the resulting formation mechanisms and chemical make-up. The following paragraphs provide further descriptions of deposit types based on temperature range.

Carbon and High Temperature Deposits From the standpoint of engine lubrication there are two regions where high temperature deposits are important—ring lands and grooves, and valve stems and guides. In both locations these deposits can lead to both abrasive wear and component sticking. A distinction is sometimes made between cold sticking and hot sticking. Especially in the earlier stages of sticking the component may stick at cold temperatures but move freely as the components warm up. A temperature of 200 °C is generally considered as the threshold. Cold sticking is a precursor to hot sticking.

These deposits are made up primarily of carbon (65–75 %) and oxygen (25–35 %), with trace amounts of nitrogen, sulfur and hydrogen. In these high temperature regions the lubricant reacts quickly with oxygen forming soluble and insoluble products that adhere to clean metal surfaces. They act as a trap to collect the carbon-based combustion products. Coking creates the hard carbon particles or deposit layers. Temperature control, and limiting the amount of oil entering these regions are the primary control techniques.

Varnish and Medium Temperature Deposits On medium temperature surfaces a thin varnish or lacquer coating can form, varying in color from yellow to light and dark brown to black. The colors are fuel and lubricant chemistry dependent but generally darker with increased severity. The medium temperature distinction can be further described as below temperatures at which coking would occur, and above temperatures at which the deposit will contain water. These deposits are thin, and in most cases have little or no impact on engine operation.

Varnish formation begins with oxidation—the reactions of hydrocarbons with oxygen, and the acceleration of these reaction rates with increased temperature, as discussed earlier. The by-products of oxidation combine to create longer molecules in polymerization processes. The polymerized molecules are held in solution in the lubricant until the fluid becomes saturated, after which point they begin precipitating out. A drop in lubricant temperature reduces solubility, increasing the quantity of molecules that precipitate out. Agglomeration occurs as the sub-micron insoluble particles bond to form larger particles. These particles, now on the order of one micron diameter remain insoluble and polarized, and have a higher molecular weight than the fluid. The agglomerated particles collect on metal surfaces, especially in cooler, low flow regions.

Sludge and Low Temperature Deposits As with both the carbon and varnish deposits, sludge begins with oxidation of fuel or lubricant molecules, creating highly reactive hydrocarbon fragments. These fragments polymerize and agglomerate to form heavier molecules. In the case of sludge the low temperature molecules act as binders, collecting other components including metal particles, dirt, and water, significantly increasing the sludge mass. Sludge ranges in color from light grey to black, and in consistency from paste to semi-solid. Both the color and consistency are dependent on the specific chemical make-up. Because of its rapid build-up in low temperature locations it can result in reduced oil flow. It may impede flow through the oil pick-up tube and screen, and through the oil filter. It may also block drain-back passages, causing oil to pool. Appearance and customer reactions are a final, important concern.

Sludge is often associated with running too cool—extended light load and idle operation, and overcooling. The move from rich to soichiometric air-fuel ratios in spark-ignition engines, for three-way catalyst operation, has increased combustion temperatures and significantly reduces sludge precursor formation. Other variables that reduce sludge build-up include reduced blow-by, increased oil or coolant temperature, increased load factors, and improved PCV systems. Oxidation inhibitors, dispersants and detergents in the lubricant additive package also serve to reduce sludge formation.

12.3   Lubrication Circuits and Systems

With the various lubricant functions in mind the stage is set to begin looking at approaches to lubrication system design. The foremost need is to adequately distribute lubricant to all of the locations where it is needed. This requires sufficient lubricant volume (system capacity), and a sufficient distribution mechanism. Various considerations in determining system capacity will be taken up later, but this is the functional consideration—how much lubricant is needed to ensure that distribution is not compromised?

The simplest approach to lubricant distribution is the splash system, used in lawn and garden, and small industrial engines. Movement of components within the crankcase is relied upon to fling the lubricant around, getting it to all of the important sliding and rolling interfaces. This may be aided by an oil slinger—typically a simple paddle or spoon attached to the connecting rod and designed to dip through the sump and distribute oil as the crankshaft rotates. Distribution will also be aided by bearing geometries designed to draw oil into the contact regions, via slots in bearing journals, and slots or holes in connecting rods near bearings. The splash lubrication system is remarkably effective, and commonly used in small, lightly loaded engines.

From the list of functions described in the first section, the one that is most difficult to accomplish with a splash lubrication system is particle removal. While the lubricant additive package provides detergent and dispersant capabilities a filter is required in order to remove the particles held in the oil, and a pump is required to send lubricant through a filter. As design complexity is added to an engine lubrication system the next step is to add an oil pump and filter circuit. Oil is drawn from the sump, through an oil filter canister, and discharged back to the sump.

The next step in lubrication system complexity is to add pressurized supply to various critical sliding joints. An oil pump supplies a drilled or cast passage referred to as the main gallery or rifle, and from this passage further passages are drilled to supply pressurized oil to the sliding joints of interest. Listed below are possible pressurized oil feeds, in approximate ranking. Note that the decision to include or neglect any of the items listed is based on the trade-off between added cost and design complexity versus improved durability, reduced wear, and reduced frictional losses.

- Main bearings
- Cam bearings
- Rod bearings
- Rocker shafts and roller follower shafts
- Piston pins
- Rocker lever ends
- If the engine is turbocharged, pressurized shaft bearing lubrication is mandatory
- Piston cooling nozzles
- Hydraulic lifters
- Hydraulic chain tensioners
- Hydraulic actuators for cam phasing or variable valve actuation systems

In laying out the lubrication system further design decisions include whether the system will have an oil cooler, whether the system will use a wet or dry sump, and whether supplemental filtration will be included along with the full-flow filter. Each of these questions will be covered in sections to follow.

Customer expectations regarding engine life necessitate the use of pressurized lubrication systems in virtually all vehicular engine applications. An oil pump is used to draw oil from a sump or oil reservoir, and supply it through pressurized passages to the majority of component interfaces susceptible to wear.

When laying out the lubrication system one must consider all of the possible paths the oil can take in traveling through the engine. These include the enclosed volumes of the crankcase and overhead as well as the pressurized oil drillings supplying oil to the various locations described in the previous section. An engine-driven oil pump draws oil from the sump or an oil reservoir, and feeds it through the oil filter and oil cooler, and then through various parallel and series paths lubricating the sliding interfaces. From each of these locations oil then drains back by gravity to the oil pan or sump beneath the crankshaft. Oil fed into the cranktrain (main bearings, rod bearings, piston pins, piston cooling nozzles) drains directly back into the sump. It should be noted that even with pressurized lube systems the cylinder walls and rings are lubricated by "splash," sometimes aided by oil squirter drillings in the connecting rods: this oil too drains directly back into the sump. Oil fed under pressure to the overhead (overhead cams, rockers levers, hydraulic lifters in

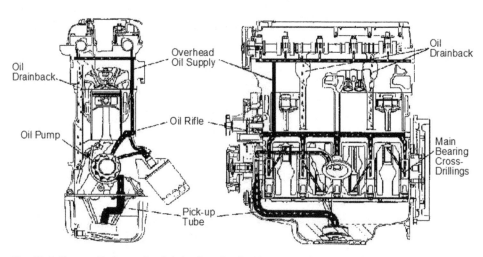

Fig. 12.2 Four-cylinder engine lubrication circuit. (Courtesy of Ford Motor Company)

overhead cam engines) drains to the oil deck of the cylinder head, and then through drain-back passages through the cylinder head and block into the sump.

Lubrication system layout is next determined. Where is the oil pump to be placed, and how is it to be driven? Where are the oil filter and oil cooler to be placed, considering plumbing complexity and joint/seal minimization, and also considering serviceability? The pressurized lubricant passages are designed to minimize machining cost and complexity and to minimize restrictions and pressure drop. The oil deck and drainback passages are design to ensure against pooling of oil in the overhead, and to ensure that the drainback oil returns directly to the sump without contacting the spinning crankshaft. Each of these considerations will be further addressed in sections to come.

Typical lubrication circuits for automobile and heavy duty engines are shown in Figs. 12.2 and 12.3 respectively. Looking first at the automobile engine, the oil pump draws oil from the sump through a pick-up tube, and sends it through a filter to the oil rifle—a pressurized drilling running the length of the block. A pressure relief valve at the oil pump exit sends a portion of the oil directly back to the pump inlet or to the sump, thus controlling rifle pressure at a nearly constant level over the engine operating speed range. Another pressure relief valve allows the filter to be bypassed if the filter inlet pressure becomes too high. This typically occurs when the oil is cold, and would also occur if the filter were to become plugged. From the oil rifle, cross-drillings supply oil to each of the main bearings and also to the camshaft bearings. Drillings within the crankshaft supply oil from the main bearings to the connecting rod bearings. Drillings within the connecting rods are often used to spray cooling oil toward the cylinder walls, the underside of the pistons, and the piston pins. Further passages carry pressurized oil to supply the hydraulic lifters, and then either through the push rods or further passages to the valve train. From each lubricated surface the oil drains into the surrounding engine cavities, and is then gravity fed back to the sump.

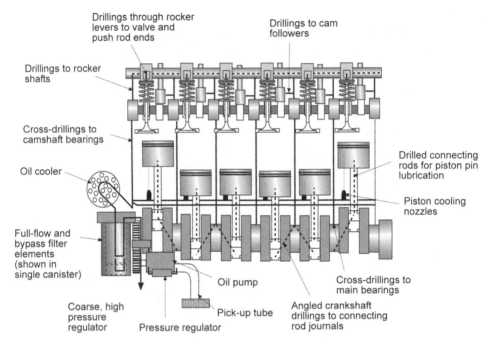

Drillings through rocker
levers to valve and
push rod ends

Drillings to cam
followers

Drillings to rocker
shafts

Cross-drillings to
camshaft bearings

Oil cooler

Drilled connecting
rods for piston pin
lubrication

Piston cooling
nozzles

Full-flow and
bypass filter
elements
(shown in
single canister)

Oil pump

Cross-drillings to
main bearings

Coarse, high
pressure
regulator

Pick-up tube

Angled crankshaft
drillings to connecting
rod journals

Pressure regulator

Fig. 12.3 Typical heavy-duty diesel engine lubrication circuit

While the sump in most engines is contained in the oil pan beneath the crankshaft *dry sump* systems are sometimes used. The dry sump system is most common in very high performance engines but is sometimes chosen for packaging reasons. With the dry sump system a second oil pump is used to immediately evacuate the returning oil from the oil pan, and send it to a separate reservoir from where it is again supplied to the engine. The resulting crankcase vacuum reduces pumping losses in the crankcase associated with piston movement and may reduce ring flutter, leading to better piston ring sealing. Crankcase vacuum is limited by the oil vapor pressure. Crankcase vacuum also reduces the oil mist and further losses associated with fluid friction between the crankshaft assembly and the air and oil vapor in the crankcase. Among the attractions for high performance applications are lower engine placement and resulting lower vehicle center of gravity, and eliminating crankshaft dipping or pickup starvation under hard cornering, acceleration, and braking. This also reduces oil aeration as it reduces churning of oil from the crankshaft and guarantees first-in-first-out of the oil in the reservoir, increasing oil residence time in the sump.

Referring now to Fig. 12.3 each of the features previously discussed with reference to the automobile engine can also be found in the heavy-duty engine. Several additional features will now be noted. First, a two-stage pressure regulator is commonly seen at the pump outlet. One stage provides coarse control when a large quantity of oil is to be bypassed. The second provides closer control of rifle pressure, and generally includes a feedback passage from the rifle. Next the oil passes through an oil cooler, where engine coolant is used to reduce the oil temperature through energy transfer to the coolant. In some cases an oil thermostat may be used to bypass the cooler until the oil reaches a threshold temperature. Where the automobile engine typically uses a single, full-flow oil filter the heavy-duty engine uses two filter stages. In addition to the full-flow filter a portion of the oil is taken through a finer filter to improve the filtration efficiency and remove smaller particles. In some cases two separate filter canisters are used, while in others both filter elements are incorporated in a single canister. Most heavy-duty engines include a second oil rifle supplying jets of oil to aid in piston cooling. This is discussed in greater detail from the perspective of piston temperature control in Chap. 15. Finally, pressurized oil is carried to a greater number of interfaces in the heavy-duty engine. Examples include pressurized oil supply to each rocker lever; through the rocker levers to the ball and socket joints; and in some cases through the connecting rods to the piston pins.

12.4   Oil Pumps

The oil pumps used in vehicular engines are positive displacement pumps driven directly or indirectly from the engine crankshaft. Two designs are commonly used—the gear pump depicted in Fig. 12.4 and the internal gear pump of Fig. 12.5. Sliding vane pumps (not shown) are increasingly seen; they are more complex and expensive than the designs shown here, but the sliding vane design lends itself well to variable displacement for pump power reduction. In both the gear and internal gear pumps one gear is driven by the engine, and it in turn drives the other gear as indicated in the figures. In either pump the separation of the gear teeth increases the volume of the inlet cavity, reducing its pressure and creating suction. The oil is carried in volumes between the gear teeth and either the pump body (gear pump) or outer rotor (internal gear pump). The meshing of the gears creates a volume reduction and corresponding pressure increase at the pump outlet. The

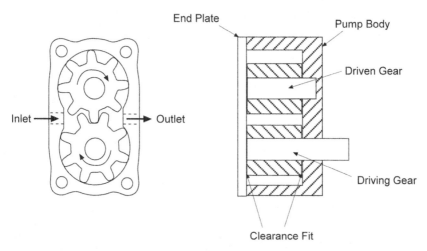

Fig. 12.4 Gear type oil pump schematic

Fig. 12.5 Internal gear pump. (Courtesy of Ford Motor Company)

pumps are constructed with a clearance fit between the parts that allows the pump to be self-lubricated by oil leakage from the high pressure to the low pressure side.

The gears are machined from iron alloys or powdered metal. The pump bodies are either sand cast from grey iron or die cast from aluminum alloys. The die cast aluminum provides lower cost in high volume engines. However the aluminum body grows with increased temperature at a greater rate than the iron alloy gears, and the leakage rate increases markedly at engine operating temperatures. The resulting efficiency reduction requires that the pump capacity be increased to compensate. Another alternative is to add end plates with Teflon seals, but this is seldom done as it defeats the cost reduction objective.

Another common practice in new engines is to design an internal gear pump around the nose of the crankshaft. The pump rotor is keyed into the crankshaft, and the housing is die cast into the front cover. This approach holds several important packaging attractions, and typically reduces the plumbing and sealing complexity. The disadvantages include the necessity of running the pump at crankshaft speed, and making the internal gear larger diameter and narrower than would otherwise be selected. Each of these design requirements penalizes efficiency.

The pump outlet pressure characteristics are shown versus speed in Fig. 12.6. The positive displacement results in an approximately linear pressure increase with speed as shown by the dashed line. In order to maintain control of flow and leakage rates, and to cap the pressure for which components such as the oil filter and oil cooler must be designed, a pressure regulator is used downstream of the pump. Regulator operation was shown previously in Fig. 8.4. The regulator is a simple device in which pump outlet pressure acts against a spring and a portion of the oil is returned to the sump or pump inlet once the desired pressure has been reached. The result is the nearly constant pressure at higher engine speeds as shown by the solid line in Fig. 12.6. The pressure continues to climb slightly with speed due to the increased force required as the regulator spring is further compressed. The oil pump is typically sized such that the regulated pressure is reached at an engine speed below that of peak torque to ensure that the design oil pressure

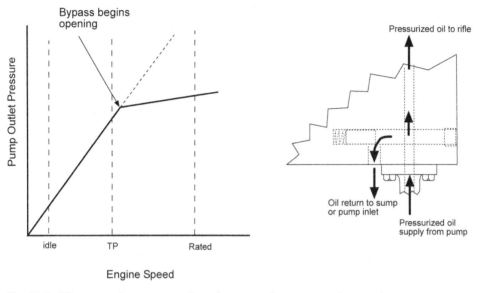

Fig. 12.6 Oil pump outlet pressure and regulator operation versus engine speed

is available over the entire full load operating speed range. With increasing demand for improved fuel efficiency more attention is being given to the oil that is pressurized and then immediately returned to the sump or pump inlet through the pressure regulator. This is not useful work, and is an additional work expenditure that increases with engine speed. A better approach than dumping oil to the sump, is to recirculate the oil to the pump inlet to reduce pumping losses. Variable displacement oil pumps are now in production on several automobile engines. A hydraulic switch changes the relative position of the pump body relative to the internal gear or vaned shaft, thus increasing pump capacity only at higher engine speeds.

Oil pump performance can be characterized in detail through the use of a pump stand in which the pump is driven with a variable speed electric motor, and inlet restriction and outlet pressure are independently controlled. The power required to drive the pump is shown over its range of operating speeds, with various outlets pressures, in Fig. 12.7a. Pump efficiency is plotted over the same range in Fig. 12.7b. This data has been collected using the same pick-up tube used on the engine, thus determining inlet restriction. If the outlet pressure is measured on the engine the engine operating curve can be overlaid. Instead of using the production pick-up tube the rig test can be modified to allow the inlet restriction to be varied. In this way the onset of cavitation can also be identified. It is important to ensure that the inlet restriction is low enough to eliminate cavitation, both to avoid pump damage and to avoid vapor entrainment in the oil rifle. Oil pickup tube diameter is not the only potential restriction, as the oil pickup screen can cause additional restriction. The pick-up screen usually flares out to large area, because of pressure drop over the screen, and the possibility of debris blocking the pickup and reducing capacity.

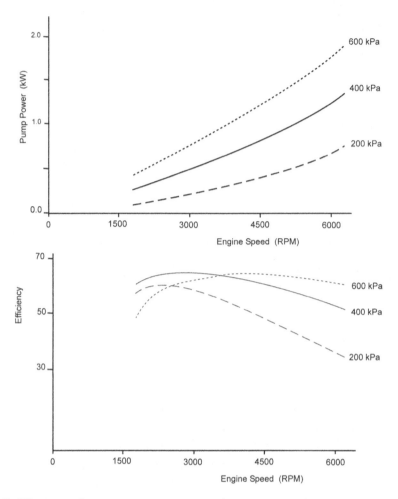

Fig. 12.7 Oil pump performance parameters measured on a pump test rig

12.5   Oil Pans, Sumps, and Windage

Most vehicular engines utilize wet sump systems, where the oil sump is contained in the oil pan as shown in Fig. 12.2 and all of the example engines in Figs. 1.6 and 1.7. Packaging constraints generally result in locating the sump at either the front or rear of the engine. Because the same engine is often used in several vehicle applications both front- and rear-sump oil pan designs may be required. In such cases it remains desirable to maintain a single pump location, and use different pick-up tube designs. Costs can be further controlled by using a symmetric oil pan design so the same stamping can be used for both oil pans. This consideration must be traded off versus other resulting design compromises pertaining to front and rear seals and the interface between the oil pan and front gear cover.

The oil pans used in automobiles are most often drawn from sheet steel alloys. They are cold formed over a progression of stamping dies, and often heat treated to reduce the resulting residual stresses. Die cast aluminum is sometimes used—especially with aluminum block engines, where the oil pan may be called upon to contribute to crankcase stiffness. Reinforced plastic oil pans are becoming increasingly common.

Design considerations include the required sump capacity, the design and location of the pick-up tube, drain plug location for various vehicle installations, and ensuring that the oil pick-up remains fully submerged under both side-to-side and fore and aft tilt. This creates a zone for ideal oil pickup that is shaped like a pyramid. Some engines have a localized depression or deep sump in the pan to aid in keeping oil around the pickup if vehicle packaging allows. In some engines threaded fittings for oil temperature sensing or an oil pre-heater must be included.

Capacity is determined by a combination of three requirements. The first is the directly functional requirement of providing sufficient lubricant for system operation. At any engine speed sufficient lubricant must remain in the sump to keep the pick-up tube submerged while oil is supplied throughout the lubrication system—pressurized lubrication and drain-back need both to be considered. The second consideration is customer or market expectations regarding oil change interval. Increasing the oil capacity slows down the rates of acid build-up, oxidation breakdown, and in the case of diesel engines soot build-up. While more oil is replaced at each oil change the vehicle service intervals are increased, still offering a customer attraction. The third system capacity consideration is direct customer expectations for the particular market. If an engine has a lower oil capacity than other engines serving the same market customers may perceive lower durability. If the engine has a greater capacity than the other engines they may perceive higher durability but will also note increased maintenance costs. Environmental impact of oil disposal is also a growing consideration.

Once the capacity has been determined the required sump dimensions can be finalized; this section concludes with a few further considerations. As was identified earlier, lubrication system performance requires that the pick-up tube remains submerged under all operating conditions. This includes the range of engine speeds, and in vehicular applications it also includes tilt, acceleration, braking, and cornering. In virtually all vehicular installations fore, aft, and side-to-side tilt must be considered, again requiring that the pick-up tube stays submerged under engine operation. Margin must be included for low-oil conditions due to low initial fill or oil consumption past the rings and valve seals in service. In high performance applications the demands must be expanded to include acceleration, braking, and cornering, all of which can push oil away from the pick-up tube. Finally, crankshaft dipping under each of these operating conditions needs to be addressed. It is important that the spinning crankshaft never dip into the oil in the pan or sump. Dipping significantly increases friction losses, and contributes to oil aeration (addressed in Sect. 12.7). Sump geometry and baffles within the oil pan can be used to address these design considerations.

The oil pan is subjected to high-cycle fatigue loading resulting from block vibration and the resulting vibration of the oil pan surfaces. Vibration of the oil pan surfaces is a significant contributor to engine noise as was discussed in Chap. 10. Accelerometer measurements over the engine speed range, coupled with servo-hydraulic rig testing constrained to duplicate the measured vibration, can be used effectively to validate oil pan durability. The oil pan is also subject to corrosive damage from water and road salt, and should be validated through corrosion testing. Finally, in most vehicle applications the oil pan is subject to impact damage from debris over which the vehicle is driven. This is most often addressed by placing a structural member of the vehicle frame ahead of the oil pan and maintaining slightly higher ground clearance at the oil pan than that of the vehicle frame member. When this cannot be done a skid plate is mounted beneath the pan, and affixed to the frame, or sometimes welded to the pan.

Another important issue for oil pan and sump design is that of windage and the minimization of crankcase pumping losses. In a multi-cylinder engine the overall crankcase volume remains relatively constant as the pistons reciprocate. However, the volume contained between each bulkhead is continuously changing, requiring the crankcase gases to be pumped back and forth across the bulkheads separating each of the cylinder bays. As engine speed increases, these pumping losses can become significant. The rapidly changing volume between bulkheads may also pump oil back and forth in the oil pan; this effect is sometimes addressed through addition of a *windage tray*—a stamped steel tray placed immediately below the crankshaft to minimize the pressure changes at the oil surface. In high performance or competition engines baffles may be included to ensure that the pick-up tube remains submerged under rapid acceleration. These may further include a one-way door to allow oil flow back into the sump. As has been introduced earlier, the challenges identified here are minimized with the dry sump system, but at a significant increase in cost and complexity.

Before leaving the topic of oil pan design a few further considerations need to be addressed:

Dipstick Design and Placement A location must be chosen that allows the customer to easily view and read the dipstick in any engine installation. This must be considered for longitudinal and transverse installations of the same engine in different chassis. While the operator is instructed to read the dipstick only when the vehicle is level, failsafe design requires that a location is selected that minimizes sensitivity to engine tilt. An automotive rule of thumb is a one quart capacity difference between the high and low marks on a dipstick, and the dipstick should be easy to read to get a clear indication of the oil level. To prevent the dipstick blade from wetting against the dipstick tube, occasionally the blade is twisted, or nubs are added to the measurement region of the stick. If plastic, the color should contrast with both clean and dirty oil for viewing. The owner's manual should have a clear indication of whether the oil should be checked hot or cold to account for expansion. Keeping the dipstick away from the spinning crankshaft and rod assemblies is important. Mounting the assembly in such a way that resonant vibration frequencies will not be encountered over the engine speed range or due to application-induced loads is important for durability. Both the tube extending out of the engine, and the dipstick extending into the oil sump must be considered in this regard. Depending on crankcase pressure during engine operation a positive seal may be required at the upper end of the tube. This is most often simply a synthetic gasket or seal ring but in some cases requires a more positive lock.

Oil Drain Location and Design An oil drain plug location needs to be selected allowing ease of drain in each application. The drain location must also ensure that the drain is complete, not leaving any pools of oil elsewhere in the pan that won't completely drain. Pan vibration and stress concentration at the threaded boss must also be considered, recognizing that in many cases the drain plug is significantly heavier than surrounding material.

Other Oil Pan Bosses Depending on engine application further items may be mounted in the oil pan. Examples include an oil heater, oil temperature sensor, oil quality sensor, or level sensor. The same considerations regarding thread stress concentration and vibration loads apply at each of these bosses.

12.6   Filtration and Cooling

The lubricant is formulated to include additives that lift deposits from engine component surfaces, and hold the particles and wear metals in suspension. A filter is then included in the system to capture these particles where they can be removed from the engine when the filter is changed. A filter's efficiency is characterized not with a single number but by the percentage of particles captured in each of a number of ranges of particle size. It is important to recognize that filters collect most but not all particles larger than the mesh of the filtration medium, and they capture some fraction of the particles smaller than the mesh. With this in mind, curves can be generated plotting filtration efficiency versus particle size. Two measures are commonly used to provide a numeric filtration capability. The nominal efficiency is the particle size at which 50 % trapping efficiency is measured. The absolute efficiency is the particle size at which 98.7 % trapping efficiency is measured.

Because the full-flow filter must accept the entire lubricant flow of the engine without excessive pressure drop the filtration media must be rather coarse. The filtration efficiency is high for large particles but quite low for small particles. Unfortunately many of the particles large enough to be damaging to the engine over time are small enough to slip through the full-flow filter. A bypass filter with a much finer mesh is often used in conjunction with the full-flow filter, especially on heavy-duty engines where the expected life-to-overhaul is long. Both filter elements may be contained in a single filter canister, or separate full-flow and bypass canisters may be used. The full-flow filters used in engines typically have nominal efficiencies of 15–20 microns and absolute efficiencies of about 30 microns. Good bypass filters have nominal efficiencies of about 5 microns and absolute efficiencies on the order of 10 microns.

The first question pertaining to engine oil coolers is whether one is to be included in the lubrication system at all. One of the functions of the lubricant is that of an energy carrier. It plays an important role in controlling piston temperature, and it carries away the majority of the energy lost in overcoming friction. Nevertheless, in many engines passive heat transfer through the crankcase and oil pan walls is sufficient to maintain the oil temperature at acceptable levels, and no further oil cooler is needed. The exceptions are the highest performance or highest load factor applications. As specific output is increased oil coolers become necessary.

The oil coolers used in engine lubrication systems fall into two general categories: oil-to-air coolers and oil-to-water coolers. Oil-to-air coolers are generally lighter, and require additional plumbing and sealed joints for a single fluid but require ram air or an additional cooling fan. Oil-to-water coolers hold the attraction of closer control of the oil temperature to the coolant temperature regardless of ambient temperature. Oil-to-water coolers allow faster heating of the oil, and due to the higher thermal capacity of coolant allow for a much smaller cooler package size. A limitation is the requirement of placing the oil cooler near a coolant passage.

Oil-to-air coolers are fin and tube heat exchangers typically manufactured of brazed aluminum. The air flows past dimpled or sometimes louvered fins, and the oil flows through tubes typically extruded with internal turbulators.

There are a variety of oil-to-water heat exchanger designs, the most common of which are tube bundle or finned tube heat exchangers. In either of these designs a "bundle" of copper tubes runs the length of a cylindrical housing. The coolant flows through the tubes in either a single- or two-pass configuration. The oil flows around the outside of the tubes, and within the cylindrical canister. A series of baffles directs the oil to make several passes (typically between five and seven) perpendicular to the tubes as it traverses the length of the heat exchanger. In the finned tube design a series of stacked aluminum fins surround the tubes, and the oil flows between the fins. The fins significantly increase oil-side pressure drop, but increase the effectiveness and reduce the package size of the heat exchanger.

Another commonly used oil-to-water heat exchanger is the brazed plate cooler. In this design each plate consists of two stamped sheets sandwiching an oil turbulator. The stampings are brazed together creating a sealed oil passage. The coolant flows around each sandwiched plate, and between the plates as they are stacked together. The assembly can be fit directly into the cylinder block cooling jacket, or placed in a separate coolant box. Increased effectiveness is accomplished by increasing the number of plates, reducing oil-side pressure drop but increasing the size of the required coolant cavity.

A third type of oil-to-water heat exchanger is the "donut" cooler. In construction it is similar to the brazed plate cooler, but is designed to fit between the oil filter head and filter canister. Oil supplied to the filter passes through the center of the donut. Returning oil passes through the turbulated oil passages in the cooler. Engine coolant is fed through a canister surrounding the oil passages.

An important oil cooler design consideration is that of what happens to the oil on engine shut-down. Does the cooler remain full, or does it drain back to the crankcase? If the oil cooler must refill when the engine is started pressurized oil supply to the main gallery is delayed, so steps are often taken to ensure that the oil cooler remains full. This unfortunately results in a quantity of oil remaining in the system when the oil is changed. In larger engines a separate oil cooler drain is often included, and the oil cooler is drained during the oil change.

Another design option that is sometimes included is an oil thermostat that bypasses the cooler until the oil reaches a predetermined operating temperature. An oil thermostat is added primarily for fuel economy improvement, ensuring that the oil always runs at a design temperature and viscosity. If the oil is cooled under all conditions its viscosity at part-load engine operation will be higher than necessary. Oil thermostats are most commonly seen in heavy-duty truck engines.

Oil cooler performance measures include heat exchange effectiveness, oil- and water-side pressure drop, and its effect on the time required for rifle pressurization under both cold and hot start-up conditions. Oil cooler structural design validation includes consideration of pressure cycling, vibration, and thermal cycling.

12.7 Lubrication System Performance Analysis

The general layout and objectives of the lubrication system were discussed in earlier sections. At this point those objectives are further reviewed with an eye toward component sizing and ensuring that performance expectations are met. This discussion begins with a look at the critical components that must be lubricated, and some general considerations regarding component location and system layout. Lube pump sizing is then covered, with an emphasis on determining bearing flow requirements. Hot and cold time to pressure is then addressed, and measurement and analysis techniques are covered. The section concludes with a detailed look at crankcase design considerations.

Pressure and Flow Requirements In laying out the lubrication system it is important to begin from a clear understanding of the key pressure and flow requirements throughout the system. The starting point, though often the hardest to accurately determine, is the flow rate requirement to each bearing—both hydrodynamic and mixed film. The combination of bearing supply pressure, geometry and clearance, and loading versus crank angle makes the flow rates extremely difficult to predict. For now it is important to recognize this challenge, and that the necessary flow rate and supply pressure will need to be identified throughout the engine; the topic will be further discussed in later paragraphs.

Many engines today use piston cooling nozzles. These are relatively high flow devices, placing heavy demands on the oil pump capacity. Especially in smaller engines a shut-off valve in the piston cooling nozzle supply is used to close off this oil flow path at low engine speeds. This ensures that sufficient oil flows throughout the rest of the system, and reduces the oil pump capacity requirements.

Hydraulic valve lash adjusters, or lifters, are included on many engines. Their flow rate requirements are relatively low, but the supply pressure at the inlet of each lifter must meet lifter design specifications.

Turbocharged engines require high oil flow rates through the turbocharger bearing assemblies due to the very high shaft speeds and floating plain bearing design. Most turbochargers use a plain hydrodynamic bearing with two oil films—one on the inside and one on the outside of the bearing. The bearing spins at an intermediate speed, with a hydrodynamic film between the bearing and the housing, and another between the shaft and the bearing. Two such bearings are used, with one inboard of each impeller. They are fed with pressurized engine oil, which is then returned by gravity through a line back to the crankcase or oil pan. The bearings must be above the level of oil in the sump to ensure drain back by gravity.

The lubrication system is then laid out as a network of drilled and sometimes cast passages, supplying oil from the pump, through the filter(s) and oil cooler, and supplying each of the locations just described.

From the initial layout oil pump size (flow rate) requirements must be determined. The pump is sized based on maximum clearances and oil temperature—the conditions under which the flow rate is at its maximum. The flow rates and system requirements must be looked at over the range of engine speeds, considering not only the full-load requirements from rated power to peak torque, but also to avoid oil starvation under extended idle conditions.

Oil flow demands (and supply pressure) to the turbocharger, piston cooling nozzles, and hydraulic lifters are most often specified by the suppliers, and can be determined experimentally through bench tests of the components. Flow rate versus supply pressure is measured, and fed into the pump sizing requirements. The most difficult aspect of pump sizing is determining the bearing flow and leakage rates. Oil flow through each bearing is dependent on flow geometry and magnified by hydraulic loading due to the changing pressure within the bearing. There are various flow rate models that use empirical relationships to estimate the contribution of each; these can be adjusted to match models to measurement but have a poor track record for correctly sizing pump flow requirements in a new engine. The largest flow demand in most engines is that through the main bearings, followed by rod bearings and cam bearings—typically in that order. The non-hydrodynamic bearings such as rocker levers, cam followers, and piston pins require much less flow; they have little impact on pump flow requirements but may impact sizing in ensuring that adequate pressure is maintained at the supply to each of these joints.

Pump sizing is then an iterative process in which initial circuit modeling is done, and pump flow calculations are made. The calculations are followed by measurements, model validation, and if necessary, pump resizing. The alternative of simply oversizing the pump holds penalties of both increased package size and increased pump power and parasitic losses.

Flow circuit modeling is an effective tool throughout this process. It does require iteration on the bearing flows, as just discussed, but once the model has been validated it becomes a very effective tool for assessing flow rates and pressures throughout the system. Circuit optimization and component placement decisions are aided during the engine development process, and if the model is maintained it will be valuable for rapidly assessing design changes or specialized engine applications once the engine is in production. With further set-up and validation the lubrication system model can be used under transient conditions as well, supporting the time-to-pressure studies of the next section.

Time-to-Pressure A critical aspect of lubrication system performance is the time it takes to build pressure at any point in the system upon engine start-up. An example of measured time-to-pressure data is shown in Fig. 12.8. Critical locations such as main and rod bearings and turbocharger bearings are designed to run under hydrodynamic conditions, and wear only occurs immediately after start-up, before oil is supplied to the bearings. While the time-to-pressure is considerably longer under cold-start conditions many applications see many more hot-starts than cold-starts, making hot time-to-pressure an important design criterion as well.

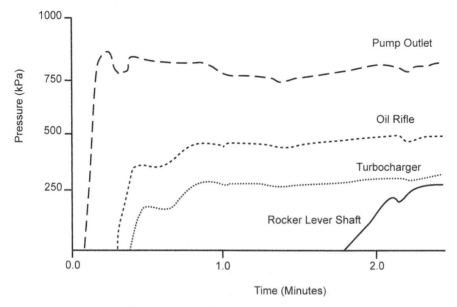

Fig. 12.8 Time to pressure at various locations in a heavy-duty diesel engine

At least initially most time-to-pressure assessment is done experimentally. Especially under cold conditions repeatability is the most challenging consideration. The engine is cold-soaked in a refrigerated environment to ensure thermal equilibrium at the cold ambient temperature. At time equals zero the engine begins cranking and then fires, with pressure monitored versus time at critical locations throughout the circuit. Engine speed during cranking and firing must also be measured versus time. Oil begins pumping during cranking, and the cranking speed and time-to-fire are critical variables in the time-to-pressure measurements. It is recommended that the test be run at least three times, ensuring that starter speed and the time between engaging the starter and the engine firing are the same in each test. Variation in these parameters undermines the ability to quantify time to pressure. Similar hot time-to-pressure tests are subject to far less variability.

The data from the cold and hot time-to-pressure tests allow validation of the transient aspects of the lubrication circuit models described in the previous section. While the resulting tool is empirically based it provides a valuable means to quickly assess the effects of design modifications on time-to-pressure.

Another important consideration in time-to-pressure assessment is the question of oil drain back time from the main oil gallery and the smaller passages. Because of the relatively tight clearances and leakage paths oil does not immediately empty from any of these passages. The time required for the main gallery to empty is different from that for a crankshaft drilling. The actual drainback time is difficult to quantify without careful experiments. Keep in mind that compressed test times between hot re-starts, whether for durability tests or repeated time-to-pressure tests, will result in different time-to-pressure results. The recommended approach is to base all restarts on an initially empty rifle. Injecting compressed air to empty the rifle between tests is recommended.

Crankcase Development Previously the role of the crankshaft assembly spinning in the oil mist of the crankshaft was discussed as regards its role in friction power loss. Bay-to-bay breathing was also discussed. Crankshaft dipping was identified in the discussion of oil pan design. In this section further crankcase performance considerations are covered.

This discussion begins with a further look at what is going on in the crankcase. Beginning with the oil, there is always a pool of oil remaining in the sump at the base of the crankcase. In a dry sump engine this pool is being continuously evacuated. The oil pool in the wet sump experiences engine vibration forces and the acceleration, braking, cornering, and tilt discussed in the pan design section. All of these forces move the oil pool around, and it is important to avoid crankshaft dipping and minimize any further contact between this oil and the crankshaft.

The oil pump is continuously drawing oil from the sump, and all of this oil must return to the sump. Depending on pressure regulator design there may be an oil flow from the regulator back to the sump (unless it has been plumbed back to the pump inlet). A significant portion of the oil is fed to the piston cooling nozzles, and another important portion is fed through the main and rod bearings. All of this oil drains back into the path of the crankshaft assembly. It is necessary for some oil to contact the crankshaft, as this mechanism is used to coat the cylinder walls and provide piston and ring lubrication. However, more than enough oil is typically available for this purpose, and steps are taken in drain-back design to send oil from the camshaft bearings and valve train, and the turbocharger return along the crankcase walls, and directly to the sump, with little or no contact with the crankshaft. In some cases enclosed passages in the skirt walls feed this oil directly down to the oil pan mounting flange, where it can run down the pan walls.

Positive Crankcase Ventilation (PCV) Superimposed on the crankcase oil flows is a continuous flow of combustion gas into and back out of the crankcase. Due to ring end gaps and axial ring motion the combustion gases are not completely sealed from the crankcase. At the elevated in-cylinder pressures of the compression and expansion strokes a small gas flow, referred to as blow-by, is intermittently entering the crankcase from each cylinder. The net result in a multi-cylinder engine is a nearly steady flow through the crankcase. If the crankcase were completely sealed this would result in pressure build-up in the crank-case and significant further crankcase pumping work. The gases must therefore be allowed to exit the crankcase. Historically this was done using a breather tube through which gas could escape to the atmosphere. Because the gas contains combustion products, and may also include some oil mist, emission regulations in vehicular applications require a sealed path. The gases are fed back into the engine intake system, through a positive crankcase ventilation (PCV) valve. In spark-ignition engines the PCV valve closes at idle to avoid intake charge dilution and misfire. It also closes at wide-open-throttle to maximize the amount of fresh charge entering the engine. Under both of these conditions a separate breather vents the crankcase into the intake system, typically upstream of the intake filter. Whether it be a breather tube or a PCV system the gas is drawn from the crankcase at the highest location possible, to minimize collecting oil in the flow. A mesh filter or centrifu-gal flow passage may be used to further capture oil.

The turbocharged or supercharged engine is an important special case for positive crankcase ventilation. Because the compressor outlet pressure is well above crankcase pressure the crankcase vapor must be fed into the intake system upstream of the compres-sor. Oil droplets are damaging to the compressor wheel, and oil is a relatively high cetane fuel. The latter presents the possibility of a runaway engine. The end result is that a far more positive oil mist separation system must be incorporated when applying PCV sys-tems to turbocharged and supercharged engines.

Oil Aeration Returning to the oil in the crankcase a further concern is oil aeration. Air is soluble in oil up to about nine percent by volume at atmospheric pressure. The solubility increases linearly as pressure is increased, and decreases linearly in a vacuum. Once the oil is saturated with air additional air in the oil will form gas bubbles. Pressurized oil that is saturated with air will emit bubbles as its pressure drops.

Oil aeration impacts lubrication system performance in several ways. It causes oil film collapse in loaded bearings, leading to reduced oil film thickness and the possibility of metal-to-metal contact. Air entraining in the piston cooling flow adversely impacts the heat transfer effectiveness. Air entrainment contributes to cavitation damage in the oil pump and on bearing surfaces. Finally as air bubbles build up they lead to oil starvation at the furthest downstream locations in the pressurized lubrication circuit. As the pressure drops more and more air comes out of the oil, and these bubbles are pushed downstream until they replace much of the liquid lubricant.

There are a number of factors impacting oil aeration, including oil age, the additive package, oil viscosity, engine speed, and design considerations such as oil mass and level (oil residence time in the sump or the number of turns of oil at peak engine speed). The problem is accentuated by crankshaft dipping, by a restrictive oil pump inlet or pick-up tube, or by a pressure bypass outlet near the pump inlet. Deep oil sumps make it more difficult for air to escape from the oil due to reduced surface area. Another important consideration is the inertia effect on oil pressure in the crankshaft drillings feeding the connecting rod bearings. The spinning crankshaft pushes oil toward the rod journal, and the restrictive inlet draws the pressure down. As the crankshaft spins faster the oil pressure within the crankshaft drilling is drawn down lower than both that at the inlet and outlet. Entrained air will come out of the oil if the pressure drops below the saturation pressure, and even though the pressure again climbs the air won't re-entrain, and the bubbles are fed into the rod bearing.

Oil aeration is closely monitored during new engine development. Steps are taken to address design problems contributing to aeration, the key one of which is the spinning crankshaft assembly "stirring" the oil and air in the crankcase. Drain-back, windage, and oil pan baffles are examples of design variables impacting aeration. Windage trays mounted below the crankshaft and above the sump are sometimes used to minimize the oil's participation and bay-to-bay breathing.

12.8   Recommendations for Further Reading

The following text, written in German, provides a detailed look at bearing and lubrication system design (see Affenzeller and Gläser 1996).

The following papers provide more detailed descriptions of oil pump design. The first of these two papers by the same authors begins with an analysis of internal gear pumps and also discusses sliding vane pumps. The second focuses on sliding vane pumps (see Manco et al. 2004a and 2004b).

System analysis and time to pressure under cold-start testing is presented in the following paper (see Kluck et al. 1986).

Especially in high speed engines another important consideration in crankcase design is ventilation. This paper provides a recent discussion (see Koch et al. 2002).

The following paper provides an excellent discussion of crankcase flow and windage tray design (see Iqbal and Arora 2013).

Examples of new engines in which variable displacement oil pumps are well described are as follows (see Konigstedt et al. 2012 and Neußer et al. 2012).

The following paper includes a description of the design considerations that went into a high-performance dry-sump system to ensure oil pick-up under acceleration, braking, and cornering (see Wasserbach et al. 2012).

References

Affenzeller, J., Gläser, H.: Lagerung und Schmierung von Verbrennungsmotoren. Springer-Verlag, Wien (1996)

Iqbal, O., Arora, K.: Windage Tray Design Comparison Using Crankcase Breathing Simulation. SAE 2013-01-0580 (2013)

Kluck, C.E., Olsen, P.W., Wickland Skriba, S.: Lubrication System Design Considerations For Heavy-Duty Diesel Engines. SAE 861224 (1986)

Koch, F., Haubner, F.G., Orlowsky, K.: Lubrication and Ventilation System Of Modern Engines—Measurements, Calculations And Analysis. SAE 2002-01-1315 (2002)

Konigstedt, J., Assmann, M., Brinkmann, C., Eiser, A., Grob, A., Jablonski, J., Muller, R.: The New 4.0-l V8 TFSI Engines from Audi. International Vienna Motor Symposium, 2012

Manco, S., Nervegna, N., Rundo, M., Armenio, G.: Displacement vs flow control in IC engines lubricating pumps. SAE 2004-01-1602 (2004)

Manco, S., Nervegna, N., Rundo, M., Armenio, G.: Modeling and Simulation of Variable Displacement Vane Pumps For IC Engine Lubrication. SAE 2004-01-1601 (2004)

Neußer, H-J., Kahrstedt, J., Jelden, H., Engler, H-J., Dorenkamp, R., Jauns-Seyfried, S., Krause, A.: Volkswagen's New Modular TDI® Generation. International Vienna Motor Symposium, 2012

Wasserbach, T., Kerkau, M., Bofinger, G., Baumann, M., Kerner, J., Neuser, H-J.: Performance and Efficiency—The Flat-Six Engines In the New Porsche 911 Carrera. International Vienna Motor Symposium, 2012

Engine Cooling

13

13.1 Tracking the Energy Transfers

In turning to the subject of engine heat transfer it is instructive to begin with a detailed look at where all of the fuel energy goes. If the entire engine is treated as a thermodynamic system, energy enters the system with the fuel. Air enters the system at ambient conditions and therefore, by convention, with zero energy. Under steady-state conditions, and averaged over an operating cycle, no energy is stored, and energy exits the system at the same rate it enters. In other words, all of the energy entering the engine in the fuel exits the engine at the same rate. The energy exits the system as either work, heat transfer, or with the exhaust flow.

At first look, one would think that studying engine heat transfer and cooling is simply a matter of quantifying the energy transfer out of the engine as heat transfer. In reality this simplified look carries very little engineering value. In managing engine temperatures and developing effective cooling systems it will be important to understand the engine heat transfer terms on a much more detailed level. Toward this objective, the energy transfers within a turbocharged and aftercooled engine with cooled, high-pressure exhaust gas recirculation (EGR) are depicted in Fig. 13.1. The terms enclosed in boxes are the transfers in and out of the engine system, while the remaining terms identify energy transfers within the system.

Looking only at the boxed terms in Fig. 13.1, one can already see engine heat transfer becoming more complicated, as the heat rejection is now distributed among three boxes: heat rejection through the charge cooler; heat rejection through the radiator; and heat rejection from engine surfaces.

Various thermodynamic sub-systems can be selected within the overall engine system. Looking first at a sub-system consisting solely of the gas within the combustion chamber, energy is transferred into the system with not only the fuel but also the warm compressed

© Springer Vienna 2016
K. Hoag, B. Dondlinger, *Vehicular Engine Design*, Powertrain,
DOI 10.1007/978-3-7091-1859-7_13

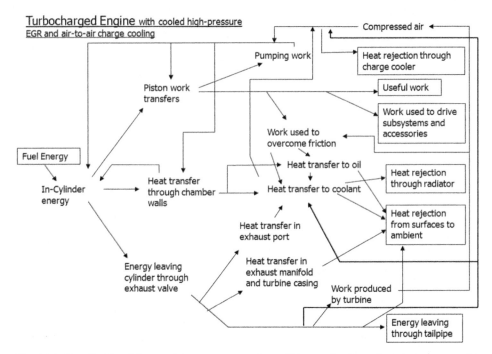

Fig. 13.1 Detailed breakdown of the energy transfers within a turbocharged engine system, with cooled high-pressure EGR and air-to-air charge cooling

air. The air that had entered the turbocharger compressor at ambient conditions (zero energy) has been compressed, cooled, and then warmed by the intake port walls and intake valves. There are piston work transfers in and out of the cylinder; at full load the net work transfer might be on the order of 40% of the fuel energy. The heat transfer is also shown into and out of the cylinder. The net heat transfer out of the cylinder is on the order of 10–15% of the fuel energy at full load. Finally, there is the exhaust energy leaving the cylinder with exhaust mass flow. At this point in the system the exhaust flow at full load carries on the order of 45–50% of the fuel energy. Each of these numbers vary a great deal with engine speed and load, and depending on the combustion system—diesel versus spark-ignition; stoichiometric versus lean burn operation as examples. The key points to keep in mind in considering engine heat transfer and cooling are that in-cylinder heat rejection is lower than one might expect, and the energy in the exhaust is higher.

Significant energy transfers occur throughout the exhaust system. As stated in the preceding paragraph on the order of 50% of the fuel energy may be leaving the cylinder with the exhaust mass. That entering the turbine might be closer to 40%, with the majority of the difference transferred to the coolant through heat rejection in the exhaust ports. Another portion is transferred directly to the ambient from the exhaust manifold and turbine casing. The key point from these observations is that exhaust port heat rejection makes a significant contribution to the cooling system heat load.

It is important to look next at the role of the lubricating oil as an energy transfer medium. Of the work expended in overcoming mechanical friction almost all of it is transferred to the lubricating oil. A significant portion of the heat rejection from the cylinder goes through the piston to the oil. When the in-cylinder temperatures are highest (during combustion) the exposed combustion chamber surfaces are primarily the cylinder head and piston crown. A large portion of the cylinder head surface area is actually the valve faces, and they present a highly resistant path to heat transfer. It follows that the majority of the combustion heat rejection is into the piston crown, from where it can pass through the ring pack to the cylinder wall, or through the undercrown to the oil. The typical split between these two paths is on the order of 50/50 although this can be measurably manipulated through cylinder cooling (jacket length), and enhanced piston oil cooling (piston cooling nozzles, gallery cooled pistons). In engines with oil-to-water oil coolers (such as the engine diagramed in Fig. 13.1) heat rejection to the oil is then transferred to the coolant. In other engines it is transferred directly to the ambient, either through an oil-to-air heat exchanger or as engine surface-to-ambient heat transfer.

Reviewing the discussion of the preceding paragraphs, and including further terms identified in Fig. 13.1, the cooling system loads are as follows:

- In-cylinder heat rejection—both directly into the coolant as well as from the oil if an oil-to-water oil cooler is used
- Exhaust port heat rejection
- Friction—a small portion through the coolant, and most of the friction energy if an oil-to-water oil cooler is used
- EGR cooler
- Charge air cooler if an air-to-water charge cooler is used

Finally it is important to recall that an important portion of the fuel energy is rejected through surface losses directly to the surrounding air. This portion is difficult to quantify but plays an important role in overall temperature management. In many automobile engines oil temperature control is almost entirely dependent on surface heat rejection. This is the case in any engine not incorporating an oil cooler.

13.2   Critical Issues in Temperature Control

As the cooling system is being developed and optimized it is important to zero in on the specific temperature control objectives. In doing this it is also important to work hand-in-hand with lubrication system development, recognizing both that oil temperature control will be of critical importance, and also that the oil serves as an important cooling medium for regions not accessible to the coolant. The reasons for temperature control are summarized in the following paragraphs:

Component Fatigue Each of the materials used in the engine have temperature-dependent fatigue lives. Understanding those limits for each engine component, and ensuring that the limits are not exceeded, is an important aspect of cooling system development. Fatigue assessment includes not only peak temperature control but control of temperature gradients as well. Alloy modification and surface treatments can also be considered, allowing the temperature limits to be increased and required cooling to be reduced. The impact of alloy changes on the component's thermal conductivity must also be considered; if the thermal conductivity is reduced the benefit of the alloy might be offset by the resulting higher operating temperature. In addition to enhancing component cooling temperature control can often be achieved by incorporating thermal barriers in the design. For example, air gaps are commonly used to reduce exhaust port and manifold heat rejection.

Knock Margin In spark-ignition engines an important consideration is the effect of combustion chamber surface temperature on knock initiation. Peak temperatures on the cylinder head, valve faces, and piston crown play a role in determining when knock will occur. Reducing the surface temperatures can improve the knock margin, allowing power and efficiency improvements.

Gasket Performance Thermal growth of the cylinder head, head bolts, and exhaust manifold impact gasket and sealing requirements at the cylinder head gasket and exhaust manifold gasket. The higher thermal growth of the cylinder head relative to that of the head bolts increases the clamping force on the head gasket as the engine temperatures increase. Thermal growth of the exhaust manifold increases the manifold gasket clamping force, and causes the exhaust manifold to "slide" relative to the cylinder head. These phenomena are further covered in Chap. 14.

Distortion and Dimensional Control Thermal growth and distortion must be taken into consideration in several areas of the engine. In addition to the gasketed joints just discussed, thermal growth is important in piston-to-cylinder clearance, cylinder wall distortion, and the press fit of valve seats into the cylinder head. Cylinder wall distortion is impacted by several variables including initial machining, head bolt clamping, cylinder pressure, and temperature variation. Cooling system development must strive to minimize cylinder wall circumferential temperature variation, taking into account inherent differences between the exhaust and intake sides of the cylinder.

Deposit Formation The hydrocarbon blends used in engine oil have a propensity to react with oxygen that increases exponentially with increased temperature. Controlling the maximum temperature seen by the lubricant in locations such as the top piston ring and land, turbocharger bearings, and valve guides is a critical aspect of engine cooling.

Oil Life Again recognizing that oil oxidation increases exponentially with increased temperature, controlling the oil temperature in the sump is an important parameter in determining the oil change interval. As specific engine output increases this becomes a decision factor in whether to add an oil cooler to the system.

13.3 Engine Cooling Circuits

The cooling circuit was previously introduced in the discussion of cylinder block and head design (Chaps. 8 and 9). In this chapter elements of that discussion are integrated to provide an overall look at the engine cooling system. While many engines in motorcycle and small industrial applications are air cooled this approach is rarely used in automobile and other, larger vehicular installations. The majority of this chapter will discuss liquid cooled systems, with the final section providing a brief overview of air cooled engines.

The liquid used in the cooling system must perform without significant change of phase over an extremely wide temperature range—from very cold ambient temperature to well past the temperature of boiling water. The coolant generally selected is a 50/50 mixture by volume of water and either ethylene glycol or propylene glycol. This mixture remains liquid to a temperature of $-57\,°C$ without freezing. Its boiling point is dependent on the degree to which the cooling system is pressurized—$125\,°C$ or more is typical. Two important points should be noted. First, as is the case with many fluid mixtures the freezing point of the mixture is below that of either fluid by itself. The minimum freezing point of ethylene glycol and water actually occurs at approximately a 60% glycol concentration, but the 50/50 mixture provides quite sufficient cold weather protection and is far easier to specify in the field. Second, the heat transfer performance of the water/glycol mixture is lower than that of pure water. Under exactly the same conditions the convective heat transfer coefficient for the mixture is about 80% of that for pure water. This means that component temperatures are higher with the glycol solution (40–50 °C in high heat flux regions), and the measured heat rejection is lower. This must be considered during engine testing, and the glycol mixture must be used in any engineering tests that would be affected by these differences.

Another important coolant consideration is stability over time. While most engine and vehicle manufacturers specify periodic cooling system flushes and coolant replacement, a significant fraction of vehicle owners ignore these service recommendations, and do not replace coolant unless a cooling system component is replaced, or a leak repaired. Coolants may contain additives that improve chemical stability, and to further protect against corrosion of iron or aluminum surfaces. In heavy-duty or longer life engines coolant filtration and supplemental additives may be used, and deionized water is specified prior to mixing with the glycol coolant.

The most commonly seen coolant flow path in automobile engines, referred to as *series cooling*, is depicted in Fig. 13.2. Coolant is supplied from the water pump to the front of the cylinder block. The majority of the coolant flows around the cylinders from the front to the back of the engine, and is then transferred into the cylinder head where it flows from the back cylinder to the front. The coolant temperature increases as it travels through

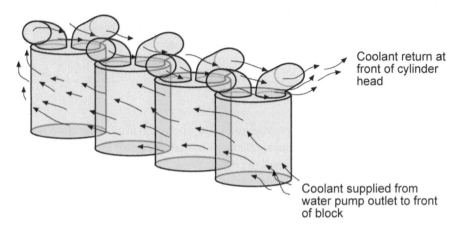

Coolant return at front of cylinder head

Coolant supplied from water pump outlet to front of block

Fig. 13.2 Representation of series flow cooling circuit for in-line four-cylinder engine

the engine, so the back cylinder wall sees higher temperature coolant than does the front cylinder, and the cylinder head sees higher coolant temperature at the front cylinder than at the back. This temperature imbalance is partially offset by transferring small portions of the coolant from the block to the head around the perimeter of each cylinder. This design holds the advantage of relatively simple jacket cores, and allows the coolant to enter and exit the engine at the front, lending itself well to front-engine, rear-drive installations. The problem of coolant flow balance was previously mentioned, although adequate balance can generally be achieved through the use of progressive transfer passage diameters—as one moves from the front to the back of the engine the transfer passage diameters are incrementally increased. A further challenge with this flow path is that of achieving adequate flow in the regions between any two cylinders.

With the move to front-wheel-drive installations and transverse engine mounting the *parallel flow* circuit of Fig. 13.3 is sometimes used. With this flow path the coolant enters the block at one end of the engine and a portion is immediately transferred to the cylinder head. The coolant then flows by two parallel paths from one end of the engine to the other, and all of the coolant exits through the cylinder head at the opposite end from its entrance. In addition to the reduced external plumbing required for this circuit in transverse installations it holds the advantage of reducing the coolant temperature in the cylinder head. The circuit must be designed to achieve the appropriate flow balance between the cylinder block and head; this poses an inherent difficulty in that the block cooling jacket tends to have lower restriction than the head while the heat rejection from the cylinder head is higher.

For reduced temperature variation between cylinders, but at the price of far more complicated castings, a *cross-flow* cooling circuit such as that shown in Fig. 13.4 may be considered. The water pump supplies coolant to a high pressure water header cast along the length of the cylinder block. From this header coolant is supplied to the separate cooling jacket surrounding each cylinder. The coolant is then transferred to the cylinder head, again from the passages surrounding each cylinder. From the cylinder head the coolant is

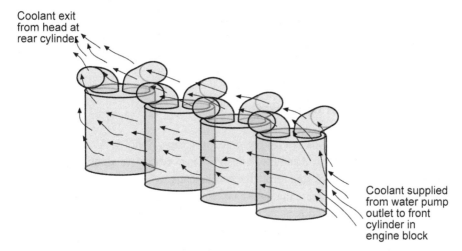

Coolant exit
from head at
rear cylinder

Coolant supplied
from water pump
outlet to front
cylinder in
engine block

Fig. 13.3 Representation of parallel flow cooling circuit for in-line four-cylinder engine

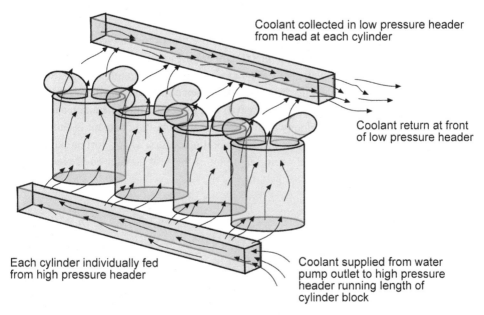

Coolant collected in low pressure header
from head at each cylinder

Coolant return at front
of low pressure header

Each cylinder individually fed
from high pressure header

Coolant supplied from water
pump outlet to high pressure
header running length of
cylinder block

Fig. 13.4 Representation of cross-flow cooling circuit for in-line four-cylinder engine

collected in a low pressure water manifold—either a separate manifold above the cylinder head or a header cast into the head or the top of the block. As compared to either of the flow circuits described previously this circuit results in far less cylinder-to-cylinder variation. In larger truck and off-highway engines the cross-flow circuit is the norm.

It should be noted that each of these circuits has been discussed relative to a single bank of cylinders. Engines having more than one cylinder bank do not open up additional flow circuit options, but duplicate one of the circuits described here in each bank.

13.4   Cooling Jacket Optimization

Cooling jacket placement was briefly presented in Chaps. 8 and 9. As was discussed in those chapters on cylinder head and block design it is necessary to identify the critical regions for temperature control, and locate the cooling jackets for maximum thermal fatigue life. It is also necessary to ensure that lubricant temperature limits are not exceeded as this would lead to lubricant breakdown and deposit formation. It is important to recognize that these objectives require not only the control of peak temperatures, but avoidance of overcooling in regions where the overcooling needlessly increases temperature gradients that penalize fatigue life. In the cylinder block overcooling results in lower than necessary oil temperatures and increased viscosity, thus penalizing fuel efficiency. The designer must also avoid stagnant regions where even though the heat flux is relatively low the lack of flow results in local coolant boiling.

As one approaches cooling jacket optimization it is important to recognize that in the regions of highest thermal loading the metal temperatures at high engine load at the surfaces exposed to coolant are well above the coolant boiling temperature. From knowledge of fluid mechanics and the velocity boundary layer the coolant in immediate contact with the wall is at zero velocity, and is thus subject to boiling. The critical parameter for jacket design is to keep the coolant velocity high enough that the boiling coolant is entrained in the bulk flow and again condensed as fast as the boiling occurs. This is further depicted in Fig. 13.5 in a pictorial representation and in a plot of the heat transfer coefficient versus

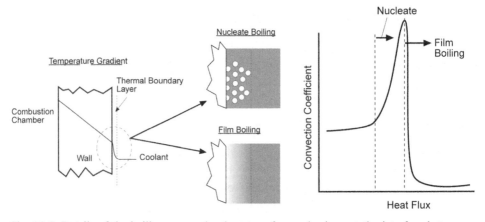

Fig. 13.5 Details of the boiling convection heat transfer mechanisms at the interface between a combustion chamber wall and the coolant

heat flux rate. At low loads the wall temperature is below the coolant boiling temperature, and heat transfer to the coolant occurs purely through forced convection to the liquid. As the heat flux rate increases the surface temperature climbs above the coolant boiling point and vapor bubbles form at the surface. As each bubble grows it breaks away from the surface and is entrained in the flowing bulk coolant where its temperature drops and it returns to the liquid phase. Because of the high heat of vaporization of the coolant this process of *nucleate boiling* provides a significant enhancement to the convection process and the heat transfer coefficient increases by an order of magnitude or more. This slows the rate at which the component temperature climbs with further increase in the heat flux. However, as the heat flux rate continues to increase a point may be reached where the coolant boils at a rate faster than it can be entrained and re-condensed. A layer of superheated coolant now covers the surface. This is referred to as *film boiling*. The heat transfer coefficient drops to a level well below that of pure convection, and the component temperature climbs rapidly. In summary, while nucleate boiling is expected and enhances component temperature control, the velocity must be maintained high enough to avoid any regions of film boiling.

An important design issue in many engines is ensuring that each cylinder receives approximately the same cooling. Variations in both coolant temperature and coolant flow rate between cylinders must be minimized. The example of flow balance in the series cooling circuit was identified in the previous section. With the series circuit the coolant exits from the water pump directly to the front cylinder. This results in very high velocities, along with the lowest coolant temperature, around the front cylinder. Jacket restriction further results in a propensity for coolant to flow directly from the front cylinder into the front of the cylinder head and back out of the engine. Such imbalances can be considerably reduced by varying the flow area of each transfer passage between the block and the head. If the front transfer passages are made very small relative to those in the back of the engine flow around the back cylinders is encouraged as depicted in Fig. 13.6.

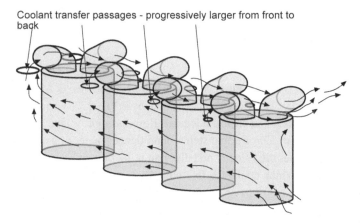

Coolant transfer passages - progressively larger from front to back

Fig. 13.6 Use of progressively increasing diameter coolant transfer passages for flow balance between cylinders

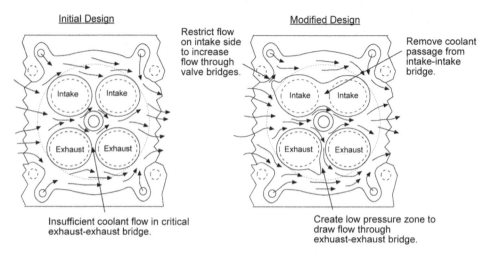

Fig. 13.7 Cylinder head cooling jacket modification for flow improvement in critical regions

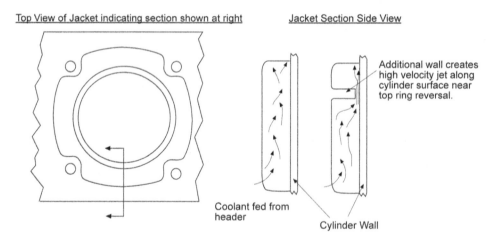

Fig. 13.8 Cylinder cooling jacket modification for improved coolant velocity and reduced temperature variation around perimeter of cylinder

The passage restriction is varied until acceptable balance is achieved. Another example of steps taken to improve coolant flow balance is shown in Fig. 13.7 where walls are added in the cylinder head jackets to ensure adequate flow through critical passages. Finally, Fig. 13.8 shows the use of a protruding step around the perimeter of a cylinder wall. This added restriction increases the velocity and distributes the flow more uniformly around the cylinders in engines in which coolant is fed to the base of each cylinder, and thus reduces temperature variation.

Experimental flow visualization is very effective in characterizing and optimizing coolant flow. A full-scale model of the cooling jacket is machined from transparent plastic. Water seeded with neutral buoyancy particles is pumped through the jacket. Because of the very high flow velocities it is difficult to make direct observations but through the use of high speed photography replayed at lower speeds the flow can be characterized in detail. Adding a scribed grid to the plastic walls allows the local coolant velocities to be quantified. Computational fluid modeling offer a further effective tool increasingly used for cooling jacket development.

13.5   Thermal Mapping

An important aspect of cooling system and engine development is thermal mapping. This effort, requiring complementary use of measurements and computational tools, provides engine component temperature maps versus relevant operating variables. In most cases the temperature maps are as a function of speed and load, but depending on the calibration and application of the engine other variables may be included as well. Examples include changes in air-fuel ratio or shifts in spark or injector timing under different operating conditions.

A complete thermal mapping process includes all of the parts that see temperatures elevated above the bulk coolant or oil temperatures, and especially parts that see large temperature gradients. The parts most often considered in thermal mapping are the cylinder head, valves and seats, piston, and the upper portion of the cylinder wall (cylinder block or liner). A complete thermal map should also include the firedeck region of the block; the head gasket itself; the head bolts; the exhaust manifold, bolts and towers; and the spark plug or fuel injector.

On a new engine project, initial thermal mapping provides the first measured check of earlier 1-D and then 3-D predicted temperatures. The experimental work of thermal mapping is used in conjunction with the models, both to "fill in" the maps in regions where measurements couldn't be made, and to refine and adjust the models. It is important to view the process as iterative. Initial thermal maps will contribute information required for further refinement of the engine (controlling critical temperatures and improving durability). As engine development continues with both performance and mechanical development, rechecking of critical temperatures will be required. In some cases the entire thermal mapping exercise may be repeated; in others select temperatures will be monitored as engine development continues.

The same principles are applied in refinement of existing engines. This may include new component development programs as well as new ratings or engine performance recipes. Thermal mapping serves several important functions. The first point is that heat transfer remains highly empirical. Thermal boundary conditions as used in models remain inexact, and will well into the future. With each mapping exercise information is gathered that improves the database of boundary conditions, but as long as the models remain empirical, each new or different engine requires boundary condition adjustment. Another important point is that complementary use of thermal mapping and 3-D models throughout the engine development process keeps an accurate model always at the ready for rapid design change evaluations.

Thermal mapping improves one's understanding of design margins relative to critical temperatures. It is always important to identify the most critical temperature regions, measure as close to those regions as possible, and develop approaches to accurately assess regions where direct measurement wasn't possible. Heat flux paths throughout the engine structure are extremely complex. Using thermal mapping to better understand and quantify those paths holds the further advantages of improved engine efficiency, improved heat exchanger sizing, and improved engine compartment and surface temperature control.

There are two distinct approaches that are taken to experimental thermal mapping, and both need to be used in most cases. The first is referred to as continuous thermal mapping. It involves the use of thermocouples to be able to monitor temperatures over a range of operating conditions—in other words, it maps temperatures versus speed, load, and other variables of interest. But the number of measurements is limited based on available data channels and limits regarding the number of drillings that can be made. The continuous thermal mapping is then complemented by detailed component thermal mapping at a specific operating condition. Templugs or other temperature indicating materials are used to provide a more detailed array of temperatures on any given component, but only a single operating condition can be assessed. It is typical to use the continuous thermal mapping measurements to identify the worst case conditions (highest temperature). The detailed component thermal mapping is then done at the selected condition.

13.6   Water Pump Design

Centrifugal water pumps are used almost universally. The pump is typically mounted at the front of the engine and gear or belt driven. The volute surrounding the pump impeller is often machined into the block or front cover of the engine although separate pump assemblies are also common. The pump impeller may be cast iron, stamped steel, or a plastic resin. Casted and stamped impeller examples are shown in the photographs of Figs. 13.9 and 13.10 respectively. In most automobile engines the shaft spins in a bushing impregnated with a solid lubricant or heavy grease intended to provide sufficient lubrication over the life of the pump. The expectation is similar in heavy duty engines although the plain bushing is often replaced with a sealed ball or roller bearing assembly.

In some engines the pump may be driven by an electric motor instead of being driven directly by the engine. This has the fuel efficiency advantage of reduced power expenditures pumping coolant when it isn't needed. It also allows faster engine warm-up, and can be used to control after-boil if the pump stays on for a specified time after engine shutdown. An electric water pump may also be used as a supplemental pump, allowing the engine-driven pump capacity to be reduced, or to pump coolant through a portion of the circuit such as that to supply a charge air cooler or EGR cooler.

The performance characteristics of a centrifugal water pump are presented in Fig. 13.11 where pump outlet pressure is shown versus volumetric flow rate over a range of engine speeds. A fundamental characteristic is the marked drop in flow rate as the pump head is

Fig. 13.9 Example water pump with cast iron impeller

Fig. 13.10 Example water pump with stamped steel impeller

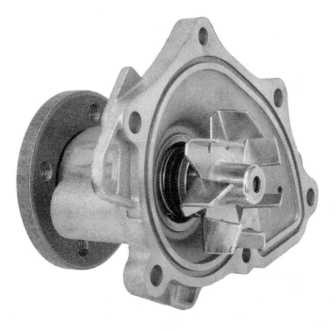

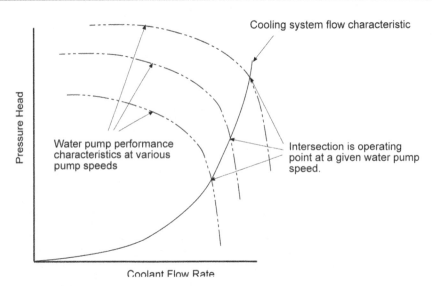

Fig. 13.11 Water pump performance characteristics, and cooling system operating point

increased beyond some threshold. The engine's cooling system will have a characteristic curve of pressure drop through the system versus flow rate as is also shown in the figure. The intersection between the pump head curve and the system characteristic at any given engine speed is the pump operating point at that speed. It is very important that the pump be sized to operate in the steeply sloped region of its performance curve. In this region changes in system restriction have little impact on the coolant flow rate. If the pump is undersized and causes operation in the flatter region of the curve slight increases in system restriction (due to water-side fouling of the radiator for example) result in a large drop of the coolant flow rate.

Cooling system design changes or specific installation requirements will shift the system characteristic curve and thus the pump operating point. For example, adding a component such as a cabin heater that draws a portion of the coolant from the pump outlet and returns it downstream of the engine reduces the overall system restriction, shifting the characteristic curve to the right. However, the flow through the cylinder block and head has been reduced and must be included in the cooling jacket flow design margin. Adding any components in series with the existing cooling circuit increases the restriction and shifts the characteristic curve to the left.

Water pump performance can be characterized through bench testing. The pump may be driven by an electric motor in which speed and power can be monitored. Inlet restriction and head pressure may be independently varied to map the flow rate and efficiency of the pump. Measuring the inlet and outlet pressure on the operating engine allows the operating point to be identified and overlaid on the performance map to determine pump power requirements and efficiency. Including a sight glass at the pump outlet on the rig allows the onset of cavitation to be identified as the inlet restriction is increased, thus ensuring cavitation margin on the operating engine.

13.7   The Cooling System

Discussion of the engine's cooling system is completed with a description of the vehicle installation. From the engine coolant is sent to the radiator—a water-to-air heat exchanger generally mounted at the front of the vehicle to take advantage of the ram air effect as vehicle speed increases. It should be noted at this point that the fan used to aid in drawing air through the radiator would actually be more effective if it could push instead of pull the air. However this would be counter to the direction of the ram air. With the electric fans used in transverse engine installations the fans are sometimes placed in front of the radiators, especially in racing applications. This however exposes the fans more directly to adverse weather so is seldom seen in production installations.

As the coolant temperature increases the pressure throughout the cooling circuit climbs. The maximum pressure is controlled with a pressure relief valve in the radiator cap. When the pressure reaches the design level determined by the spring constant of the relief valve air or coolant is vented from the system. Most engines include an overflow tank to which the system vents. Air escapes from this tank, and coolant is drawn from the tank back into the system as needed.

The engine's cooling circuit is completed with a thermostat that controls how much coolant flows to the radiator under any given combination of engine operating and ambient conditions, thus maintaining a nearly constant engine coolant temperature. When the coolant is below the desired temperature the thermostat is closed, and coolant is sent from the engine outlet directly back to the water pump inlet. As temperature increases the thermostat begins opening, sending a portion of the coolant to the radiator. The thermostat will cycle open and closed, maintaining a nearly constant engine coolant temperature under most operating conditions. At high engine load and increasing ambient temperature the percentage of time during which the thermostat is open increases, eventually staying open. At this point further ambient temperature increases result in a coolant temperature increase. The ambient temperature at which the thermostat remains open and coolant temperature climbs is determined by the effectiveness of the radiator. Increasing the heat exchanger effectiveness (determined by frontal area, core thickness, and air-side fin density), or increasing the air flow (determined by fan and shroud efficiency, and air-side restriction) allow the maximum ambient temperature at which the vehicle can be operated at load to increase.

The need to reduce engine warm-up time for improved emission control, and the opportunities for fuel economy improvement through closer control of cylinder wall temperatures, have resulted in recent changes in approach to the thermostat. One such change is to place the thermostat at the engine inlet, immediately downstream of the water pump, as opposed to its traditional placement at the engine outlet. This holds the advantage of not pumping the coolant through the engine when the thermostat is closed, in turn reducing engine warm-up time. Careful development is required since hot spots can quickly occur during rapid transients. Coolant pressure in the block and heads is also reduced, making the coolant more susceptible to boiling in high heat flux regions. Other changes in thermostat design include electronic controls and using separate valves with parallel flow systems to independently control coolant flow in the block and head.

The vehicle installation may include various other heat exchangers that draw a portion of the coolant from the engine. Each of these heat exchangers draw coolant immediately downstream of the water pump outlet, and return it downstream of the engine cooling jackets. Most installations include a passenger compartment heater core. Some engines may have a small heat exchanger for cooling the recirculated exhaust gas (EGR cooler) used for emission control. An oil-to-water heat exchanger may be used to control engine oil temperature—typical in heavy duty engines. Turbocharged or supercharged engines may use a charge air cooler in which engine coolant is used to cool the intake charge after compression. Most charge air coolers used in on-highway vehicles today use ambient air instead of engine coolant as the cooling medium, in order to directly take advantage of the lower "coolant" temperature. In highly boosted automobile engines, a separate, lower temperature cooling circuit may be used for the charge air cooler.

13.8   Venting and Deaeration

A final concern in cooling system design is that of venting the air during initial fill and removing air that may be entrained in the coolant during engine operation. Air in the coolant reduces its effectiveness in controlling engine temperature, and if the amount of air in the system is large the water pump may be unable to maintain the coolant flow. In filling the engine's cooling system coolant is supplied at the top of the radiator. As the radiator is filled coolant enters the engine through the lower radiator hose to the water pump inlet. The cooling jackets are filled from the bottom, displacing air to the top. The cooling jackets must be designed such that air can escape from the top of each jacket to the thermostat housing. Because the thermostat is closed at this time a small hole is sometimes placed adjacent to the thermostat to allow the air to escape back to the top of the radiator. If this vent is not included the system may not completely fill until the engine is operated and the coolant heats up sufficiently to open the thermostat.

Heavy duty engine installations generally specify an active deaeration. A vent line runs from the top of the engine back to the top tank of the radiator or to a separate surge tank. This upper tank is separated from the high coolant flow portion of the system (using a baffle in the radiator top tank, or the separate surge tank). During engine operation a small portion of the coolant continuously flows through the vent line to this separate region where the much lower flow rate allows air to dissipate to the surface. As the air collects it will escape the system through the pressure relief valve in the fill cap. In this way air is continuously removed from the system. The cooling system must meet deaeration specifications in which a known quantity of air is injected into the system and must be removed in a specified amount of time. The air removal is tracked visually by adding a clear section of tube to the lower radiator hose.

13.9   Trends in Cooling System Requirements

Further trends and considerations in cooling system development are summarized here. Each of these trends place further demands on the system, and result in innovative approaches increasingly seen on new engines.

- One of the levers being increasingly utilized in improving engine efficiency is the combination of down-sizing, down-speeding, and boosting. Smaller displacement engines, operated at lower speeds allow the engines to run at higher load factors, reducing pumping and friction losses. The higher load factors result in higher component and underhood temperatures that the cooling system must manage.
- Along with the added superchargers or turbochargers comes the addition of charge coolers. In many cases these are air-to-air heat exchangers that don't contribute to the cooling system load but require an additional heat exchanger at the front of the vehicle. For closer charge temperature control, and to reduce the volume of the charge cooler and plumbing, several new vehicles are incorporating air-to-water charge coolers. These systems most often use a separate, lower temperature coolant circuit, and a separate electric water pump.
- Cooled exhaust gas recirculation (EGR) is sometimes used with spark-ignition engines, and is typical on new diesel engines. Engine coolant, using either the main circuit, or a lower temperature circuit also serving the charge cooler is used in these exhaust gas-to-water heat exchangers. An increasing number of applications are using two EGR circuits—a high pressure circuit, and a low pressure circuit—and both of these may utilize EGR coolers.
- Rapid catalyst light-off requires rapid engine warm-up and minimizing both the thermal inertia and heat rejection between the engine and close-coupled catalyst.
- Allowing the coolant and especially the lubricant temperature to run warmer reduces oil viscosity and friction. This must be closely balanced versus its impact on oil change interval, component durability, and knock margin. Separate flow circuits and thermostats for coolant supplied to the cylinder jackets and the head jackets are now sometimes seen to aid in optimizing these trade-offs.
- Reducing water pump and oil pump power requirements is another consideration for fuel efficiency improvement. Water pump clutches, variable speed drives, and electric primary or supplemental pumps are among the design options now being considered.
- Vehicle aerodynamics plays an important role in fuel efficiency, but creates additional challenges in cooling system development and under-hood temperature management. Reduced radiator frontal area is the first challenge, and leads to the need to increase system pressure and temperature. Further aerodynamic refinements are now including undercarriage panels that further restrict flow through the engine compartment.
- Demands throughout the vehicle often increase the number and size of heat exchangers competing for space in the already limited frontal area. In addition to engine radiators, oil coolers, and charge air coolers these include air conditioning condensers,

transmission coolers, and power steering coolers. Heat exchanger, shroud, and electric- or engine-driven fan design has become increasingly complex.

13.10   Air-Cooled Engines

An alternative to liquid cooling is to utilize ambient air along with the engine's lubricant to control critical temperatures. Fins are cast into the cylinder head and block in the same critical regions where cooling jackets would otherwise be placed, and air is forced between the fins to enhance the convective heat transfer coefficient, and provide the needed temperature control. In motorcycle applications it is typical to rely on the ram effect of air directed at the exposed engine to force flow between the cooling fins. In other vehicular and stationary applications shrouds are placed around the cooling fins, and an engine-driven fan supplies air through the resulting passages.

Fin spacing and placement is driven both by casting constraints and the need to keep the fins from plugging with debris during operation. Within these constraints the fin and shroud design process has historically been almost entirely experimental—optimization based on measured temperatures. Computational fluid dynamics plays an important role today, allowing air flow rates, air temperature rise, and convective heat transfer coefficients to be spatially evaluated. Experience and validation make the tool quite effective, and widely used in air-cooled engine development today.

As specific output and engine life demands continue to increase, hybrid systems that use enhanced oil cooling or partial water cooling to supplement air cooling are also being seen. Oil cooling passages may be added to circulate pressurized lubricant into critical regions of the cylinder head firedeck, or near top ring reversal around the cylinder wall. An oil-to-air heat exchanger is then used to maintain acceptable oil temperature. Some motorcycle engines use a small liquid cooling circuit, water pump, and radiator to cool portions of the engine, with air cooling fins controlling the temperature in other regions.

13.11   Recommendations for Further Reading

The following papers focus on cooling jacket design and optimization. Examples using both experimental and computational approaches are presented (see Kruger et al. 2008).

C.C.J. French at Ricardo has long been recognized as a leader in engine thermal management. Although a few years have passed since it was written we recommend his paper (see French 1970; Sandford and Postlethwaite 1993; Brasmer and Hoag 1989).

There have been a number of studies that specifically address boiling heat transfer in cooling jackets. The first paper listed is a computational study of the effect of including boiling in thermal analysis of engine components (see Bo 2004; Finlay and Parks 1985; Norris et al. 1989, 1993, 1994; Campbell et al. 1999; Finlay et al. 1987).

The following paper provides a comprehensive study of water pump design and optimization (see Zoz et al. 2001).

These two papers look at future trends in automobile cooling systems. The first looks at advanced, demand response systems while the second provides a more general look at cooling system component development (see Melzer et al. 2001; Rocklage et al. 2001).

Recent new engine papers demonstrating advances in cooling systems are as follows (see Hadler et al. 2012; Heiduk et al. 2011; Bauder et al. 2011):

References

Bauder, R., Eiglmeier, C., Eiser, A., Marckwardt, H.: The new high-performance diesel engine from Audi, the 3.0l V6 TDI with dual-stage turbocharging. International Vienna Motor Symposium (2011)

Bo, T.: CFD homogeneous mixing flow modelling to simulate subcooled nucleate boiling flow. SAE 2004-01-1512 (2004)

Brasmer, S., Hoag, K.: The use of flow visualization and computational fluid mechanics in cylinder head cooling jacket development. SAE 891897 (1989)

Campbell, N.A.F., et al.: Nucleate boiling investigations and the effects of surface roughness. SAE 1999-01-0577 (1999)

Finlay, I.C., Parks, B.A.: Factors influencing combustion chamber wall temperatures in a liquid-cooled, automotive, spark-ignition engine. Proc. Instit. Mech. Eng. **199**(D3) (1985)

Finlay, I.C., Boyle, R.J., Pirault, J.P., Biddulph, T.: Nucleate and film boiling of engine coolants flowing in a uniformly heated duct of small cross section. SAE 870032 (1987)

French, C.C.J.: Problems arising from the water cooling of engine components. IMechE P29/70 (1970)

Heiduk, T., Dornhofer, R., Eiser, A., Grigo, M., Pelzer, A., Wurms, R.: The new generation of the R4 TFSI engine from Audi. International Vienna Motor Symposium (2011)

Hadler, J., Neuser, H.-J., Szengel, R., Middendorf, H., Theobald, J., Moller, N.: The new TSI. International Vienna Motor Symposium (2012)

Kruger, M., Kessler, M.P., Ataides, R., de la Rosa Siqueira, C., dos Reise, M.V., Mendes, A.S., Tomoyose, R., Argachoy, C.: Numerical analysis of flow at water jacket of an internal combustion engine. SAE 2008-01-0393 (2008)

Melzer, F., Hesse, U., Rocklage, G., Schmitt, M.: Thermomanagement. SAE 1999-01-0238 (2001)

Norris, P., Wepfer, W., Hoag, K., Courtine-White, D.: Experimental and analytical studies of cylinder head cooling. SAE 931122 (1993)

Norris, P., Hastings, M., Hoag, K.: An experimental investigation of liquid coolant heat transfer in a diesel engine. SAE 891898 (1989)

Norris, P., Hoag, K., Wepfer, W.: Heat transfer regimes in the coolant passages of a diesel engine cylinder head. Exp. Heat Transf. **7**, 43–53 (1994)

Rocklage, G., Riehl, G., Vogt, R.: Requirements on new components for future cooling systems. SAE 2001-01-1767 (2001)

Sandford, M., Postlethwaite, I.: Engine coolant flow simulation—a correlation study. SAE 930068 (1993)

Zoz, S., Thelen, W., Alcenius, T., Wiseman, M.: Validation of methods for rapid design and performance prediction of water pumps. SAE 2001-01-1715 (2001)

Gaskets and Seals

14

14.1 Gasketed Joint Fundamentals

An important challenge in engine design is that of providing leak-free joints at each of the component mating surfaces exposed to one or more of the working fluids. Minimizing the number and complexity of joints is one design goal. Another is that of providing durable, effective seals at each of the remaining joints. The working fluids include fuel, lubricant, coolant, intake air, and combustion products. Some joints must maintain separation between two or more of the engine's working fluids, while others seal one of the working fluids from the atmosphere. Some joints are stationary, while others include the need for one component to move relative to the other. These moving joints include the interface between the pistons and cylinder walls, those between the valve stems and guides, and those associated with spinning shafts protruding from the engine (the crankshaft and the water pump are examples). A categorization of seal types is presented in Fig. 14.1. Piston rings and valve stem seals will be covered in Chaps. 15 and 17 respectively.

Looking first at the static joints Fig. 14.2 provides a general summary of the characteristics of a gasketed joint. Sufficient clamping force must be applied to the gasket to ensure an effective seal. Any given combination of gasket material and surface finish will result in some minimum sealing load as shown in the figure. The sealing load is provided by screws or bolts located at various points along the gasketed surface. These fasteners provide a clamp force over the gasket area which results in an average sealing contact stress, or unit sealing stress, as shown in Eq. 14.1.

$$\text{Average Sealing Contact Stress} = \frac{\# \, \text{Fasteners} \cdot \text{Clampload Each}}{\text{Total Gasket Area}} \quad (14.1)$$

© Springer Vienna 2016
K. Hoag, B. Dondlinger, *Vehicular Engine Design*, Powertrain,
DOI 10.1007/978-3-7091-1859-7_14

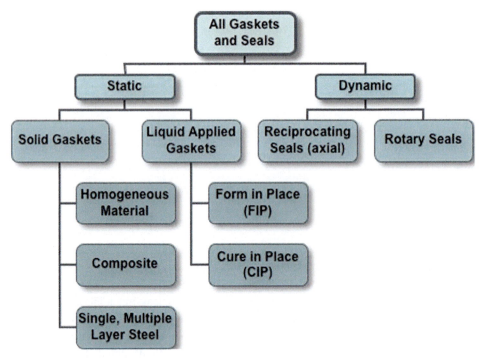

Fig. 14.1 Categorization of seal types

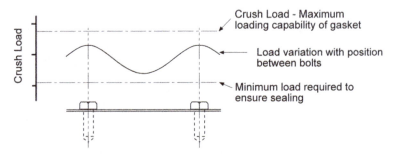

Fig. 14.2 General characteristics of the gasketed joint, showing *upper* and *lower* load limits

Assuming flat surfaces, the clamping force will be highest at each of the fasteners, and will drop to a local minimum between each of them. If the stiffness of both surfaces is constant along its length the local minimum clamping force will occur midway between the two surrounding cap screws. This can be represented by a beam fixed at both ends with a distributed load as shown in Fig. 14.3, and modeled in Eq. 14.2:

$$y_{\max} = \frac{W \cdot L^3}{384 \cdot E \cdot I} \qquad (14.2)$$

Fig. 14.3 Fixed-Fixed Beam
with uniform load

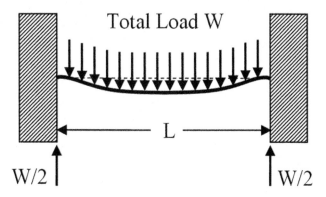

Where:

y_{max} =   Maximum deflection at midspan of beam
W =   Total load
L =   Length of span between fastener centerlines
E =   Modulus of Elasticity of flange
I =   Moment of inertia of flange

A general guideline is to limit the maximum deflection, at midspan of the flange, to less than 0.034 mm of deflection per 1.0 MPa of sealing contact stress for flat gaskets and liquid sealants as shown in Eq. 14.3. This generally results in a fastener span of seven to ten times the fastener diameter as a first approximation, to be validated by FEA or component testing.

$$\text{Flange deflection} = \frac{0.034 \text{ mm}}{1.0 \text{ MPa}} \quad \text{or} \quad \frac{< 0.001''}{100 \text{ psi}} \quad (14.3)$$

$$\text{Fastener span} = 7 - 10x \text{ fastener diameter}$$

Another important gasket property is the maximum clamping force it can accommodate without being damaged—generally referred to as the maximum *crush* load. The design variables of bolt placement and surface stiffness determine the range of clamping loads seen along the gasket. This range must be maintained above the minimum sealing load and below the maximum crush load over the entire gasket as shown in Fig. 14.2 previously. Minimum sealing and maximum crush load are determined primarily by the properties of the gasket material itself, and are also impacted by the characteristics of the two mating surfaces (flange width, flatness, and surface finish). Operating conditions such as temperature changes and pressurization further impact the clamping loads as will be discussed in Sect. 14.5. Relative motion (sliding or fretting) between the two surfaces may also be important, and will further impact material requirements and the clamping load range.

Gaskets may be manufactured from a variety of materials to meet the particular requirements of sealing a given joint. There are many considerations when choosing a gasket material as listed here:

- Cost
- Operating temperature and pressure
- Flange geometry
- Storage temperature
- Fluid compatibility
- Assembly method, gasket retention, and serviceability
- Required durability and reliability
- NVH and transmissibility
- Styling requirements

A few common families of gasket materials are summarized in the following paragraphs.

Cork Cork material is generally ground and combined with elastomers and adhesives to provide the required characteristics. It is a highly compressible material that requires very low clamping loads to provide an effective seal. Its flexibility allows it to conform well regardless of surface finish, and it is often used to seal rough "as cast" surfaces.

Paper or Reinforced Fiber Plain paper gaskets, made from heavy stock such as tag board or similar weight materials are common in the aftermarket for low temperature and pressure joints. Paper is sometimes treated with cellulose, glycerin and animal glues, or saturated with elastomer binders. Such treatments greatly improve the gaskets' ability to provide a seal and their resistance to oils and fuels. Slurries made from a variety of materials provide the raw materials for a wide variety of gaskets. Materials such as wood pulp,

Fig. 14.4 Fiber gasket

synthetics, and glass fibers are mixed in slurries with elastomer or starch binders. The gaskets can be made in a wide range of thicknesses, with a variety of properties needed for specific applications. An example is shown in Fig. 14.4.

Elastomer A wide variety of gaskets with a wide range of properties are made from elastomers, which are a subset of polymers. Common engine elastomers are made from synthetic rubber or urethane and include Nitrile (NBR), Hydrogenated Nitrile (HNBR), Polyacrylate (ACM), Ethylene Propylene Diene (EPDM), Silicone (VMO), Fluorosilicone (FVMO), and Fluorocarbon (FKM, Viton). These include die cut, molded, and extruded materials, coatings on other materials, and liquid applied anaerobic or room temperature vulcanized (RTV) silicone gaskets. An example of a molded gasket is shown in Fig. 14.5.

Graphite In high temperature applications a graphite foil is sometimes used on a steel carrier. The seal between an exhaust manifold and cylinder head is an example joint as shown in Fig. 14.6. The graphite also has good sliding contact characteristics, so maintains a seal under the relative motion seen between the head and manifold under thermal cycling. It is an expensive material and quite brittle so susceptible to handling damage.

Metal Steel is often used as a stand-alone gasket material for high pressure joints; an example is the seal ring for a spark plug. Steel may be crimped onto a reinforced fiber or

Fig. 14.5 Molded elastomer, engineered X-section

Fig. 14.6 Graphite exhaust gasket with stainless steel sealing rings

Fig. 14.7 Elastomer coated, embossed steel head gasket

graphite gasket as is often done to provide the combustion seal in head gaskets. It is also commonly used as a carrier to which other materials are bonded, such as Multiple Layer Steel (MLS) and Metal-Elastomer head gaskets as shown in Fig. 14.7. Softer than steel, copper and aluminum conform very well to flange surfaces, and provide high strength under high pressure loads. Spark plug seal rings and the seal between diesel fuel injectors and their cylinder head mating surfaces are often made from copper. Flat metal o-rings are occasionally used to seal the cylinder head to block, and can be made of copper, aluminum, and soft steel. Embossments are often added to lower the local spring rate to make the gasket more tolerant of varying sealing contact stress in the operating environment.

Material properties vary widely depending on composition of the gasket for elastomers and fiber gaskets. Performance also varies as a function of material and joint design. Some representative values are shown in Tables 14.1, 14.2.

14.2   The Gasket Operating Environment

Gaskets have a variety of fluids and operating environments to contend with. The following Table 14.3 summarizes typical gasket working environments.

14.3   Flange Sealing Types

There are three basic types of flange gaskets: flat gaskets, o-rings, and liquid applied gaskets. Compression types of flat gaskets are made from cork, paper, elastomer, or metal and are used where an inexpensive and serviceable seal is required. O-rings are circular in cross-section, are made of elastomers, and are typically applied in a groove where the compression of the o-ring is limited. An o-ring type seal is often bonded to a plastic or aluminum carrier, to ease installation and limit the compression height.

Table 14.1 Material properties of gasket types

Material	Cork	Paper or reinforced fiber	Elastomer	Graphite/Steel Carrier	Metal
Sealing contact stress, minimum	2.75–3.5 MPa	5–45 MPa Depending on material and thickness	*Solid elastomer* 1 MPa *Liquid sealant* 0.5–2.5 MPa	Bulk = 28–55 MPa For Head Gasket combustion seal = 175 N/linear mm	*Flat x-section* Cu 100–310 MPa Al 70–140 MPa Soft steel 210–480 MPa Stainless steel 240–650 MPa *Coated MLS or Metal-Elastomer* Half emboss = 28 N/linear mm Full emboss = 105 N/linear mm
Surface finish required	>3.2 µ-m	0.2–3.2 µ-m	*Solid elastomer* 0.8–3.2 µ-m (static) 0.4 µ-m (dynamic) *Liquid Sealant* 0.8–6.3 µ-m	0.4–1.6 µ-m for bimetallic joints to accommodate thermal expansion 0.4–3.2 µ-m for common joint materials	*Coated MLS or metal-elastomer* 2.0 µ-m max for combustion seal
Flange flatness required	n/a	0.025–0.127 mm/mm 0.25 mm total	*Liquid sealant* 1.27 mm total	0.025 mm/mm 0.1 mm total	0.025 mm/mm 0.1 mm total
Operating temperature, max or range	120°C max	150°C max	−60–315°C range, depending on material	150–370°C max	*Flat x-section* Cu 315°C max Al 430°C max Soft Steel 540°C max S. Steel 870°C max *Coated MLS or metal-elastomer* 200°C continuous, 290°C max intermittent
Operating pressure, maximum	<7 MPa	<7 MPa	*Solid elastomer* 8.6 MPa for o-ring *Liquid sealant* Flange dependent	<17 MPa	<17 MPa
Flange width recommendation	6–9 mm	6–9 mm	*Liquid sealant* 5 mm for liquid applied sealants	5 mm min around comb chamber 3 mm min. around other sealing features	*Coated MLS or metal-elastomer* 8 mm min. around comb chamber 2.5–5 mm min. around other sealing features

Table 14.2 Relative performance of gasket types

Material	Cork	Paper or reinforced fiber	Elastomer	Graphite/Steel carrier	Metal
Fretting Tolerance	Medium	High	*Solid elastomer* Medium *Liquid sealant* Low (0.025 mm)	High	Medium
Relative cost	$	$	$-$$$	$$$	$$-$$$$$
Pros/Cons	Inexpensive High compressibility enables high tolerance for rough surfaces, but low torque retention is a disadvantage Seals surface porosity Prone to wicking of fluid Gasket is not reusable	Stiffer than cork, enables higher torque retention Seals surface porosity Prone to wicking of fluid Gasket is not reusable	*Solid elastomer* Vulnerable to surface porosity in line contact application (o-ring). Compression limiting feature adds cost (groove or boss). *Liquid Sealant* Intolerant of fretting, Seals surface porosity	High fretting resistance |High temperature resistance Higher surface finish requirement adds cost	*Flat x-section* High operating temperature resistance for uncoated metal, high stiffness is an advantage *Coated MLS or metal-elastomer* Vulnerable to surface porosity, high stiffness is an advantage
Other			*Solid elastomer* Within the subset of elastomers, relative cost varies significantly ($ 1 $\times$-$ 60 $\times$) *Liquid sealant* RTV skins over in 3–7 min, assemble immediately	High fretting and temperature resistance enables it to be used in applications with relative movement due to differential thermal growth, such as head gasket and exhaust manifold	*Coated MLS or metal-elastomer* Typically have a lower sealing contact stress requirement due to embossed beads and elastomer coating. When used as a head gasket requires lower fastener clampload, which reduces cylinder bore distortion Primary failure modes are bead fatigue and fretting of the elastomer off the substrate

Table 14.3 Typical gasket environment

System	Temperature range	Pressure range	Gasket loading
Engine cooling	−34–138°C	0.2 MPa max	Thermal differential, vibration
Engine, Trans Oil Pan	−34–149°C	Vacuum to 0.04 MPa	Thermal differential, vibration
Engine oil pump	−34–149°C	Vacuum to 0.7 MPa	Thermal differential, vibration
Engine air intake (turbocharged)	−34–149°C	Vacuum to 0.3 MPa	Thermal differential, vibration
In-cylinder combustion gas	980 °C max	Vacuum to 19,300 kPa	Thermal differential, vibration, dynamic deflections
Engine exhaust (turbocharged)	815 °C max	Vacuum to 0.2 MPa	Thermal differential, vibration, dynamic deflections

Liquid applied gaskets are applied to the flange surfaces and fill gaps and scratches. They are either cured in the joint after assembly (Form-in-place, FIP) or cured by ultraviolet (UV) light prior to assembly (Cured-in-place, CIP).

14.4 Engine Cover Design

Many gasketed joints in engines involve the attachment of relatively light weight, low cost covers to stiff components such as the cylinder block or cylinder head. Examples of such covers include the valve cover, the front cover over the camshaft drive, and the oil pan. Stamped steel covers are often used, and their design considerations for sealing are summarized in Fig. 14.8. These covers are flexible compared to their mating component, and each fastener that must be added to reduce the sealing contact stress variation adds cost to the engine. Turning up the flange edge, designing back-draft into the stamping, and

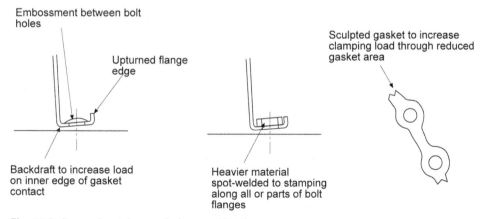

Fig. 14.8 Stamped metal cover design considerations

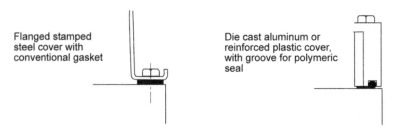

Fig. 14.9 Comparative characteristics of stamped and cast covers and approaches to sealing

embossing the gasket flange between each bolt are important design enhancements. When sufficient stiffness cannot be achieved through this combination of techniques, heavier steel reinforcement strips can be spot welded to the flange. Because the seal is dependent on gasket contact stress, and not gasket area, sculpted gaskets are sometimes used. Sculpted gaskets are used to increase the gasket contact stress achieved with a given screw clampload by reducing the area over which the bolt load acts at mid-span. Additionally, the flange may be pre-curved toward the mating component, between fasteners, to attempt to achieve uniform sealing contact stress in the assembled state.

Die cast aluminum covers are also commonly used on engines. They are typically stiffer than the stamped steel covers due to their greater wall thicknesses. The gasketed surface can often be incorporated in the wall edge, eliminating the cantilevered flange, resulting in a stiffer joint as depicted in Fig. 14.9. Reinforced plastic covers are becoming more widely used, and provide important advantages for sealing. Using either die cast aluminum or reinforced plastic a seal groove can be easily incorporated in the cover. This allows an elastomeric seal to be positively fitted into the cover, and the sealing contact stress to be controlled by limiting the gasket compression. The sealing contact stress is now constant around the entire perimeter of the cover, and is no longer dependent on controlling the force applied at each fastener. The risk of assembly problems is also significantly reduced. Each of these cover materials requires high tooling costs, but carries with it a low piece price. As a result they are commonly seen in high volume engines.

As can be surmised from Figs. 14.2 and 14.10, bolt placement is a critical variable in gasketed joint design. One design guideline is to ensure that whenever possible straight lines drawn between adjacent bolts will pass over the sealing interface. The arrows point to a region of low sealing contact stress, which would be improved by the addition of a fastener at location "A". When this is not possible the stiffness of the mating components in that particular region must be significantly increased to ensure that the minimum sealing force is maintained.

Another important consideration in cover design is its role in engine noise. The resulting panel stiffness determines the natural frequencies at which their contribution to noise will be especially high. Changing the cover material will change the natural frequencies, and may either reduce or increase engine noise. In some cases the gasket must be designed to isolate the cover from the engine surface to which it mates. This is generally done by using a relatively thick, compliant gasket material to dampen the high frequency motion of

Fig. 14.10 Fastener line of
action

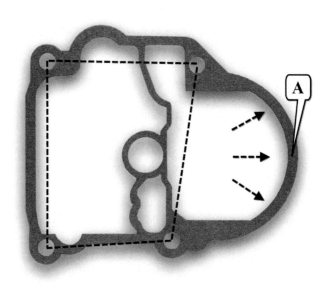

the engine surface, and minimize its transmission to the cover. Such an approach increases
the cost of the gasket but may significantly reduce engine noise. Another approach is to
add sound deadening material, glued to the panel surfaces—this is sometimes seen on oil
pans and front covers.

An alternative to fasteners piercing through the sealing flange, is to have a single fastener in the center of the cover, independent of the sealing edge, as shown in Fig. 14.11. If
the cover is stiff enough and the sealing contact stress required is low enough, the clamp-
load of a single fastener can be spread to the outer seal like an umbrella.

A final consideration for engine covers is disassembly. When an engine is to be inspected or repaired, the first parts to come off are the covers. The use of liquid applied
gaskets means that the cover will be adhered to its mating part. Over time, an elastomer
seal may also adhere to the mating component. A conscientious designer will add tabs or

Fig. 14.11 Cover with three centrally located fasteners

undercuts to a cover to allow pry points for removal during service. As an alternative, additional threaded holes may be added to a cover, so that a screw can be treaded in to push against a mating flange and separate the cover.

14.5   Clamping Load Parameters

In the opening section, the gasketed joint designer was presented with the need to provide clamping force sufficient to meet the minimum sealing contact stress, while further ensuring that the maximum crush load was not exceeded. Several phenomena further complicate the design of these joints. The first is that of creep relaxation in the gasket. Most gasket materials are permanently deformed under the initial clamping loads, and over time the materials relax, reducing their resistance to the clamping force, and thus causing the force to drop. This is depicted in Fig. 14.12, where deflection is plotted versus the applied force. As the fasteners are loaded they elongate (positive deflection or stretch). Because the bolts do not reach their yield strength this elongation is fully elastic, and linear as shown in the plot. The gasket begins at its un-deformed thickness shown as the 'x' intercept in the figure. As the clampload is applied the gasket is compressed non-linearly as shown (negative deflection or crush). The intersection between the two plots identifies the initial gasket loading. Creep relaxation of the gasket over time causes its deflection curve to shift to the left as shown, resulting in a drop in the resulting sealing load. Fortunately the majority of this relaxation occurs very quickly—typically just a few minutes. Over time it is common to have gasket relaxation due to thermal cycling, and a graphite gasket may only retain

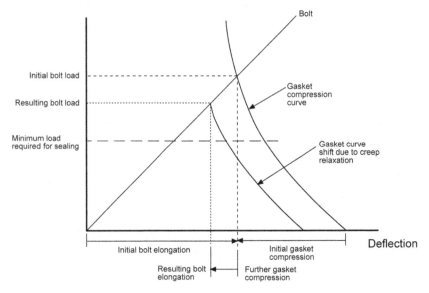

Fig. 14.12 The effect of bolt load and gasket creep on resulting clamping load

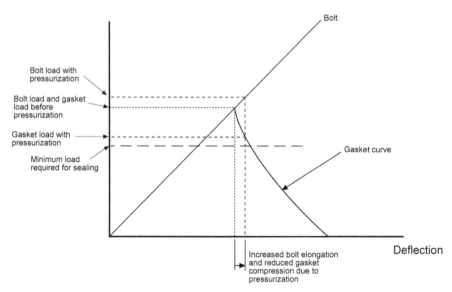

Fig. 14.13 The effect of cylinder pressure on gasket clamping load

70 % of its initial thickness, while an MLS gasket will retain 90 %. In some circumstances, as when a stiff cover with many fasteners is assembled, the actual clampload on the first fastener torqued changes as subsequent fasteners are torqued, which is called cross-talk. It is because of these characteristics that specifications often call for repeating the torque sequence on critical gasketed joints. Also shown in the figure is the minimum load required for sealing. It must be remembered that this figure applies everywhere along the gasketed joint, and that the resulting load and deflection will be lower between the bolts than at a given bolt location; the minimum force required for sealing must be achieved at all locations. While not specifically addressed in this section, it is common practice for hoses to be sealed with a spring clamp, rather than a fixed-distance clamp such as a worm gear clamp. With the fixed-distance clamp, as the rubber creeps over time, sealing contact stress is lost for the same reasons listed previously for a gasketed joint.

Some gasketed joints must maintain their seal under high internal pressure. The head gasket is an important example of such a joint. The effect of internal pressurization is depicted in Fig. 14.13. The increased applied firing pressure attempts to separate the mated components, and in doing so further elongates the fasteners. The new load experienced by the fastener is shown in the figure. As a result of the further bolt elongation, the gasket crush is reduced as shown. Fastener length and the deflection of the mating components in response to the increased pressure are important further parameters in the joint design. If fastener yield is exceeded, low cycle fatigue will result. If the minimum sealing contact stress is violated, leaks will occur.

Still another important consideration in some of the gasketed joints of an engine is that of temperature change in the mated components and screws. Especially important is the case where the bolted component experiences significant temperature increases during

Fig. 14.14 EGR transfer manifold with bellows

operation. Perhaps the most important case is that of the exhaust manifold. As the manifold temperature increases and experiences thermal growth the screw loads increase significantly. This is due to the combination of the lower thermal expansion coefficient of the high strength screw material, and the fact that the screw temperatures remain below that of the manifold (due both to thermal resistance and the direct contact of the screws with the cooler cylinder head). Increasing the length of the screws significantly reduces the load change due to manifold growth, and is done both to control the gasket load and limit the stress in the screws themselves. Another important design consideration is the need for the manifold to slide at the gasket as it elongates with temperature. The screw hole diameters must be enlarged to account for thermal growth and relative movement. Especially long exhaust manifolds are sometimes constructed in pieces, with sliding joints or bellows to reduce dimensional shift and loading as shown in Fig. 14.14.

As can be seen from Figs. 14.12 and 14.13, a less stiff fastener with a shallower slope force-deflection curve will be less sensitive to gasket creep over time. A short, stiff fastener will be compared with a long slender fastener under the same load. Both fasteners are made of steel, and both are under the same load of 15 kN. The first fastener is M10 × 1.5 with a 10 mm grip length (1:1), and the second fastener is M6 × 1 with a grip length of 24 mm (1:4). The equation for elongation of a slender rod (fastener) is given as Eq. 14.4.

$$\Delta = \frac{F \cdot L}{A \cdot E} \tag{14.4}$$

Table 14.4 Comparison of different fasteners in same application

	F = 15 kN	
Fastener	#1	#2
Diameter	M10 × 1.5	M6 × 1
Grip length	10 mm	24 mm
Elongation at load	$\Delta_1 = 0.009$ mm	$\Delta_2 = 0.062$ mm

Where:

F =   Fastener clamp load
L =   Length between head and first tread of engagement (grip length, or clamped length of joint)
A =   Cross sectional area of fastener
E =   Modulus of elasticity of fastener

Assuming the same load (F) on each fastener, and the same material (E), this equation simplifies to Eq. 14.5:

$$\Delta_1 \approx \frac{L_1}{A_1} \text{ and } \Delta_2 \approx \frac{L_2}{A_2} \tag{14.5}$$

Calculating the actual elongation of each fastener produces very different results as displayed in Table 14.4. The second fastener stretches almost seven times further, yielding a joint design that is less sensitive to gasket creep relaxation.

Thermal loading must also be taken into account in designing the cylinder head gasket joint. While cylinder head temperatures do not change nearly as much as exhaust manifolds, their thermal growth has an important impact on head gasket loading. It should be noted that while increasing the screw length reduces the impact of thermal loads it increases the impact of the cylinder pressure loads discussed earlier. The relative impact of these two effects must be compared as the number, length, and diameter of the screws is chosen for cylinder head clamping. An additional impact on the number of fasteners used per head gasket is cylinder bore distortion with thermal loads. As the engine heats up, the cylinder bore will distort to match the bolt pattern, due to the added stiffness at each fastener location. A four-bolt head gasket will produce a deformed cylinder bore shape similar to a four-leaf clover. While increasing the number of fasteners leads to a more robust joint, this will reduce the span between fasteners and increase the number of lobes in the deflected shape of the cylinder bore. This makes it more difficult for the piston ring to conform to the cylinder bore.

The importance of computational methods in designing critical seals such as the combustion seal at the cylinder head firedeck should be apparent from the preceding paragraphs. However, such analysis presents its own challenges as well. The most difficult aspect of this analysis is properly modeling the plasticity of the head gasket itself. Accounting directly for the plastic deformation significantly increases computational intensity.

Various approaches have been taken to simplify the computations, with the most accepted method approximating the already crushed gasket with an offset elastic deformation curve. If increased confidence is desired in the analysis, the plastic deformation of the head gasket can be measured by compressing it between two thick plates on a tensile test machine, or by other methods of direct measurement in service.

14.6   Bolt Torque and Sealing Load Control

The objective in assembling a gasketed joint is to apply a force exceeding the minimum sealing contact stress required along the entire gasket, while not exceeding the maximum crush load. In most cases a bolt torque specification is provided for assembly, and used in an attempt to provide the desired load. An approximate relationship between bolt torque and bolt load is given by Eq. 14.6.

$$T = k \cdot F \cdot d \text{ or } F = \frac{T}{k \cdot d} \qquad (14.6)$$

Where:

T = Applied torque on fastener
k = 'Nut factor' or friction factor
F = Clamp load
d = Fastener nominal diameter

The nut factor (k) is a catch all factor, and varies significantly depending on fastener material, coatings, fastener diameter, and many other parameters. The nut factor is the slope of the torque-clampload (torque-tension) relationship. Typical ranges for common applications are shown in Table 14.5. After the initial design is complete and prototype hardware is available, this number should be experimentally validated for each application.

Over 80 % of the applied torque is used to overcome friction between the threads and the contact area under the head of the fastener. At best—a well-calibrated torque wrench pulled smoothly to ensure a constant friction coefficient and clean threads—the resulting clampload can be expected to vary over a range of ±25 %. This unfortunate variability

Table 14.5 Nut factor (k) for common engine applications

Fastener or coating	K factor range
Uncoated steel	0.18–0.20
Zinc coated steel	0.18–0.22
Zinc coated steel w/oil	0.12–0.21
Zinc coated steel w/moly or copper based lubricant	0.08–0.17
Corroded steel fastener	0.36–0.39

must be taken into account when designing a bolted joint. Alternative methods can be considered for especially critical joints.

One such method is referred to as **Torque-Plus-Turn**, or **Turn-of-the-Nut**. In this method, the torque wrench is used to bring the bolt to some portion of the desired load, usually this is the torque required to overcome the rundown friction and that which starts actually stretching the fastener and clamping the joint (prevailing torque or snug torque). The bolt is then turned through a further turn angle in the linear potion of bolt stretch. As the fastener turns, the inclined angle of the thread advances the threaded portion of the fastener a linear amount. Since the tension in the fastener is a function of this stretch, the tension can be more closely controlled. The variability due to friction is reduced and confined to the prevailing torque set point. It is typically estimated that the clampload can be controlled within $\pm 10\%$ with this method.

Another method that results in very close control of the resulting clamp load is that referred to as **Torque-to-Yield**. This method requires a more sophisticated tightening tool that senses the change in the relationship between applied force and elongation (sensed as twist) that occurs as the bolt begins to yield. The bolt is sized to yield at the desired bolt clamping force. This method provides more accurate clampload control of $\pm 8\%$, but the sophistication of the required tools limits its application to initial assembly. Fasteners are often discarded after one use, or maximum length specifications are given for reuse.

The final method used to assemble joints is direct measurement of fastener elongation. This is only done when both ends of the fastener are exposed, as in a fastener and nut clamping two plates or on connecting rod caps. A micrometer is used to measure the initial length of the fastener, then final length, and the difference can be correlated to tension. It is typically estimated that the clampload can be controlled within $\pm 3\text{--}5\%$ with this method.

During development of gasketed joints, a more accurate method to measure fastener stretch is needed. This is obtained by strain gauging of the fasteners, which usually requires modifying prototype components to allow places for the strain gauge wires to exit the assembly. Or, the change in length of the fastener can be measured ultrasonically. The fastener is prepared by grinding both ends flat. An ultrasonic probe is attached to one end, which measures the time of flight of a sound wave from the head of the bolt to the tip, and then back again. As the bolt is stretched, its length increases and so does the time of flight. Both the strain gauging and ultrasonic bolt measurements require careful calibration in the lab, but accuracy can be $\pm 1\%$.

14.7   Shaft Seal Design

At each end of the engine the crankshaft must protrude from the crankcase, requiring an oil seal to be maintained at these spinning shaft interfaces. The water pump shaft is another example, in this case requiring a coolant seal. In the case of the rear of the crankshaft, the seal is sometimes split and incorporated in the rear main bearing journal and cap. Three types of seal are commonly used. The isomeric seal depicted on the left in Fig. 14.15 can

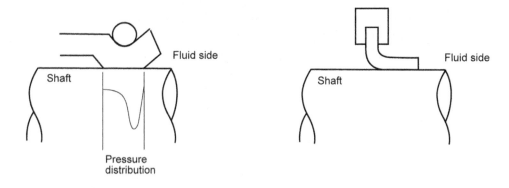

Fig. 14.15 Elastomeric and Teflon shaft seal designs

Fig. 14.16 Radial shaft
seals

be either continuous or split. The seal is shaped so that in its installed state it provides the pressure gradient shown in the figure. A steep pressure gradient at the inner end pushes oil that seeps between the shaft and seal back toward the crankcase. At the outer end a shallow pressure gradient minimizes the amount of oil that seeps past the shaft to the environment. A small film of oil is needed between the shaft and the seal during operation to prevent excessive wear. Another option used in split seal applications is the rope seal. Finally the Teflon seal shown at right in Fig. 14.15 can be used in continuous seal applications. A gaiter spring is sometimes used to ensure consistent seal contact pressure over the life of the engine as shown in Fig. 14.16. Also shown in this figure is a seal surface that is designed with very small spiral grooves which continue to push oil back toward the crankcase.

Important design considerations with any of these seals is concentricity and dynamic runout between the shaft and seal. Positive locators such as dowel pins are required for the front cover in order to ensure concentricity, and stiff bearing support of the shaft will minimize dynamic runout. Engine designs demanding especially long life typically require

Fig. 14.17 Integrated seal
carrier

continuous seals, and so the rear crankcase seal will also be mounted in a separate cover, again designed for positive location as shown in Fig. 14.17.

The direction that a shaft is ground may leave a very shallow helical spiral or lead on the shaft, if the grinding wheel transverses the shaft axially. This helical spiral, or grinding lay, may act as a pump to move oil past the seal. It is important to consider shaft rotation direction when specifying the manufacturing method, or alternately to plunge the grinding wheel to the shaft purely radially to eliminate any lead. Shaft surface finish is critical to ensure that a small film of oil is present under the sealing lip to enhance durability. Typical surface finish is 0.25–0.5 µ-m Ra, with close control of further machining parameters.

14.8   Example: Cylinder Head Gasket Joint Design

The cylinder head gasket joint is statically indeterminate, which means that there are more variables than equations to solve the sum of the forces in equilibrium. The solution is to add an additional equation to describe the deformation based on geometry, or the equation for elastic deformation.

The gasketed joint used in this example will have an aluminum cylinder head and an aluminum cylinder block, with four M10 × 1.5 Class 12.9 fasteners per cylinder. The example below will analyze the clamped joint per fastener—not for the whole cylinder head. Many of the calculations have the gasket contribution divided by the number of fasteners per cylinder, to account for the gasket contribution per fastener. Peak cylinder firing pressure is 69 bar applied over a cylinder bore diameter of 89 mm. The initial design will be for a graphite head gasket as shown in Fig. 14.18, which will generally package protect for an MLS head gasket as shown in Fig. 14.19. It is generally desirable to have the thinnest head gasket possible, to reduce the crevice volume between the cylinder head and

Fig. 14.18 Example of
graphite head gasket

Fig. 14.19 Example of
multi-layer steel (MLS)
head gasket

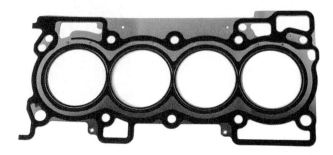

cylinder block to minimize HC emission. The two most important design considerations
are protecting the cylinder head fastener from fatigue, and ensuring the minimum gasket
pressure required for sealing is maintained under all operating conditions.

The general method for design is to:

1. Calculate external applied load and target fastener clampload
2. Calculate equivalent stiffness for the fasteners, abutment, gasket, and load factors
3. Calculate fastener and abutment deflection and loading
4. Calculate design margins for joint

14.8.1   Calculated External Applied Loads and Target Fastener Clampload

The external load on the cylinder head that is trying to separate the joint is due to peak
combustion pressure. This is calculated by multiplying peak cylinder pressure by bore
area in Eq. 14.7 to determine gas force.

$$F_{Gas} = \frac{\pi \cdot D_{cyl}^2}{4} \cdot P_{max}$$

$$F_{Gas} = \frac{\pi \cdot 89^2 \ \text{mm}^2}{4} \cdot 69 \ bar = 42.9 \ \text{kN} \tag{14.7}$$

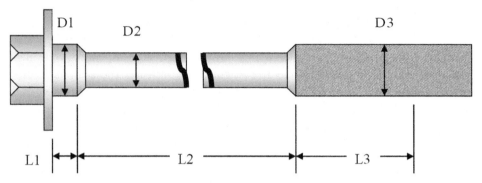

Fig. 14.20 Reduced shank cylinder head fastener

Table 14.6 Fastener dimensions

L1 = 4.5 mm	D1 = 10.00 mm
L2 = 116.5 mm	D2 = 7.78 mm
L3 = 26.0 mm	D3 = 8.34 mm (Root diameter of M10 × 1.5 threads)

Good design practice for a cylinder head fastener is to have a necked down, reduced diameter portion of the shank as shown in Fig. 14.20. This increases bolt stretch and reduces the criticality of the stress concentration in the treads (Table 14.6).

The fasteners will apply a clampload to the gasket. Typically a fastener diameter is chosen such that the initial assembly torque preloads the fastener to 60–90% of proof strength. If the type of torque method employed is very consistent, for instance yield control tightening, a target clampload closer to proof strength can be used. If a less precise method is chosen, for instance torque control only, a target clampload closer to 60% would allow greater design margin to the proof strength, which accommodates the greater clampload scatter. It is not uncommon to have a special drive feature at the head of the bolt, such as a TORX® drive, to prevent any slipping during tightening. For this example, 75% of proof strength will be chosen and shown in Eq. 14.8.

$$F_b = \% \cdot S_p \cdot \left(\frac{\pi \cdot D_{min}^2}{4} \right)$$

$$F_b = 0.75 \cdot 970 \text{ N / mm}^2 \cdot \left(\frac{\pi \cdot 7.78^2 \text{ mm}^2}{4} \right) = 34.6 \text{ kN}$$

(14.8)

Where

F_b =   Initial fastener clampload
S_p =   Proof strength of fastener, class 12.9
% =   Percent of proof strength targeted
D_{min} =   minimum diameter of fastener, either in thread root or necked down region

14.8.2   Calculated Equivalent Stiffness and Load Factors

The three different diameters of the fastener have different stiffness, and an equivalent stiffness must be determined for the grip length of the fastener. The analysis method approximates springs in series, using the different portions of the fastener as separate 'springs'. Starting with the equation for axial deflection of a constant diameter rod, this can be rearranged to get the spring rate for a constant diameter rod in Eq. 14.9.

$$\Delta = \frac{F \cdot L}{A \cdot E} \quad \text{Rearranged to:} \quad k = \frac{F}{\Delta} = \frac{A \cdot E}{L} \tag{14.9}$$

For a fastener with stepped diameters, the equivalent stiffness is approximated as springs in series as shown in Eqs. 14.10 and 14.11.

$$\frac{1}{k_b} = \frac{1}{k_1} + \frac{1}{k_2} + \frac{1}{k_3} + \cdots \frac{1}{k_n} \tag{14.10}$$

When rearranged:

$$k_b = \frac{\dfrac{\pi}{4} \cdot E_b}{\dfrac{L_1}{D_1^2} + \dfrac{L_2}{D_2^2} + \dfrac{L_3}{D_3^2} + \cdots \dfrac{L_n}{D_n^2}}$$

$$k_b = \frac{\dfrac{\pi}{4} \cdot 2.1 x 10^5 \ \text{N} / \text{mm}^2}{\dfrac{4.5 \ \text{mm}}{10^2 \ \text{mm}^2} + \dfrac{116.5 \ \text{mm}}{7.78^2 \ \text{mm}^2} + \dfrac{26 \ \text{mm}}{8.34^2 \ \text{mm}^2}} = 70.4 \ \text{kN} / \text{mm} \tag{14.11}$$

Where:

$E_b =$ Modulus of elasticity of fastener

Next the equivalent stiffness of the compressed members will be estimated. As the clamped members are loaded, the stress spreads out as it penetrates the part. The stress approximates a cone, with the small end underneath the head of the fastener and the large end at the head gasket surface. For this reason, approximations must be made for the clamped area, and actual measured stiffness of the cylinder head would be preferable.

If using a large slab head, like on an inline four cylinder, it can be assumed that the stress has an opportunity to spread to a large cone as shown in Fig. 14.21. The effective diameter (d_e), and thus the effective area (A_c), is approximated to be the average cross sectional diameter of a frustum cone. There are a number of different methods to approximate

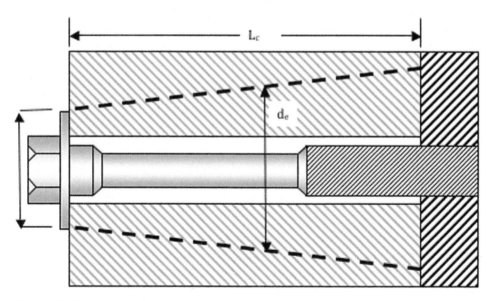

Fig. 14.21 Frustum cone dimensions

this effective area, depending on thickness of the cylinder head and the grip length of the fastener (L_c), and one method is illustrated in Eq. 14.12. It is assumed that the gasket is a small portion of the total length, and can be neglected here.

$$A_c = \frac{\pi}{4}\left[d_e^{\;2} - d_{hole}^2 \right]$$

$$A_c = \frac{\pi}{4}\left[\left(d_{bolt} + \frac{1}{10} \cdot L_c \right)^2 - d_{hole}^2 \right]$$ (14.12)

$$A_c = \frac{\pi}{4}\left[\left(20.8\;\text{mm} + \frac{1}{10} \cdot (4.5 + 116.5 + 26.0\;\text{mm}) \right)^2 - 10.75^2\;\text{mm}^2 \right] = 899\;\text{mm}^2$$

Where:

$A_c =$ Effective area for stiffness calculation
$d_e =$ Effective diameter of frustum cone
$d_{bolt} =$ Diameter for the bearing area under the head of fastener
$d_{hole} =$ Clearance hole diameter in cylinder head
$L_c =$ Grip length, or clamped length of joint

If a slab cylinder head is not used, for example on a single cylinder engine, then A_c might be much smaller. If the effective diameter (d_c) calculated in the method above is larger than the available diameter in the head, for example if the fastener hole is near an edge, this method is not valid. In this case, the largest boss diameter (d_{boss}) at the cylinder head gasket surface is used as in Eq. 14.13.

$$A_c = \frac{\pi}{4}\left(d_{boss}^2 - d_{hole}^2\right)$$
(14.13)

Once the effective area is calculated, the stiffness of the clamped members can be calculated using Eq. 14.14.

$$k_c = \frac{A_c \cdot E_c}{L_c} = \frac{899 \text{ mm}^2 \cdot 7.2x10^4 \text{ N}/\text{mm}^2}{147 \text{ mm}} = 440 \text{ kN/mm}$$
(14.14)

Next the gasket stiffness is determined via a similar method as shown in Eq. 14.15. Gasket stiffness is typically non-linear, however a simplifying assumption is made that it responds in a linear fashion. The total gasket area is divided by the number of fasteners, in order to get the gasket contribution to stiffness per fastener.

$$k_g = \frac{A_g \cdot E_g}{z \cdot L_g} = \frac{10,370 \text{ mm}^2 \cdot 344.5 \text{ N}/\text{mm}^2}{4 \cdot 0.7 \text{ mm}} = 1276 \text{ kN}/\text{mm}$$
(14.15)

Where:

k_g =   Gasket stiffness
A_g =   Clamped area of the gasket per cylinder
E_g =   Modulus of Elasticity of gasket material (consult manufacturer)
z =   Number of fasteners per cylinder
L_g =   Uncompressed gasket thickness

Since total gasket compression is typically a key design parameter, it can be calculated using Eqs. 14.16 and 14.17.

$$T_g = \frac{F_b}{k_g \cdot L_g} = \frac{34.6 \text{ kN}}{1,276 \text{ kN}/\text{mm} \cdot 0.7 \text{ mm}} = 0.039 \text{ or } 3.9\%$$
(14.16)

$$L_c = L_g(1-T_g) = 0.7 \text{ mm}(1-0.039) = 0.67 \text{ mm}$$
(14.17)

Where:

T_g =   % gasket compression
L_c =   Compressed gasket thickness

Now the stiffness of the clamped members and the stiffness of the head gasket can be combined to determine an equivalent stiffness in Eq. 14.18. This uses the same springs-in-series approach as the fastener:

$$\frac{1}{k_{c \cdot g}} = \frac{1}{k_c} + \frac{1}{k_g} = \frac{1}{440 \text{ kN / mm}} + \frac{1}{1276 \text{ kN / mm}} = 0.003 \qquad (14.18)$$

$$k_{c \cdot g} = 333 \text{ kN / mm}$$

Where:

k_{cg} =   Combined stiffness of the abutment (head + gasket)

14.8.3   Calculate Fastener and Abutment Deflection and Loading

The next step is to calculate the fastener extension, and abutment (head + gasket) compression due to tightening using Eqs. 14.19 and 14.20.

$$\Delta_b = \frac{F_b}{k_b} = \frac{34.6 \text{ kN}}{70.4 \text{ kN / mm}} = 0.49 \text{ mm} \qquad (14.19)$$

$$\Delta_{c \cdot g} = \frac{F_b}{k_{c \cdot g}} = \frac{34.6 \text{ kN}}{333 \text{ kN / mm}} = 0.10 \text{ mm} \qquad (14.20)$$

Now the contribution of the fastener and abutment to the overall joint stiffness can be calculated. The fastener and abutment Load Factors are calculated using Eqs. 14.21 and 14.22.

$$L_b = \frac{k_b}{k_b + k_{c \cdot g}}$$

$$L_b = \frac{70.4 \text{ kN / mm}}{70.4 \text{ kN / mm} + 333 \text{ kN / mm}} = 0.175 \qquad (14.21)$$

$$L_{c \cdot g} = \frac{k_{c \cdot g}}{k_b + k_{c \cdot g}}$$

$$L_{c \cdot g} = \frac{333 \text{ kN / mm}}{70.4 \text{ kN / mm} + 333 \text{ kN / mm}} = 0.825 \qquad (14.22)$$

Note the different contribution to joint stiffness from the fastener and abutment. The fastener is less stiff than the cylinder head and gasket combination. Now the portions of the gas load taken by the fastener and by the abutment are calculated using Eqs. 14.23 and 14.24.

$$F_{b,gas} = \frac{L_b \cdot F_{gas}}{z}$$

$$F_{b,gas} = \frac{0.175 \cdot 42.9 \text{ kN}}{4} = 1.88 \text{ kN}$$

(14.23)

$$F_{c \cdot g,gas} = \frac{L_{c \cdot g} \cdot F_{gas}}{z}$$

$$F_{c \cdot g,gas} = \frac{0.825 \cdot 42.9 \text{ kN}}{4} = 8.85 \text{ kN}$$

(14.24)

It can be seen that less load is taken up by the fastener, improving fatigue life. The next calculations are the total load on the fastener and abutment with gas force addition using Eqs. 14.25 and 14.26, and total extension of the fastener and abutment using Eqs. 14.27 and 14.28. It can be seen that fastener load increases with gas force as it tries to lift the cylinder head off of the cylinder block, and the clamped member load decreases with gas force.

$$F_{b,total} = F_b + F_{b,gas}$$

$$F_{b,total} = 34.6 + 1.88 = 36.5 \text{ kN}$$

(14.25)

$$F_{c \cdot g,total} = F_b - F_{c \cdot g,gas}$$

$$F_{c \cdot g,total} = 34.6 - 8.85 = 25.8 \text{ kN}$$

(14.26)

$$\Delta_b = \frac{F_{b,total}}{k_b}$$

$$\Delta_b = \frac{36.5 \text{ kN}}{70.4 \text{ kN} / \text{mm}} = 0.52 \text{ mm}$$

(14.27)

$$\Delta_{c \cdot g} = \frac{F_{c \cdot g,total}}{k_{c \cdot g}}$$

$$\Delta_{c \cdot g} = \frac{25.8 \text{ kN}}{333 \text{ kN} / \text{mm}} = 0.08 \text{ mm}$$

(14.28)

14.8.4   Calculate Thermal Affects

The previous calculations were for the condition of initial fastener tightening at room temperature, and engine loading at startup. In reality, this is a small percentage of engine operating life. The engine spends the majority of its operating life at elevated temperatures, and these elevated temperatures must be accounted for. Using Eq. 14.29 for thermal growth, the force due to differential thermal growth of dissimilar materials can be accounted for with Eq. 14.30, assuming both members have a uniform temperature rise. Conversely, at low operating temperatures, bolt clampload decreases from the room temperature value. The gasket has a small contribution to these calculations, and is not considered here.

$$\Delta L = \alpha L_o (T_2 - T_1) \tag{14.29}$$

$$F_{b,therm} = \frac{(\alpha_c - \alpha_b) L_c (T_2 - T_1)}{\dfrac{1}{k_c} + \dfrac{1}{k_b}} \tag{14.30}$$

$$F_{b,therm} = \frac{\left(20.86x10^{-6} - 11.6x10^{-6}\ \frac{1}{°C}\right) 147\ \text{mm}\,(107 - 21°C)}{\dfrac{1}{440\ \text{kN}\,/\,\text{mm}} + \dfrac{1}{70.4\ \text{kN}\,/\,\text{mm}}} = +7.1\ \text{kN at } 107°C$$

$$F_{b,therm} = \frac{\left(20.86x10^{-6} - 11.6x10^{-6}\ \frac{1}{°C}\right) 147\ \text{mm}\,(-30 - 21°C)}{\dfrac{1}{440\ \text{kN}\,/\,\text{mm}} + \dfrac{1}{70.4\ \text{kN}\,/\,\text{mm}}} = -4.2\ \text{kN at } -30°C$$

Where:

$\Delta L =$ Change in length of feature
$\alpha =$ Coefficient of linear expansion for material
$L_c =$ Clamped length of joint
$T_1 =$ Initial temperature of joint
$T_2 =$ Final temperature of joint

In the example, a steel fastener with an aluminum head is analyzed. As the temperature increases, the aluminum head will grow further than the steel fastener; this will increase the initial fastener clampload. This will also beneficially increase the clampload on the cylinder head and gasket. Care must be taken not to exceed the maximum gasket contact stress. When the operating temperature decreases from the assembly temperature, clampload decreases. Care must be taken to ensure the minimum gasket contact stress is maintained to prevent gasket blow out, when starting the engine at low temperatures. The load of initial assembly, thermal affects, and firing force are calculated in Eqs. 14.31 and 14.32.

$$F_{b,total,therm} = (F_b + F_{b,therm}) + F_{b,gas}$$

$$F_{b,total,therm} = 34.6 + 7.1 + 1.88 = 43.6 \text{ kN at } 107\,^{\circ}\text{C} \tag{14.31}$$

$$F_{c\cdot g,total,therm} = (F_b + F_{b,therm}) - F_{c\cdot g,gas}$$

$$F_{c\cdot g,total,therm} = 34.6 + 7.1 - 8.85 = 32.9 \text{ kN at } 107\,^{\circ}\text{C} \tag{14.32}$$

14.8.5   Design Margins

As discussed in the previous section, the head gasket must be evaluated at multiple operating conditions. The goal is to avoid "cold pop", a situation that occurs when the engine does not have enough sealing contact stress at startup in cold conditions, and also to avoid yielding components during extended engine operation at high temperatures. Design margins must be evaluated at three conditions: cold storage startup, initial assembly at room temperature, and peak operating temperatures.

14.8.5.1   Minimum Gasket Contact Pressure and Joint Separation

A common recommendation for a graphite head gasket with a combustion seal ring is 175 N/linear mm. An MLS head gasket may only require 28 N/linear mm for a half emboss, and 105 N/linear mm for a full emboss design. The clamp force required to achieve this gasket pressure per cylinder is calculated using all of the fasteners per cylinder. Beyond exceeding the minimum gasket contact pressure, joint separation occurs when gas force equals the total clamp force on the joint. First the required clamp force for the gasket will be calculated using Eq. 14.33, and then the design margin is calculated for the gasket at cold storage temperature using Eq. 14.34.

$$F_{min,gasket} = \frac{(\pi \cdot D_{cyl}) \cdot \left(\dfrac{\text{Force}}{\text{Linear Distance}} \right)}{z}$$

$$F_{min,gasket} = \frac{\pi \cdot 89 \text{ mm} \cdot 175 \text{ N / mm}}{4} = 12.2 \text{ kN} \tag{14.33}$$

$$\text{Sealing Design Margin} = \frac{F_{c\cdot g,total,therm}}{F_{min,gasket}} = \frac{34.6 - 4.2 - 8.85 \text{ kN}}{12.2 \text{ kN}} = 1.8 \text{ at } -30\,^{\circ}\text{C} \tag{14.34}$$

14.8.5.2   Fatigue Life of Fastener

The cylinder head fastener must be designed to last the life of the engine. The key to designing a fastener for fatigue resistance is to maximize the initial tension as close to the proof strength as possible. Increasing initial tension of the fastener by selecting a higher strength material of smaller diameter reduces the relative stiffness of the fastener, which diminishes the bolt fluctuating stress and improves fatigue life.

The minimum and maximum stresses on the fastener in operation are calculated using Eqs. 14.35 and 14.36.

$$\sigma_{min} = \frac{F_b + F_{b,therm}}{\left(\dfrac{\pi \cdot D_{min}^2}{4}\right)} = \frac{34.6 + 7.1 \text{ kN}}{\left(\dfrac{\pi \cdot 7.78^2 \text{ mm}^2}{4}\right)} = 877 \text{ MPa} \qquad (14.35)$$

$$\sigma_{max} = \frac{F_{b,total,therm}}{\left(\dfrac{\pi \cdot D_{min}^2}{4}\right)} = \frac{34.6 + 7.1 + 1.88 \text{ kN}}{\left(\dfrac{\pi \cdot 7.78^2 \text{ mm}^2}{4}\right)} = 917 \text{ MPa} \qquad (14.36)$$

The alternating and mean stresses are calculated using Eqs. 14.37 and 14.38.

$$\sigma_{alt} = \frac{\sigma_{max} - \sigma_{min}}{2} = \frac{917 - 877 \text{ MPa}}{2} = 20 \text{ MPa} \qquad (14.37)$$

$$\sigma_{mean} = \frac{\sigma_{max} + \sigma_{min}}{2} = \frac{917 + 877 \text{ MPa}}{2} = 897 \text{ MPa} \qquad (14.38)$$

Using the modified Goodman line, the design margin for fastener fatigue is calculated using Eq. 14.39.

$$\text{Fatigue Design Margin} = \frac{S_e}{\sigma_{eq}} = \frac{S_e}{\sigma_{alt} + \left(\dfrac{S_e}{S_{uts}}\right) \cdot \sigma_{mean}}$$

$$\text{Fatigue Design Margin} = \frac{162 \text{ MPa}}{20 \text{ MPa} + \left(\dfrac{162 \text{ MPa}}{1240 \text{ MPa}}\right) \cdot 897 \text{ MPa}} = 1.2 \qquad (14.39)$$

Where:

S_e =   Endurance strength for material
S_{uts} =   Ultimate tensile strength for material

14.8.5.3   Thread Stripping of Nut Member

Another important consideration in the head gasket bolted joint design is stripping of the nut member. Typically for powertrain design, a fastener is threaded into a blind hole in the cylinder block. A conservative guideline is that thread engagement should be 2.5 times the fastener diameter. Typically, the nut material is not as strong as the fastener material, and is the limiting condition. In this example it is assumed the cylinder block is aluminum, and care must be taken as the yield strength may drop at elevated temperatures. A triangular thread will fail in shear, and the strength of the threaded joint is the strength of each thread multiplied by the total number of threads. First, the cross-sectional area of the threads engaged is calculated using Eq. 14.40, and then the design margin can be calculated using Eq. 14.41.

$$A_t = \frac{\pi \cdot L_e \cdot D_m}{p} \cdot [0.5p + (D_m - D_p)\tan 30^\circ]$$

$$A_t = \frac{\pi \cdot 25 \text{ mm} \cdot 9.732 \text{ mm}}{1.5 \text{ mm}} \cdot [(0.5 \cdot 1.5) + (9.732 - 9.206)\tan 30^\circ]$$

$$A_t = 537 \text{ mm}^2$$

(14.40)

Where:

A_t =   Thread stripping cross-sectional area
L_e =   Length of engaged threads (# of thds × Thd Pitch)
D_m =   Major diameter of external thread
p =   Thread pitch or width of one thread (1/n where n=threads/distance)
D_p =   Pitch diameter of internal thread

$$\text{Thread Strip Design Margin} = \frac{0.58 \cdot S_y}{\left(\dfrac{F_{b,total,therm}}{A_t}\right)}$$

$$\text{Thread Strip Design Margin} = \frac{0.58 \cdot 225 \text{ MPa}}{\left(\dfrac{43.6 \text{ kN}}{537 \text{ mm}^2}\right)} = 1.6$$

(14.41)

Where:

S_y =   Yield strength of material

14.8.5.4   Embedment of the Fastener in the Cylinder Head

The area of maximum compressive stress usually occurs under the head of the fastener, or between the fastener washer and clamped member. When the compressive yield strength of the clamped material is exceeded, embedment occurs. This is usually not a concern for the clamped member, as the stress spreads out due to the frustum cone and ultimately stops yielding. However, when embedment occurs, a major concern with the fastener is the loss of stretched length and the associated loss of clampload. For this reason, a large fastener head or a hardened ground washer is often used. Attention must be paid to the hot hardness of materials such as aluminum, since the compressive yield strength will decrease at elevated temperatures. Compressive stress can be calculated using Eq. 14.42, and design margin using Eq. 14.43.

$$\sigma_{comp} = \frac{F_{b,total,therm}}{A} = \frac{F_{b,total,therm}}{\frac{\pi}{4}\left(d_{bolt}^2 - d_{hole}^2\right)}$$

$$\sigma_{comp} = \frac{43.6 \text{ kN}}{\frac{\pi}{4}(20.8^2 \text{ mm}^2 - 10.75^2 \text{ mm}^2)} = 175 \text{ MPa}$$

(14.42)

$$\text{Embedment Design Margin} = \frac{S_{y,comp}}{\sigma_{comp}}$$

$$\text{Embedment Design Margin} = \frac{225 \text{ MPa}}{175 \text{ MPa}} = 1.3$$

(14.43)

Where:

d_{bolt} =   Diameter for the bearing area under the head of fastener
d_{hole} =   Clearance hole diameter in cylinder head
$S_{y,\ comp}$ =   Compressive yield strength of clamped material

Summary of Design Margins
It is now possible to summarize the gasketed joint design margins as shown in Fig. 14.22. These results can be summarized in a table or plotted in a graph to pictorially show robustness of the cylinder head gasket joint. It is also recommended that each of the design margins calculated previously be evaluate at limits of tolerance of all values.

14.9   Test Methods for Gaskets

There are many tools for accelerating and validating gasket designs. Initial seal design is completed using previous knowledge, design calculations, and finite element analysis. Pressure tests may be run on a servo-hydraulic rig to simulate working pressures. Vibration tests can be run on electro- or hydro-dynamic shakers. Hot gas cycle tests can be run to simulate elevated temperatures at the head gasket and exhaust manifold gasket. However, it is difficult to accelerate the aging of a gasket due to UV and atmospheric exposure, and this is best accomplished with vehicle testing at environmental extremes.

Since gasket behavior is difficult to model due to its visco-elastic nature, there are many methods available for direct measurement. The oldest method is the lead pellet test. This method involves cutting a small hole in a flat gasket and placing a lead pellet

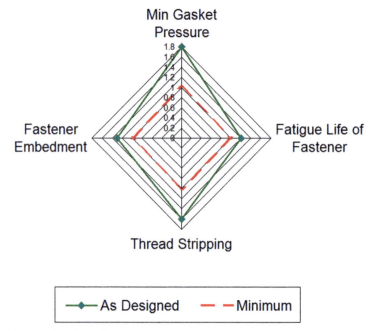

Fig. 14.22 Target plot of results

within. As the components are assembled, the pellet is compressed, and the thickness can be measured upon disassembly. Another direct measurement method is to employ pressure sensitive film, known by the trade name of Fujifilm Prescale® by Sensor Products Incorporated. This film has microencapsulated beads of dye that break when a certain threshold pressure is achieved. Fujifilm Prescale® gives a nice pictorial representation of the quality of seal pressure distribution. Quantitative measurement is slightly more difficult, as color must be compared to a calibrated scale. Different ranges of film are available with different resolutions. Both the lead pellet and Fujifilm Prescale® methods have the limitation of only capturing the peak displacement or pressure, and cross-talk of the bolted joint during assembly may result in a final pressure less than the peak value. If a dynamic pressure map is required, a pressure sensitive strip made by Tekscan® may be inserted with active sensors. However, the pressure map will be unique to each application, due to flange geometry and the need to pass screws through it. Real time gap sensors can be used for cylinder head liftoff testing. Finally, unit gasket stiffness curves can be developed on a tensile test machine between two thick platens. But this has the limitation of not replicating actual component flange stiffness.

There are three different ways to classify a failed gasket: *seep*, *weep*, and *leak*. A seep is the least severe form of failure, when dry dust gathers on the external surface of the gasket. This can be the case in a cork or fiber gasket where the fluid penetrates the body of the seal via wicking, and not by a leak at the sealing surface, and is wet enough to gather dust on the gasket only. A weep is a more significant type of failure where the surface of the mating components is wetted, and fluid is wicking away from the gasket. Finally, a leak is a total failure of the gasket, where significant quantities of fluid are present, and possibly dripping off of the cover. Depending on the engine application, different levels of gasket failures can be tolerated and may allow a less expensive type of gasket to be used. A cork gasket may be acceptable for the oil pan in applications where the presence of a weep will be hidden from view and covered with an NVH shield. At the opposite end of the spectrum is when the engine is on display and highly cosmetic, such as on a motorcycle. The customer expectation may be that even a seep is not acceptable.

Once working prototypes are produced there are several methods of detecting effectiveness. For leak detection on a running engine, fluorescent dye may be added to the oil and leaks found using ultraviolet light. Dry powder can be sprinkled on the surface of the sealed components, as passive leak detection. A soapy water solution can be squirted on the joint surface, and bubbles will indicate locations of pressurized air leaks. Vacuum or pressure can be applied to a sealed cavity and monitored for flow or pressure decay. This method is generally used for gross leaks, as air temperature of the expanding gas can affect the measurement significantly. Hydrogen gas can be injected into a working cavity, and a gas detector can be employed. Hydrogen gas leak detection, when mixed with nitrogen gas to lower flammability, is effective because of the small molecules that diffuse rapidly. However, hydrogen gas leak detection can be limited due to low resolution of where the leak is actually occurring.

14.10   Recommendations for Further Reading

The following references provide important general considerations about gasketed joint design: Bickford (1995), Czernik (1996) and AE-13 (1988).

For an analysis of flange design and consideration of flange bending the following paper is recommended: Widder et al. (1993).

During the 1980s focus was placed on elimination of asbestos in engines, and alternative gasket materials were sought. These papers provide summaries of those efforts: Majewski et al. (1988), Percival and Williams (1985).

Silicon-based Room Temperature Vulcanized (RTV) gaskets continue to see increased application. The following papers provide an overview of RTV joint design and analysis: Roberts (1990) and Littmann (1990).

The cylinder head gasket is an especially challenging seal, and it has received much attention in the literature. The first two papers listed below discuss structural analysis of the head gasket joint. This is followed by a paper on metal head gasket evaluation, and a series of recent papers on Multi-Layer Steel (MLS) head gasket development (Raub 1992; Graves et al. 1993; Ishigaki et al. 1993; Kestly et al. 2000; Popielas et al. 2000, 2003a, b).

Finally, the following paper summarizes detailed measurements of radial shaft seal operation (Sato et al. 1999).

References

AE-13: Gasket and Joint Design Manual for Engine and Transmission Systems. SAE Publication (1988)

Bickford, J.H.: An Introduction to the Design and Behavior of Bolted Joints. Marcel Dekker Inc, U S A (1995)

Czernik, D.E.: Gaskets: Design, Selection, and Testing. McGraw-Hill (1996)

Graves, S., Utley, T.L., Isikbay, N.: Nonlinear finite element analysis of diesel engine cylinder head gasket joints. SAE 932456. (1993)

Ishigaki, T., Kitagawa, J., Tanaka, A.: New evaluation method of metal head gasket. SAE 930122. (1993)

Kestly, M., Popielas, F., Grafl, D., Weiss, A.: Accelerated testing of multi-layer steel cylinder head gaskets. SAE 2000-01-1188. (2000)

Littmann, W.J.: Silicone sealing through application development. SAE 900116. (1990)

Majewski, K.-P., Zerfass, H.-R., Scislowicz, M.: Asbestos substitution in cylinder head gaskets. SAE 880144. (1988)

Percival, P., Williams, B.J.G.: Non-asbestos gasket engineering. SAE 850191. (1985)

Popielas, F., Chen, C., Obermaier, S.: CAE approach for multi-layer-steel cylinder head gaskets. SAE 2000-01-1348. (2000)

Popielas, F., Chen, C., Ramkumar, R., Rebien, H., Waldvogel, H.: CAE approach for multi-layer-steel cylinder head gaskets—part 2. SAE 2003-01-0483. (2003a)

Popielas, F., Chen, C., Mockenhaupt, M., Pietraski, J.: MLS influence on engine structure and sealing function. SAE 2003-01-0484. (2003b)

Raub, J.: Structural analysis of diesel engine cylinder head gasket joints. SAE 921725. (1992)

Roberts, K.T.: Designing cured-in-place silicone rubber engine gaskets. SAE 900119. (1990)

Sato, Y., Toda, A., Ono, S., Nakamura, K.: A study of the sealing mechanism of radial lip seal with helical ribs—measurement of the lubricant fluid behavior under sealing contact. SAE 1999-01-0878. (1999)

Widder, E., Sadowski, M., Novak, G.: Gasketed joint analysis including flange bending effects. SAE 930120 (1993)

Pistons and Rings

15

15.1 Piston Construction

The piston remains one of the most challenging components to successfully design, and is certainly quite critical to the performance and durability of the engine. The root of the challenge lies in the piston's role as the moving combustion chamber wall. It is thus directly exposed to the severe conditions of the combustion chamber, and must manage the work transfers between the combustion gases and the connecting rod. Further challenges implied by this role include the necessity of maintaining a combustion seal at this moving boundary, under a wide variety of operating conditions; the need to provide adequate lubrication and minimal friction and wear at an elevated temperature, and under continually starting and stopping conditions; and the requirement of managing the reciprocating forces as well as secondary forces resulting from the pivoting motion at its pin to the connecting rod.

Typical pistons as used in an automobile engine are depicted in Fig. 15.1a, b. The piston is either cast or forged from an aluminum alloy. It is instructive to consider the piston as consisting of the four regions shown in the figure—the crown, the ring lands, the pin boss, and the skirt. Each region will be discussed in turn in the initial sections of this chapter. Later sections will address the rings and the cylinder wall.

In high volume applications the pistons are typically die cast, with squeeze casting (a die casting process conducted under controlled pressure in a special mold) becoming more common to reduce porosity and improve strength. Sand casting is sometimes used in low volume applications. High performance engines often use forged pistons to improve strength. Most piston alloys are eutectic to hyper-eutectic alloys of aluminum and silicon (12–18 % silicon). The silicon improves strength at elevated temperature, and significantly reduces wear. Fiber-reinforced alloys using whiskers of alumina (Al_2O_3) in quantities of

© Springer Vienna 2016

K. Hoag, B. Dondlinger, *Vehicular Engine Design*, Powertrain,
DOI 10.1007/978-3-7091-1859-7_15

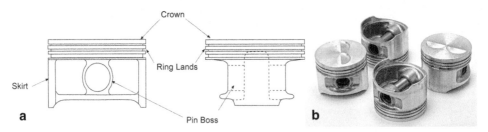

Fig. 15.1 **a** Typical window type automobile piston diagram. **b** Typical automobile pistons

10–15 % by volume are sometimes used. The alumina reduces the thermal expansion coefficient, and further increases the high temperature strength.

A number of piston coatings are commonly seen, for a wide variety of purposes. Tin, lead, or phenolic resin graphite coatings may be used on the piston skirt to improve sliding and reduce the friction work. In order to reduce the propensity to scuff it is important that the piston and cylinder wall materials are not the same. In engines using aluminum blocks it remains most common to cast an iron cylinder liner into the block. However, to further reduce cost and weight an aluminum alloy block may be used with high silicon content, either throughout the casting or in the cylinder wall material. In these engines it is necessary to coat the piston for scuff resistance. An iron coating is most typical although chromium is sometimes seen as well. Finally, a hard anodizing (zinc) may be coated on the crown of the piston to increase its high temperature capability. A 15–20 °C increase in peak crown temperature may be allowed through use of the anodized coating.

In heavy-duty diesel engines used in trucks, agricultural, construction, and marine applications various further piston design options are seen. Increasingly popular for the highest output engines is the thin-wall ductile iron or steel piston such as shown in Fig. 15.2. These pistons hold the structural advantage of significantly higher strength and high temperature capability. They also have much lower thermal expansion than does aluminum, allowing tighter running clearance and considerably less clearance variation with load. However, the much higher weight increases reciprocating loads and required rod and main

Fig. 15.2 Forged steel, single piece heavy duty diesel piston

Fig. 15.3 Articulated
piston for heavy duty diesel
engine

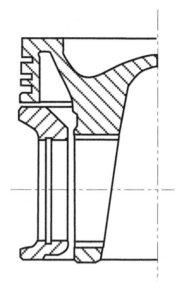

bearing area. Connecting rod cap loading is always a design challenge, and is certainly aggravated with iron pistons. Every opportunity for weight reduction is taken through the use of thin walls and minimal skirt area as can be seen in Fig. 15.2. It should be noted that as design refinement continues to reduce the weight, iron alloy pistons are gaining attention in lighter duty engines as well.

Also seen in heavy-duty diesel engines is the articulated piston shown in Fig. 15.3. The articulated piston uses an iron or steel crown and ring belt in conjunction with an aluminum skirt. The two pieces share a common pin at the connecting rod. The articulated piston allows the high temperature strength and clearance control advantages of the iron piston, but with reduced weight. However the increased cost and complexity has caused this design to rapidly lose favor as compared to the single-piece iron alloy pistons described previously. In engines larger than those addressed in this book composite pistons become common. In the composite piston an iron or steel crown is bolted to an aluminum piston base.

Single-piece aluminum pistons are also used in many heavy-duty engines. In some cases a cooling gallery is cast around the perimeter of the piston, immediately behind the ring belt. This gallery, supplied by a directed spray of pressurized oil, is very effective in controlling peak crown and top ring temperatures. A gallery cooled piston is shown in Fig. 15.4. Gallery cooled pistons are also sometimes seen in passenger car diesel and boosted spark-ignition engines. While the gallery cooled design is very effective in controlling temperature it becomes critical to positively locate the oil spray nozzles, and to ensure that the spray column does not break up before reaching the piston. Such design considerations will be further discussed in the next section.

The cross-section of the compression rings is another design variable that may affect piston construction. While many rings have a rectangular cross-section, a keystone

Fig. 15.4 Gallery cooled
aluminum piston for heavy
duty diesel engine

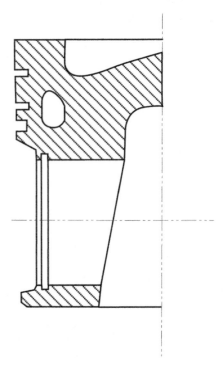

cross-section is often used as well. The keystone shape is less prone to collapse at high cylinder pressures, is less likely to stick as carbon deposits form, and its additional movement slows the rate of deposit build-up. However, keystone rings of necessity experience more vertical motion in the ring groove, and may result in significant groove wear. In order to minimize groove wear the grooves may be anodized, or a nickel-based alloy ring groove insert may be cast into the aluminum piston. The insert may be used for the top ring alone or for both compression rings. The second ring typically experiences more movement, for reasons to be discussed in Sect. 15.6, but the temperatures are considerably higher at the top ring groove.

A final piston construction feature that should be mentioned is the steel frame around which an aluminum piston is sometimes cast. The purpose of the frame is to restrict the thermal growth of the piston, thus reducing the difficulties of maintaining proper piston clearance. The cast-in frame has all but disappeared as machining process control now allows much closer specification of piston diameter variation—both axially and circumferentially. This will be further discussed in Sect. 15.4.

15.2   Piston Crown and Ring Land Development

Piston crown geometry is one of the areas where diesel and spark-ignition engines differ greatly. This follows directly from the very different criteria of combustion chamber design. The clearance volume (combustion chamber volume at TDC) is invariably contained in a piston bowl in the diesel engine, and is almost invariably cast or machined into the cylinder head in spark-ignition engines. While spark-ignition engines sometimes use piston bowls there are several reasons why this is unattractive. First, the hydrocarbon emissions from the spark-ignition engine are quite sensitive to ring placement. The high ring placement required for low hydrocarbon emissions makes ring temperature control difficult if the piston simultaneously contains a piston bowl. Second, the piston bowl results in a squish flow of air from all around the perimeter toward the center of the combustion chamber as the piston approaches TDC. This flow is generally undesirable to the spark-ignition flame front, and can be a source of miss-fire. Finally, placing the combustion chamber in a piston bowl aggravates piston temperature, creating durability challenges with the higher temperatures seen in spark-ignition engines as compared to the diesel. For the reasons just enumerated most spark-ignition engines use flat-crown pistons. In high performance engines designed to operate on higher octane fuels the crown is often domed to increase the compression ratio. Examples of piston crowns in spark-ignition engines are shown in Fig. 15.5. During the valve overlap period the piston approaches TDC while both intake and exhaust valves are partially open. In some cases this requires valve reliefs to be cut into the piston crown as shown in the figure.

A fundamental objective in diesel combustion chamber layout is to maximize the contact area between the fuel spray plumes and the air in the combustion chamber. From this objective it follows that a centrally-located fuel injector spraying fuel into a chamber whose walls mirror the shape of the spray plume will be most effective. It would be quite impossible to create such geometry on a cylinder head surface that must also contain the intake and exhaust valves. A more detailed explanation is beyond the scope of this book,

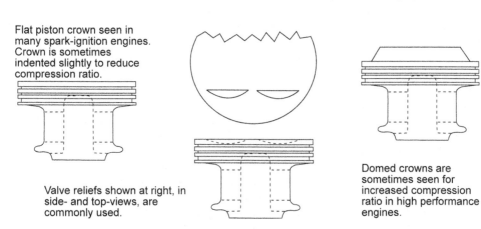

Flat piston crown seen in many spark-ignition engines. Crown is sometimes indented slightly to reduce compression ratio.

Valve reliefs shown at right, in side- and top-views, are commonly used.

Domed crowns are sometimes seen for increased compression ratio in high performance engines.

Fig. 15.5 Example piston crowns for spark-ignition engines

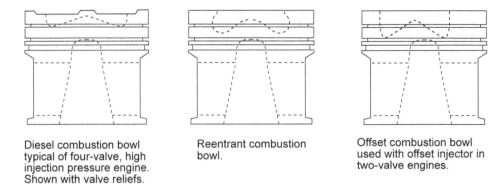

Diesel combustion bowl
typical of four-valve, high
injection pressure engine.
Shown with valve reliefs.

Reentrant combustion
bowl.

Offset combustion bowl
used with offset injector in
two-valve engines.

Fig. 15.6 Example piston crowns for diesel engines

but the result is that the diesel combustion chamber is invariably contained in a bowl in the piston. It should also be noted that the squish flow resulting from the rim around the piston bowl is quite advantageous in diesel combustion as it creates high rates of shear between the air and the injected fuel, thus enhancing mixing. Several example diesel engine piston crowns are depicted in Fig. 15.6. The four-valve configurations shown in the figure naturally result in central injector placement. The two-valve configuration necessitates offsetting the fuel injector in order to maximize valve area. In these engines it remains advantageous to center the injector in the piston bowl, so the bowl is offset to coincide with the injector offset. This design is falling out of favor, and new diesel engines almost invariably use the four-valve cylinder head and central injector. As with the spark-ignition engines valve cut-outs are sometimes required to provide sufficient clearance during overlap. However, the cut-outs in the piston rim have generally been found quite detrimental to the optimization required for emission control, and are now most often avoided. In many cases optimization has led to a reentrant lip protruding over the bowl from the rim, adding durability challenges as will be discussed in the paragraphs to follow.

The primary loads experienced by the piston crown are thermal and cylinder pressure loads. Especially with aluminum alloy pistons the peak temperatures seen on the crown are critical to piston durability. Maximum crown temperatures are limited to between 320 and 330 °C, with the higher end of this range usually requiring an anodized coating. At these temperatures the fatigue strength of the alloy is on the order of half that at ambient temperature and the strength rapidly plummets as the temperature is raised further. An important point to emphasize is that while at first glance the temperatures and stresses appear nearly symmetric around the piston the variation is in fact significant. The maximum piston temperature may vary by as much as 50 °C between that seen beneath the intake valves and that beneath the exhaust valves. While this may seem insignificant, because of the strong impact of temperature on fatigue in this temperature range it is in fact quite important. Immediately beneath the crown the piston incorporates the pin boss. The piston expands more readily along the pin axis than perpendicular to the axis, and this too has an important impact on the crown stress profile. It should also be noted that pin and boss stiffness impact crown stress. As the pin is made less stiff crown stress is reduced (although stress at the pin boss will certainly increase).

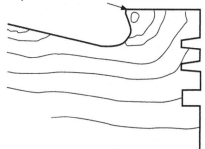

Rapid surface temperature decrease
results in tensile stress due to hotter
temperatures beneath surface

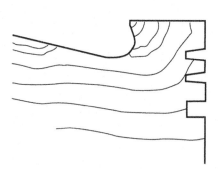

Isotherms under full load, steady-state   Isotherms immediately after sudden load
conditions   decrease

Fig. 15.7 The thermal shock mechanism resulting from sudden load change depicted on a diesel piston crown

Another important piston crown development consideration is that of rapid thermal transients. Because the high temperature crown of the piston is constrained by the lower temperature pin boss and skirt the crown normally experiences high compressive stress. However, if the engine load is rapidly reduced the surface temperature may drop more rapidly than that just beneath the surface. This situation is depicted in Fig. 15.7, and results in a momentary tensile hoop stress on the crown surface. The degree to which this occurs is greatly aggravated when the piston design results in additional area exposed to combustion gases. The bowl rim on diesel pistons is an example, and the reentrant bowl described previously makes the situation more severe. In any case the alternating loads produced by thermal transient cycles are an important consideration in piston durability validation. Analytical approaches such as finite element analysis become extremely computationally intensive when transients must be considered. Rig tests developed for rapid thermal cycling are sometimes used, but most thermal shock assessment at this writing is done with engine testing. Cycling the engine between conditions of high load and hot intake temperatures and low or zero load with refrigerated air can be quite effective. Whenever possible such testing should be correlated with field data; test confidence is much higher if one can ascertain how well the design being validated does on the particular test relative to successful designs already in production.

Clearly temperature control is among the most critical issues determining piston crown durability. Another crucial focus of temperature control is that of minimizing deposits on the ring lands and in the ring grooves. Lubricant oxidation breakdown is exponentially dependent on temperature so small temperature increases result in significant increases in the rates at which lubricants react with oxygen—a 15 °C temperature increase will approximately double the reaction rates. The oxidation reactions result in lighter fractions of the lubricant hydrocarbons burning off, and the heavier fractions forming solid deposits that collect fuel fragments, wear particles, and dirt, and adhere to the piston surfaces as deposits. Chemical analysis typically shows the deposits to consist primarily of carbon, oxygen, and silicon, with small amounts of a variety of additional substances. Maximum allowable temperatures at the top ring or on the land surface between the top and second ring are generally kept below approximately 220 °C to ensure minimal deposit build-up. This temperature limit is driven by lubricant reaction rates and not material properties so it remains the same regardless of piston material.

While it remains difficult to measure the piston temperatures of operating engines several techniques are widely used. One approach is to use a "grasshopper linkage" to physically carry thermocouple wires from the piston. If the wires are run parallel to the axis of each linkage joint (so that they twist instead of bending) many hours of data can be recorded without wire breakage. Various electronic signal transmission techniques have been developed and are gaining in popularity. Another approach that can be used on both pistons and rings is to install templugs—small, threaded plugs of special alloys whose hardness changes based on the highest temperature they have seen. Templugs are installed at the locations of interest, and the engine is operated for a specified period of time at one speed and load point. The templugs are available in small enough sizes that at least on heavy-duty engines they can be installed directly on the rings as well. The disadvantages are that the temperatures can only be obtained at one operating point per engine build, and that they are not as accurate as thermocouples.

The use of lubricant spray for piston cooling was introduced in the opening section. Most automobile engines rely on oil mist in the crankcase to aid in cooling, often supplemented with a spray through a drilling in the connecting rods. Piston undercrown cooling can be significantly enhanced by incorporating high flow, spray cooling nozzles fed from a pressurized oil rifle. These are often non-directed sprays. The pistons shown in Figs. 15.2, 15.3 and 15.4 each require using a directed nozzle. The nozzle is designed to provide an oil column that remains intact, if possible the entire distance to the piston's TDC position. If the column breaks up earlier oil flow into the gallery will drop off as the piston travels upward in the cylinder. The combination of oil flow rate, column break-up length, and drain-back drilling diameter in the piston (generally 180° from the supply drilling into the oil gallery) are used to determine the amount of oil in the gallery. It is important to keep the residence time in the gallery short to minimize oil temperature rise and avoid coking in the gallery. It is recommended that the oil gallery is not flooded, but maintained one-half to two-thirds full. This results in a "cocktail shaker" effect, providing a heat transfer coefficient significantly higher than that obtained with the forced convection heat transfer mechanism of a full gallery. Articulated pistons and single-piece ductile iron pistons typically use a stamped cover pressed in place as shown in Figs. 15.2 and 15.3 to enclose the oil gallery.

15.3 Piston Pin Boss Development

Two different options for constraining the piston pin are presented in Fig. 15.8. In the full-floating design the piston pin is clearance-fit into both the piston and connecting rod. Snap rings fitted into the pin boss at each end constrain the pin, and the pin is free to float

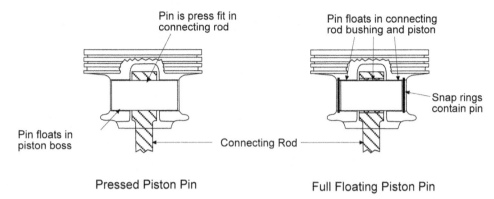

Fig. 15.8 Pressed and full-floating piston pin designs

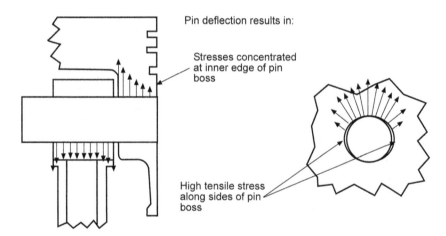

Fig. 15.9 Piston pin loads

(rotate) relative to both the piston and the connecting rod. The advantages of the full-floating pin include reduced friction and increased durability through reduced pin wear. However the snap ring grooves reduce the length of the piston pin and thus available bearing area. A bushing must be added to the connecting rod since the rod itself is an unacceptable bearing material. The rod bushing, snap rings and grooves all add material and assembly cost. Finally, snap ring breakage or misassembly lead to severe cylinder damage. In the alternative fixed pin design the pin is press fit into the connecting rod and clearance fit in the piston. The primary advantage of this design is in its simplicity, resulting in significantly lower cost. A further advantage is the ability to increase the bearing contact area between the piston and pin.

Attention now turns to managing the loads at the pin boss. Pin and boss loading is depicted in Fig. 15.9. Pin deflection results in peak loading at the inside edge of the pin boss on the piston, and the outside edge of the connecting rod journal. The piston, pin, and rod experience high-cycle alternating loads at this joint—the cylinder pressure force

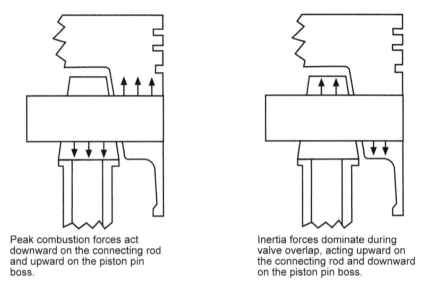

Peak combustion forces act
downward on the connecting rod
and upward on the piston pin
boss.

Inertia forces dominate during
valve overlap, acting upward on
the connecting rod and downward
on the piston pin boss.

Fig. 15.10 Keystone piston pin boss design for low speed, high cylinder pressure engines

dominates as the piston approaches TDC during the compression/expansion processes. Upward reciprocating forces dominate as the piston approaches TDC during the valve overlap period, and downward reciprocating forces dominate each time the piston approaches BDC. For highly loaded low speed engines a keystone design is often used to maximize bearing area on the upper surface of the piston pin boss and the lower surface of the connecting as shown in Fig. 15.10. In these engines the downward loading due to cylinder pressure dominates over that generated by the reciprocating forces. The keystone geometry allows bearing area to be scaled based on the relative magnitudes of the pressure and reciprocating loads. Piston pin deflection in the radial direction is also shown in Fig. 15.9. Ovalizing of the pin in the circular pin boss results in tensile stress on either side of the applied load.

Increasing the pin stiffness may be effective in managing the pin boss stresses depicted in Fig. 15.9. However it must be remembered that increasing pin stiffness will also increase stresses in the piston crown. Although the various regions of the piston are being discussed separately in this chapter it is important to remember that the different regions cannot be developed independently. Further approaches that are commonly used to improve piston durability at the pin boss are shown in Fig. 15.11. It is common to taper the piston pin boss diameter, with the magnitude of the taper selected to provide even loading with the measured or calculated pin deflection. Cold forming the pin boss results in a compressive residual stress that may be effective toward managing the tensile stress seen because of pin ovalizing under load. Ovalizing the pin bore and adding side-reliefs to keep the load from being distributed around the pin boss are also effective.

The approaches just described may be used in conjunction with one another for further benefit. Pin bore unit loading is defined as the maximum force acting at the pin bore

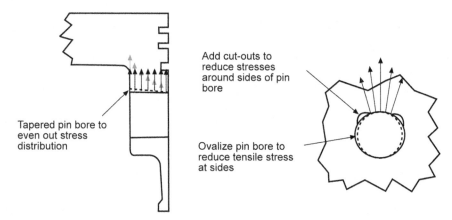

Fig. 15.11 Design modifications for increased pin boss durability

divided by the available bearing area, and is an important pin boss durability measure. As an example of the effect of the design modifications just discussed, for a given piston the maximum pin bore unit loading with a drilled, round pin bore was determined to be 35MPa. Cold forming the bore allowed the maximum unit loading to be increased to 50MPa. Adding side reliefs increased the maximum loading to 60MPa, and tapering and ovalizing the bore further increased the maximum pin bore unit loading to 75MPa. For this example, the combination of techniques more than doubled the load carrying capability of the pin boss while maintaining the required fatigue life. The techniques to be used in a given design are determined by the necessary load carrying capability, the required fatigue life, and the cost of each process. Because the loading at the piston pin is well understood, and consists of high-cycle mechanical loads at an elevated but relatively constant temperature, rig testing is very effective for design validation. Especially with aluminum pistons it is important to elevate the temperature for such testing because of the strong temperature-dependence of the fatigue life.

15.4   Piston Skirt Development

The slider-crank mechanism of the reciprocating piston engine is designed to harness the linear force generated by pressure acting on the piston crown and convert it to rotational force, or torque, at the crankshaft. As the linear force acting along the cylinder axis is transferred from the piston into the connecting rod a side force is generated, the magnitude of which at any point in time is dependent on the length and instantaneous angle of the connecting rod. As a result of the combination of pressure and side forces the piston pivots about the piston pin. This *secondary motion* of the piston varies throughout the operating cycle based on the combination of pressure and side forces acting on the piston. This is shown in Fig. 15.12, where the secondary motion for a particular engine is exaggerated throughout the operating cycle. The side forces are highest during combustion and early in the expansion process. The portion of the skirt experiencing the side force at this time is

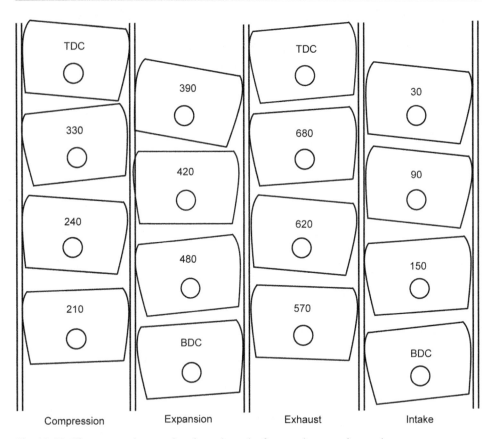

Fig. 15.12 Piston secondary motion throughout the four-stroke operating cycle

referred to as the ***major thrust*** side; its opposite is referred to as the ***minor thrust*** side. A slight offset of the piston pin is often used to create a torque that counteracts the secondary motion of the piston reducing the peak side force. It must be recognized however that the piston secondary motion is dependent on instantaneous forces and geometry, so modifications that impact the peak forces impact the forces and secondary motion throughout the operating cycle. Offsetting the crankshaft from the cylinder centerline is another option sometimes used to change the secondary motion trace made by the piston throughout the operating cycle.

Piston-to-cylinder wall clearance is an important element of skirt design. The piston sees higher operating temperatures than does the surrounding cylinder, and in the case of aluminum pistons and an iron block the pistons will have a significantly greater thermal expansion coefficient. This combination of factors results in large changes in piston-to-wall clearance with engine load. Enough clearance must be maintained to avoid scuffing at high loads, however too much clearance results in excessive secondary motion and skirt impact loading against the cylinder wall. Such impact loads contribute to engine noise, and may result in skirt cracking. In order to best optimize the piston-to-wall clearance a

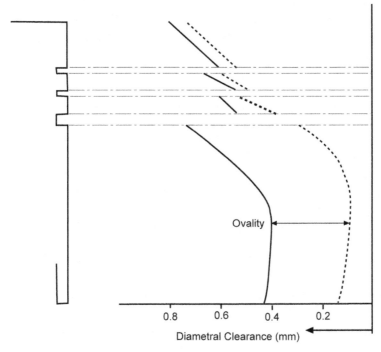

Fig. 15.13 Piston diameter variation and cold clearance

varying piston diameter is machined as the axis is traversed. This is shown in Fig. 15.13. As can be seen in the figure, the "cold" piston-to-wall clearance is much greater near the crown, where high temperatures will give rise to the greatest thermal expansion. The clearance is progressively reduced with position down the skirt, and then increased again near the base of the skirt in order to ensure an adequate oil film between the piston and cylinder wall. It should also be noted from the figure that the piston diameter is not the same along the thrust axis as it is along the piston pin axis. The piston tends to bend about the pin axis under cylinder pressure loading and thus requires greater clearance along the thrust axis. The profiles shown in the figure are often determined from experience. Increasingly finite element modeling is used to predict thermal and mechanical distortion and aid in determining the ideal cold, or machined, clearances. Validation should include visual evaluation for evidence of scuff after engine operation under high-temperature over-loaded conditions. Rig testing, again under elevated temperature conditions may be used to load and unload the piston skirts for validation against fatigue cracking.

15.5   Piston Ring Construction

The fundamental purpose of the ring pack is to provide a seal at the moving wall of the combustion chamber, keeping combustion gases from entering the crankcase and lubricating oil from entering the combustion chamber. Another, nearly equally important purpose is to control the oil film thickness on the cylinder wall, providing sufficient oil to minimize friction and wear while not providing more oil than is necessary and as a result, increasing oil consumption. Most ring packs today consist of three rings as shown in Fig. 15.14—two compression rings and an oil control ring. The compression rings may be of rectangular, keystone, or half-keystone cross-section as discussed previously, where the keystone rings are capable of handling higher cylinder pressure, less prone to deposit build-up and sticking, but result in greater ring and groove wear. The top compression ring surface in contact with the cylinder wall is most often barrel faced. This geometry allows the ring to readily lift to a hydrodynamic oil film, and provides an effective compression seal. The second compression ring typically has a wedge shaped face in contact with the cylinder wall. The wedge is downward scraping at an angle of one to two degrees from the cylinder wall, and aids in controlling the oil film thickness as well as in providing the combustion seal.

The compression rings are cast from ductile iron or steel, and their running faces are usually coated with a high hardness material such as chromium or sometimes a ceramic. Assembly of the rings into the piston grooves requires each ring to be split. Effort has long focused on minimizing the effect of these *end gaps*. However, recent findings pertaining to axial ring dynamics have demonstrated that the controlled flow of blow-by gases through the end gaps is important to stable operation. This will be explained further in the next section.

While there is a much wider variety of oil control ring construction each of the designs share the same operating principles. In each case the oil ring consists of two parallel scraping rings whose outward force controls the oil film thickness. These scrapers are separated by a flexible spacer. In some designs each of these elements is a separate piece while in others they are molded as a single piece. There may also be a separate tensioning spring controlling the outward force. In any case the spacer between the two scrapers is designed with sufficient flexibility that the spacers are not constrained to move in and out together; this is required to make both spacers contribute effectively to oil film thickness control. The spacer also consists of sufficient open space to allow oil to flow ahead of the oil ring scrapers back to the crankcase. In some cases drillings or slots behind the oil ring are used to allow oil to drain directly back to the crankcase. When slots are used this is also done to create a heat sink and reduce heat transfer into the piston skirt for cooler operation.

15.6 Dynamic Operation of the Piston Rings

While it remains extremely difficult to characterize ring pack operation in an operating engine, an understanding of the physical processes combined with detailed measurements and careful engineering judgment provides valuable insight. Returning to Fig. 15.14 a general summary of ring operation on the downward and upward strokes of the piston is depicted. As the piston approaches TDC the cylinder wall is typically flooded with more oil than is necessary. On the downward stroke the region between the piston skirt and cylinder wall is filled with oil. The oil control ring is responsible for pushing excess oil down

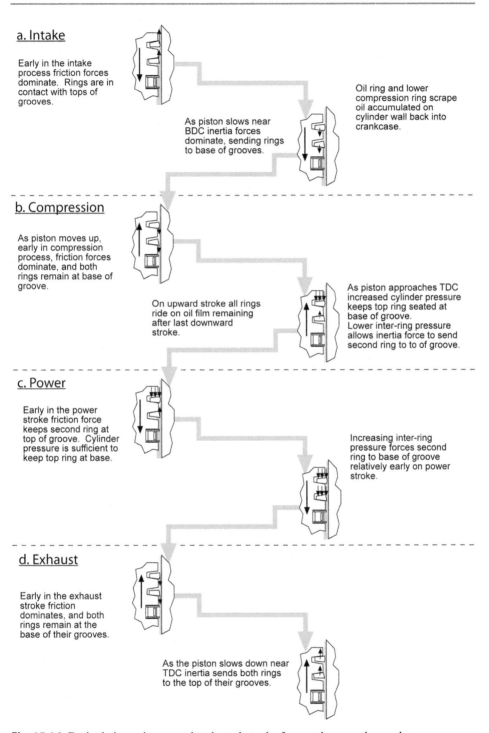

a. Intake

Early in the intake process friction forces dominate. Rings are in contact with tops of grooves.

As piston slows near BDC inertia forces dominate, sending rings to base of grooves.

Oil ring and lower compression ring scrape oil accumulated on cylinder wall back into crankcase.

b. Compression

As piston moves up, early in compression process, friction forces dominate, and both rings remain at base of groove.

On upward stroke all rings ride on oil film remaining after last downward stroke.

As piston approaches TDC increased cylinder pressure keeps top ring seated at base of groove. Lower inter-ring pressure allows inertia force to send second ring to to of groove.

c. Power

Early in the power stroke friction force keeps second ring at top of groove. Cylinder pressure is sufficient to keep top ring at base.

Increasing inter-ring pressure forces second ring to base of groove relatively early on power stroke.

d. Exhaust

Early in the exhaust stroke friction dominates, and both rings remain at the base of their grooves.

As the piston slows down near TDC inertia sends both rings to the top of their grooves.

Fig. 15.14 Desired piston ring operation throughout the four-stroke operating cycle

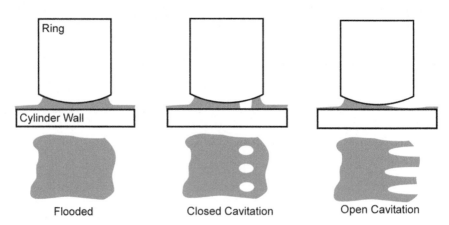

Fig. 15.15 Oil film distribution scenarios

the cylinder wall ahead of the ring pack, leaving an oil film of the desired thickness in its wake. The wedge shape of the second compression ring pushes additional oil ahead of it, and the top compression ring rides on the remaining oil film. It is important to note that the oil film available to provide hydrodynamic lubrication to the rings on the next up-stroke of the piston is only that remaining adhered to the cylinder wall after the preceding down-stroke. The oil film does not burn off the cylinder during the combustion and expansion process, and it is important that the oil adhere to the cylinder wall without running down and leaving portions of the surface uncovered.

Each ring exerts a circumferential outward force on the oil film, and it is important that the force be as uniform as possible. It is also important that circumferential cylinder distortion be minimized. These two criteria, along with conformability of the ring to any remaining cylinder wall distortion are necessary in order to ensure a complete oil film. Various scenarios are depicted in Fig. 15.15. A continuous oil film around the circumference of each ring is required for combustion sealing. Axial distortion of the cylinder wall is also important as excessive axial distortion may result in the rings momentarily jumping away from the cylinder wall, resulting in loss of the combustion seal.

Looking further at the outward circumferential force exerted by the rings, and the resulting oil film thickness it is found that the radius of the barrel face on the top compression ring governs a trade-off between how quickly the ring picks up oil and becomes fully hydrodynamic, and how quickly the film collapses at the piston approaches either end of the stroke. This trade-off is further depicted in Fig. 15.16.

It is now important to look at the axial dynamics of ring pack operation. Axial motion of the two compression rings is determined by the balance of pressure, friction, and inertia forces acting on the rings. Returning once again to Fig. 15.14, if one begins with the intake stroke shown in Fig. 15.14a, as the piston moves downward friction forces between the rings and cylinder wall are dominant, and both compression rings ride in contact with the tops of the piston grooves. As the piston slows down near the end of the intake stroke the friction force drops, and the inertia of the rings sends them both to the base of their grooves.

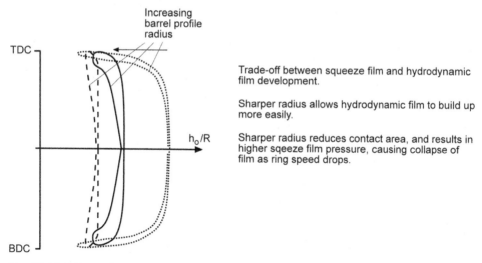

Increasing barrel profile radius

TDC

BDC

h_o/R

Trade-off between squeeze film and hydrodynamic film development.

Sharper radius allows hydrodynamic film to build up more easily.

Sharper radius reduces contact area, and results in higher sqeeze film pressure, causing collapse of film as ring speed drops.

Fig. 15.16 Effects of top compression ring barrel face geometry on oil film thickness along stroke

Early in the compression process, shown in Fig. 15.14b, the friction forces are again dominant, now holding the rings at the base of their grooves. The pressure in the combustion chamber, and thus above the top compression ring, builds throughout the compression process. The pressure is sufficient to overcome the top ring inertia force, and when the piston slows down near TDC the top ring remains firmly against the base of the groove. With the in-cylinder pressure rise the pressure begins building between the top and second rings due to flow of combustion gas through the top ring end gap. However, late in the compression stroke the pressure is not yet high enough to hold the second ring at the base of the groove, and the ring's inertia force sends it back to the top of the groove.

As the expansion stroke shown in Fig. 15.14c begins the in-cylinder pressure is high enough to overcome the friction force at the top ring and the ring remains seated at the base of the groove. Inter-ring pressure between the top and second rings continues to build, and early in the expansion stroke the pressure becomes sufficient to overcome the friction force and send the second ring to the base of its groove. As the expansion process continues the in-cylinder pressure begins to fall, but by the time the pressure drops below that necessary to keep the rings at the base of their grooves the piston is slowing down and the inertia forces increase, aiding the remaining pressure forces in keeping the rings seated.

Finally, the exhaust stroke is shown in Fig. 15.14d. Early in the exhaust stroke the friction forces between the rings and cylinder wall keep the rings at the base of their grooves. Late in the exhaust stroke the falling piston speed results in ring inertia forces sending both compression rings back to the top of the grooves.

The ring movement just described is the desired axial ring motion. One of the goals of ring pack development is to accomplish this desired motion throughout the entire range of speeds and loads over which the engine is to be operated. However, the relative

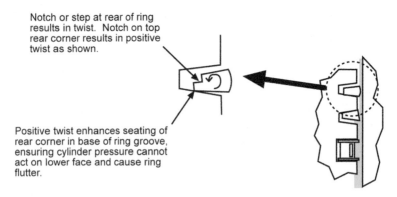

Notch or step at rear of ring
results in twist. Notch on top
rear corner results in positive
twist as shown.

Positive twist enhances seating of
rear corner in base of ring groove,
ensuring cylinder pressure cannot
act on lower face and cause ring
flutter.

Fig. 15.17 Control of top ring flutter through positive twist

magnitudes of the forces just described vary considerably. The friction forces are strongly dependent on speed and have some load dependence (due to pressure acting on the backs of the rings). The inertia forces for a given ring design are solely dependent on speed, and the pressure forces are almost entirely dependent on load. Clearly it is very difficult to obtain the desired ring pack performance over the entire engine operating map. The ring pack is developed for optimal performance over the portions of the map where the engine is expected to spend the majority of its time, and acceptable performance over remaining operating regimes.

It is now instructive to review a few important deviations from ideal operation, and approaches that are taken to address them. As the pressure in the cylinder builds during compression and combustion it is possible for the rear edge of the top ring to lift from the base of its groove. If this happens pressure begins to act underneath the ring, rapidly changing the balance of forces, and allowing the ring to jump to the top of the groove. When the ring contacts the top of the groove pressure is shut off to the underside of the ring, again changing the balance of forces and causing the ring to drop. The resulting instability is referred to as *top ring flutter*. The most common solution to top ring flutter is to add *positive twist* to the ring as depicted in Fig. 15.17. This is done by adding a notch or bevel cut to the top rear edge of the ring, causing it to preferentially twist in the direction shown in the figure. The positive twist increases the contact force along the rear edge of the ring to ensure a complete seal.

The second ring may also experience instability (*second ring flutter*). In this case, as the second ring moves to the base of its groove during the expansion stroke the volume between the top and second rings increases. This results in a drop of inter-ring pressure that may allow the top ring to jump back up, again causing the pressure to build, and repeating the cycle. In the case of the second ring either positive or *negative twist* may be used, depending on whether the objective is to place the ring at the top or bottom of the groove. Another approach sometimes taken is to cut material from the land between the top and second rings, thus reducing the rate of pressure build-up, and reducing the volume change associated with ring movement.

Visual assessment of the rings from engines under various operating conditions includes not only face wear but ring and groove wear due to axial motion, and evidence of complete circumferential contact between each compression ring and the lower groove surface. Techniques have been developed for in-situ measurement of inter-ring pressure, instantaneous axial ring motion, ring rotation, and time- and location-specific oil film thickness. However, all of these remain research techniques not routinely available for product development. Total engine blow-by measurements are routinely made with simple flow meters assessing flow from the crankcase. Oil consumption measurements typically require careful monitoring over many hours of operation. While analytical models of ring dynamics regularly appear in the literature, and several commercial models are available these must be used with great care and even then are quite suspect. Such models require very detailed geometric description (including distortion), accurate lubricant properties versus temperature, blow-by gas flow modeling, and a friction model that includes not only hydrodynamic lubrication but mixed film lubrication as well. The results are extremely sensitive to each of these models, none of which is easily validated. As a result, the models are often demonstrated to predict incorrect trends even after every possible care has been taken in their validation.

15.7 Cylinder Wall Machining

One of the requirements for a successful piston and ring pack design is that a lubricating oil film adhere to the cylinder wall behind the piston's downward stroke, and thus be available to provide a hydrodynamic film on the subsequent upward stroke. In order to meet this requirement a smooth surface finish cannot be used, even though this would be desirable to minimize friction and wear. The ideally machined cylinder wall is one with a smooth running surface for minimized friction and closely spaced grooves for oil retention. The desired surface is characterized in the plot shown in Fig. 15.18. The plot shows the percentage of any given section of the cylinder wall surface having grooves at various depths. A small percentage of the grooves will be very shallow, and will disappear during engine run-in. As shown in the figure these grooves have a depth of less than 1.5 micron, and typically make up less than ten percent of the grooves. The vast majority of the grooves are in the range from depths of 1.5 to 3 microns, and make up the running cylinder wall surface. Most of the remaining grooves range in depth from 3 to 8 microns, and determine the oil retention capability of the surface.

A cross-hatch pattern of grooves is desired, with the grooves placed at an angle of 15–35° from the horizontal. The groove angle is controlled by the ratio of hone speed to feed rate. In some cases a plateau honing process is used. An initial honing operation is followed by a finer hone that removes the peaks to create a smooth operating surface and leaves the valleys for oil retention. The resulting surface characterization is shown in Fig. 15.19. This higher cost approach reduces wear during run-in, allowing more rapid ring seating and reduced friction during run-in. The longer term effects of this process are not conclusive.

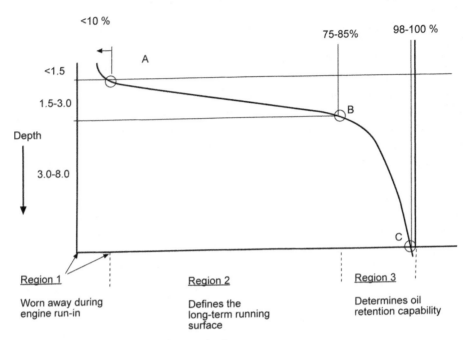

Fig. 15.18 Cylinder wall surface characterization

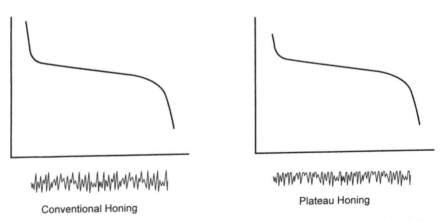

Fig. 15.19 Effects of conventional and plateau honing on cylinder wall surface characteristics

The problems posed by both axial and circumferential distortion were identified in the preceding section. Because the ring is no longer in contact around the entire circumference of the cylinder friction is actually reduced, but at a severe price. The distortion is seen to reduce minimum oil film thickness, thus increasing wear under mixed film operation and significantly increase blowby. It has been found experimentally that severe distortion results in ring rotation until the end gaps align, followed by a significant further blowby increase. This result should not be surprising. Under normal operation the rings will turn in their grooves, especially at high speeds and light loads. As load increases the rings stop turning due to the increased pressure holding them in contact with the grooves. However, because of the end gaps, the outward tension on the rings is not completely uniform, but peaks at the gaps. As the rings turn the distorted cylinder wall results in a preferred location that tends to be the same for both gaps.

Cylinder wall distortion is impacted by a combination of machining process capability, mechanical distortion primarily due to head bolt clamping loads, and thermal distortion. The first can be addressed only through improved machine and fixture capability. Mechanical distortion is sometimes addressed through the use of a torque plate during machining. The torque plate has similar stiffness to the cylinder head, and is bolted in place prior to machining. It has been successfully applied to parent bore engines, but is problematic in engines having removable liners. Thermal distortion is best addressed through careful placement of the cooling jackets and control of coolant flow. The problem is more complicated than simply minimizing temperature variation around the perimeter of each cylinder, as it involves temperature, thermal expansion, and block stiffness. A combination of temperature measurement and block distortion measurement or analysis is required.

Rig tests using small sample sections of rings and cylinder walls are commonly used for base material, coating, and manufacturing process development. Such tests can be used with varying loads applied at the ring to cylinder wall contact, and various lubricant conditions and temperatures.

15.8 Recommendations for Further Reading

These recommendations lead off with an excellent recent paper on piston loading. It includes experiment and analysis, and while it was done on a diesel engine, many of the approaches and findings are more broadly applicable. The next paper is a nice recent example of piston crown fatigue and failure analysis. The remaining papers in this section provide further examples of thermal and mechanical analysis (see Kenningly and Morgenstern 2012; Morgan et al. 2013; Mizuno et al. 2009; Reipert and Voigt 2001; Keribar et al. 1993; Keribar and Morel 1987).

An important aspect of piston design is secondary motion and piston slap. This aspect affects engine noise, piston durability, and friction and lubrication. The following five papers are recent analytical and experimental studies addressing this topic (see Takahashi et al. 2013; Offner et al. 2001; Teraguchi et al. 2001; Hoffman et al. 2003; Stout et al. 2003).

The following papers address recent approaches to piston design and materials. The first two discusses advances in aluminum alloy pistons. This is followed by a paper addressing trade-offs between aluminum and steel in automobile diesel engines. The fourth presents the current state-of-the-art in heavy-duty diesel pistons as it discusses single piece thin wall iron pistons in comparison to other approaches (see Spangenberg et al. 2013, 2014; Barnes and Lades 2002; Kemnitz et al. 2000).

The following papers discuss design optimization pertaining to the piston pin and pin boss. The first is an older paper on a topic on which very little has been published. The approaches described in this paper remain quite relevant. This is followed by a pin motion study in a full-floating design, and two companion papers that revisit pressed pin designs (see Fletcher-Jones et al. 1986; Clark et al 2009; Kondo and Ohbayashi 2012; Nishikawa 2012).

An interesting experimental study of skirt design is reported in the following paper (see Kim et al. 2009):

Each of these papers discuss aspects of ring design. The first two deal with compression rings, and the third addresses oil control rings (see Mihara and Inoue 1995; Murray et al. 1986; Reid et al. 1986).

Ring dynamics is an important and little understood aspect of engine development. The following three papers discuss ring dynamic analysis and measurement. These are followed by two papers describing measurement techniques that address oil film thickness and inter-ring pressures and blowby respectively (see Tomanik and Nigro 2001; Furuhama et al. 1979; Veettil and Shi 2011; Richardson and Borman 1991; Dursunkaya et al. 1993).

Finally, quite a bit of attention is now being placed on cylinder wall surfaces and coatings. These papers discuss the topics of cylinder wall surface finish, coatings, and bore distortion (see Urabe et al. 2014; Morawitz et al. 2013; Rejowski et al. 2012; Verpoort et al. 2008; Ghasemi 2012; Maassen et al. 2001; Lu et al. 2004; Schneider et al. 1993).

References

Barnes, S.J., Lades, K.: The Evolution of Aluminum Based Piston Alloys for Direct Injection Diesel Engines. SAE 2002-01-0493 (2002)

Clark, K., Antonevich, J., Kemppainen, D., Barna, G.: Piston Pin Dynamics and Temperature in a C.I. Engine. SAE 2009-01-0189 (2009)

Dursunkaya, Z., Keribar, R., Richardson, D.E.: Experimental and Numerical Investigation of Inter-Ring Gas Pressures and Blowby in A Diesel Engine. SAE 930792 (1993)

Fletcher-Jones, D., Adams, D.R., Barraclough, E., Avezou, J.C.: Optimization of Piston Pin Hole Shape. AE Group Symposium '86, Paper No. 31 (1986)

Furuhama, S., Hiruma, M., Tsuzita, M.: Piston Ring Motion and its Influence on Engine Tribology. SAE 790860 (1979)

Ghasemi, A.: CAE Simulations for Engine Block Bore Distortion. SAE 2012-01-1320 (2012)

Hoffman, R.M., Sudjianto, A., Du, X., Stout, J.: Robust Piston Design and Optimization Using Piston Secondary Motion Analysis. SAE 2003-01-0148 (2003)

Kemnitz, P., Maier, O., Klein, R.: Monotherm, A New Forged Steel Piston Design for Highly Loaded Diesel Engines. SAE 2000-01-0924 (2000)

Kenningly, S., Morgenstern, R.: Thermal and Mechanical Loading in The Combustion Bowl Region of Light Vehicle Diesel Alsicunimg Pistons; Reviewed with Emphasis on Advanced Finite Element Analysis and Instrumented Engine Testing Techniques. SAE 2012-01-1330 (2012)

Keribar, R., Morel, T.: Thermal Shock Calculations in I.C. Engines. SAE 870162, (1987)

Keribar, R., Dursankaya, Z., Ganapathy, V.: An Integrated Design Analysis Methodology to Address Piston Tribological Issues. SAE 930793 (1993)

Kim, K.-s., Shah, P., Takiguchi, M., Aoki, S.: Part 3: A Study of Friction and Lubrication Behavior for Gasoline Piston Skirt Profile Concepts. SAE 2009-01-0193 (2009)

Kondo, T., Ohbayashi, H.: Study of Piston Pin Noise of Semi-Floating System. SAE 2012-01-0889 (2012)

Lu, S., Iyer, K., Hu, S.J.: Functional Characterization of Surface Roughness Generated by Plateau Honing Process Using Wavelet Analysis. SAE 2004-01-1558 (2004)

Maassen, F., Koch, F., Schwaderlapp, M., Ortjohann, T., Dohmen, J.: Analytical and Empirical Methods for Optimization of Cylinder Liner Bore Distortion. SAE 2001-01-0569 (2001)

Mihara, K., Inoue, H.: Effect of Piston Top Ring Design on Oil Consumption. SAE 950937 (1995)

Mizuno, H., Ashida, K., Teraji, A., Ushijima, K., Takemura, S.: Transient Analysis of the Piston Temperature with Consideration of In-Cylinder Phenomena Using Engine Measurement and Heat Transfer Simulation Coupled with Three-Dimensional Combustion Simulation. SAE 2009-01-0187 (2009)

Morawitz, U., Mehring, J., Schramm, L.: Benefits of Thermal Spray Coatings in Internal Combustion Engines, with Specific View on Friction Reduction and Thermal Management. SAE 2013-01-0292 (2013)

Morgan, W., Barnes, S., Ryu, K.H., Jun, S., Shim, W.: A Non-Linear Finite Element Approach Applied to Diesel Piston Combustion Bowl Rim Strength Assessment. SAE 2013-01-0293 (2013)

Murray, E.J., Holt, J.W., Inwood, B.C., Revello, P.L., Cecchi, L.: The Development of Compression Ring Design and Surface Treatments for Future Automotive Engines. AE Group, Symposium '86, Paper No. 26 (1986)

Nishikawa, C.: Optimization of Semi-Floating Piston Pin Boss Formed by Using Oil-Film Simulations. SAE 2012-01-0908 (2012)

Offner, G., Herbst, H.M., Priebsch, H.H.: A Methodology to Simulate Piston Secondary Movement Under Lubricated Contact Conditions. SAE 2001-01-0565 (2001)

Reid, T.J., Haisell, O., Plant, R.: Design Features of Oil Control Rings. AE Group. Symposium '86, Paper No. 27 (1986)

Reipert, P., Voigt, M.: Simulation of the Piston/Cylinder Behavior for Diesel Engines. SAE 2001-01-0563 (2001)

Rejowski, E.D., Mordente, P., Pillis, M.F., Casserly, T.: Application of DLC Coating on Cylinder Liners for Friction Reduction. SAE 2012-01-1329 (2012)

Richardson, D.E., Borman, G.L.: Using Fiber Optics and Laser Fluorescence for Measuring Thin Oil Films with Application to Engines. SAE 912388 (1991)

Schneider, E.W., Blossfeld, D.H., Lechman, D.C., Hill, R.F., Reising, R.F., Brevick, J.E.: Effect of Cylinder Bore Out-of-Roundness on Piston Ring Rotation and Engine Oil Consumption. SAE 930796 (1993)

Spangenberg, S., Adelmann, J., Hettich, T., Hammen, A.: Lightweight Pistons for Gasoline Engines with Optimized Frictional Loss. 34th Internaional Vienna Motor Symposium (April 2013)

Spangenberg, S., Hettich, T., Lazzara, M., Schreer, K.: Pistons for Passenger Car Diesel Engines—Aluminum or Steel? 35th International Vienna Motor Symposium (April 2014)

Stout, J.L., Williams, R., Hoffman, R.: Eliminating Piston Slap Through A Design For Robustness CAE Approach. SAE 2003-01-1728 (2003)

Takahashi, M., Isarai, R., Hara, H.: Measurement of Piston Secondary Motion Using The New Digital Telemeter. SAE 2013-01-1708 (2013)

Teraguchi, S., Suzuki, W., Takiguchi, M., Sato, D.: Effects of Lubricating Oil Supply on Reductions of Piston Slap Vibration and Piston Friction. SAE 2001-01-0566 (2001)

Tomanik, E., Nigro, F.E.B.: Piston Ring Pack and Cylinder Wear Modeling. SAE 2001-01-0572 (2001)

Urabe, M., Takakura, T., Metoki, S., Yanagisawa, M., Murata, H.: Mechanism of and Fuel Effi-
    ciency Improvement by Dimple Texturing on Liner Surface for Reduction of Friction Between
    Piston Rings and Cylinder Bore. SAE 2014-01-1661 (2014)
Veettil, M.P., Shi, F.: CFD Analysis of Oil/Gas Flow in Piston Ring Pack. SAE 2011-01-1406 (2011)
Verpoort, C., Bobzin, K., Ernst, F., Richardt, K., Schlaefer, T., Schwenk, A., Cook, D., Flores, G.,
    Blume, W.: Thermal Spraying of Nano-Crystalline Coatings for Al-Cylinder Bores. SAE 2008-
    01-1050 (2008)

Cranktrain (Crankshafts, Connecting Rods, and Flywheel)

16

16.1 Definition of Cranktrain Function and Terminology

The Cranktrain is at the heart of the reciprocating piston engine, and its purpose is to translate the linear motion of the pistons into rotary motion for the purpose of extracting useful work. The cranktrain is typically composed of connecting rods, the crankshaft, and a flywheel or power takeoff device.

The crankshaft is usually composed of one or multiple throws, to which the connecting rods are attached with either fluid film journal bearings as shown in Fig. 16.1 or in some cases rolling element bearings. The crankshaft is typically cast or forged of iron or steel. In addition to its main function of translating linear to rotary motion, the crankshaft drives many of the engine accessories including the valvetrain, oil pump, water pump, and charging system. Crankshaft rotation direction can be either clockwise or counterclockwise, depending on driveline packaging and requirements. In most automobile applications the rotation direction is clockwise as viewed from the crankshaft nose and front of the engine. The crankshaft typically incorporates additional balance weights as have been discussed in Chap. 6.

The connecting rod in its simplest form is a beam with a pin joints at each end as shown in Fig. 16.2. The *small end* sees stop-and-start motion as it is connected to the piston, while the *big end*, attached to the crankshaft sees high rotation speeds. The connecting rod supports bearings or bushings at each end. It is one of the most stressed components in the engine and subjected to high tensile, compressive, and bending stress. It can be cast, forged, or sintered from powdered metal.

The flywheel serves several functions. First, it adds additional rotating inertia to the crankshaft to reduce the cyclical speed variation produced by the reciprocating pistons and intermittent combustion events. When attempting to move a vehicle from stationary, the flywheel provides stored energy to overcome the inertia of the vehicle for smooth

© Springer Vienna 2016
K. Hoag, B. Dondlinger, *Vehicular Engine Design*, Powertrain,
DOI 10.1007/978-3-7091-1859-7_16

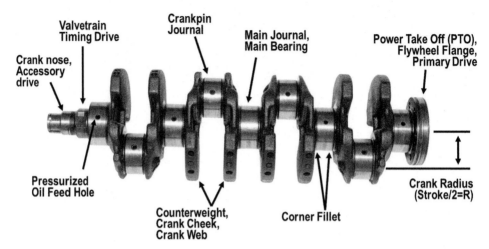

Fig. 16.1 Crankshaft terminology

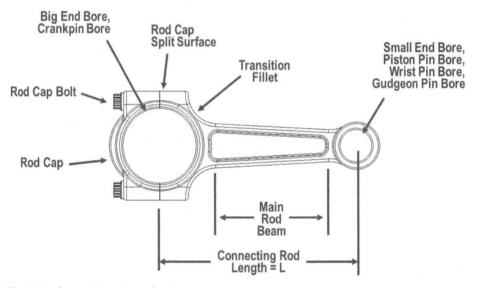

Fig. 16.2 Connecting rod terminology

starts. If the inertia of the cranktrain is too low, the engine may stall or unnecessarily high engine speed will be required to launch the vehicle from rest. A further function of the flywheel is power transfer to the vehicle drivetrain; the clutch assembly or torque converter is mounted to the flywheel, and contributes to the total flywheel mass. A final function of the flywheel is to hold the ring gear against which the starter will engage. An example flywheel bolted to the rear crankshaft flange of a large diesel engine is shown in Fig. 16.3.

Fig. 16.3 Flywheel mounted
to crankshaft

Fig. 16.4 Single-plane
crankshaft

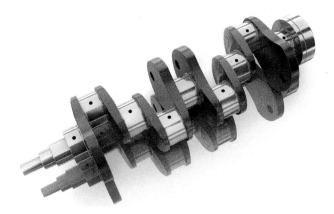

16.2   Description of Common Cranktrain Configurations and Architectures

16.2.1   Crankshaft Configurations

Crankshaft throw configurations were previously discussed in Chap. 6, as determined by engine balance and firing order requirements. Multi-throw crankshafts can be arranged in either a single plane or multiple planes. A single-plane or flat-plane crankshaft has all crankpin throws at either 0 or 180°, as shown with the in-line four cylinder example in Fig. 16.4. A multi-plane crankshaft has throws in more than one plane; the cruciform arrangement used in most V-8 engines is given as an example in Fig. 16.5.

The vast majority of engines use a single piece crankshaft that is cast or forged and then machined to achieve its final geometry. Each of the examples shown in Figs. 16.1, 16.4 and 16.5 are single piece crankshafts. In very low volumes a single-piece crankshaft might be machined from billet steel. The single-piece crankshaft requires that both the main and rod bearings are split for assembly. Connecting rod and cap design will be covered later in

Fig. 16.5 Dual-plane
crankshaft

Fig. 16.6 Assembled
crankshaft

this chapter. Various approaches for accomplishing the split main bearings were covered in Chap. 8 on cylinder block design.

An alternative to the single-piece crankshaft is the assembled crankshaft such as that shown in Fig. 16.6. The assembled crankshaft is composed of multiple pieces that were manufactured separately, then joined together to form a single crankshaft by means of a press fit, bolted, or welded joint.

An assembled crankshaft is typically used on either very small or very large engines to ease the manufacture. The manufacture of small crankshafts, typically single throw, can be made less expensive since the creation of multiple circular pieces are made on a simple lathe and are then centerless ground. The requirement for a more expensive offset crankshaft grinding machine is eliminated. This also makes manufacture of very large crankshafts for ships and stationary pumping engines easier, as they can be made of more easily managed 'small' pieces.

One of the key challenges for the assembled crankshaft is aligning all of the sub-components during assembly to a high tolerance. This usually requires the application of large force to assemble and then straighten the components to achieve the desired straightness

Fig. 16.7 Plain connecting rod

Fig. 16.8 Angled split connecting rod

Fig. 16.9 One piece connecting rod

of the crankshaft for the alignment of bearings. The key failure mode of an assembled crankshaft is the misalignment of these components during operation. The torque of the engine may cause the assembled joints to slip with respect to one another, and "scissor", leading to misalignment of the crankshaft.

16.2.2 Connecting Rod Configurations

There are four basic configurations for the big end or crankpin bore of the connecting rod as shown in Figs. 16.7, 16.8, 16.9, and 16.10.

1. Plain (most common),
2. Angled split (allowing assembly of larger diameter rod bearings),
3. One piece (for rolling element bearings), and
4. Articulated (for radial engines).

Fig. 16.10 Articulated connecting rod

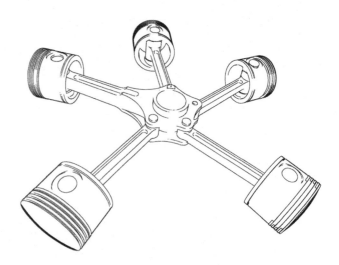

The plain connecting rod is most common in current automotive practice. An angle split
connecting rod allows a larger diameter rod bearing while maintaining capability to remove and replace the connecting rod through the top of the engine. This is common in
over the highway trucks and locomotives, which have a large crankpin and consequently
large bearing. A disadvantage of an angled split rod is non-symmetrical loading of the connecting rod crankpin bore. The rod cap is now at an angle to the loading, and it makes the
cap slide with respect to the upper rod, thus requiring additional attention to cap locating.
A one-piece connecting rod allows a rolling element bearing without a split cage to be
used to reduce friction and lower lubricant flow. However, the one-piece connecting rod
is limited to engines with an assembled crankshaft. An articulated rod is used for radial
engines where more than one connecting rod must attach to the same crank journal. This
arrangement reduces the axial length of the crankpin journal for multiple cylinders, and is
most common in air-cooled radial aircraft engines.

Regardless of crankpin bore arrangement one of the requirements of the connecting
rod is to minimize crankpin bearing deflection a to ensure an even oil film and prevent
pinching of the rod bearing relative to the crankshaft. The connecting rod cap may have
reinforcing ribs to increase stiffness, as shown in Fig. 16.11. A single rib is easiest to cast
or forge, but a double rib provides a more efficient distribution of stiffness.

There are two general configurations of connecting rod beam cross-section. The I-
Beam is shown in Figs. 16.7 through 16.10, and again in Fig. 16.13. An H-Beam connecting rod is shown in Fig. 16.12. The beam is the portion of the connecting rod that connects
the piston pin and crankpin ends, and when viewed in cross section resembles a capital
letter "I" or capital letter "H".

The I-Beam connecting rod is most common in use because it puts the most material
in the areas of maximum bending stress while reducing material where it is not needed.
The I-Beam shape also lends itself well to the casting, forging, or Powdered Metal (PM)
processes. The pull direction for the beam is in the same direction of the pin bosses, making forming easier and reducing the requirement for material removal.

Fig. 16.11 Rib on connecting
rod cap

Fig. 16.12 H-beam

The H-Beam connecting rod is more difficult to manufacture, but this geometry offers a more stable, lighter design. The H-beam rod allows a more gradual transition of stiffness between the main beam and the crankpin end of the connecting rod, reducing stress in this critical radius. It also has greater bending stiffness in the crankpin axial direction than an I-bean rod. This may lead to increased edge loading on the crankpin bearing, when two rods share the same journal. Due to the added expense of manufacture, this is usually only seen in competition engines.

Fig. 16.13 Pressed small end bushing, tapered piston pin end geometry, splash oil hole

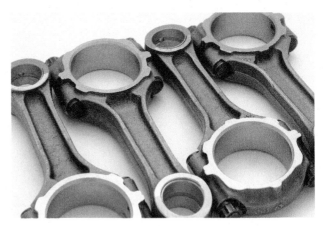

If the small-end boss is to be used with a full-floating piston pin a pressed bushing such as is shown in Fig. 16.13 is typical. If a bushing is not used, a coating is required on one of the two mating components to reduce wear. The small end of the rod sees only reversing, stop-and-start motion, resulting in mixed-film lubrication. In some engines an oil feed hole is added supplying the connecting rod piston pin bushing either as a passive splash oiling hole from the top or a pressurized lubricant drilling from the crankpin bearing.

16.2.3   Flywheel Configurations

The most common arrangement of flywheel in automotive applications is a separate disc attached to the end of the crankshaft through a bolted joint. The total cranktrain inertia is composed of the flywheel, the crankshaft, the rotating mass of the connecting rods, the accessory drive pulley and vibration dampener, and the clutch or torque converter. Occasionally the flywheel is incorporated as part of the crankshaft counterweights, the disc varying in thickness to achieve the desired balance counterweight. This is common practice in motorcycle applications and is seen in the assembled crankshaft shown in Fig. 16.6.

16.3   Detailed Design of Crankshaft Geometry

The geometry of a crankshaft is complicated, and the loading varies as a function of its rotational position. Because of the dissimilar cross-sections of the crankshaft, it will be sensitive to discontinuities in stiffness. It is at these stiffness discontinuities that stress will be concentrated. Generally, the crankshaft fillet will be the most highly stressed area and will require the most detailed design attention as the materials used to make crankshafts are typically sensitive to notch factors. Loading at the fillet alternates as shown in

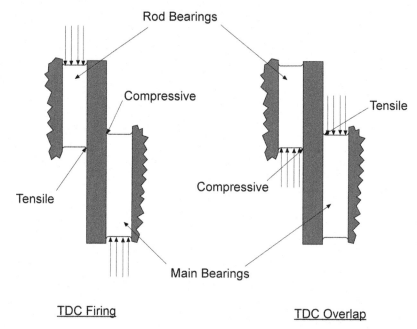

Fig. 16.14 Main and rod bearing loading near TDC firing (**a**) and TDC valve overlap (**b**)

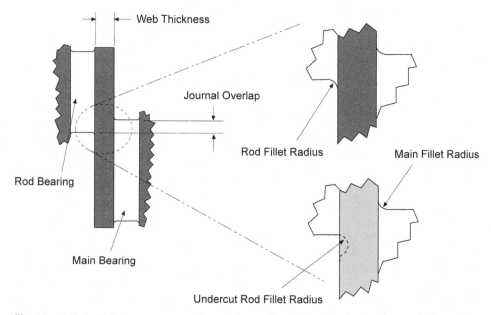

Fig. 16.15 Rod fillet stress versus crank angle for each cylinder of an in-line four-cylinder engine

Fig. 16.16 Critical crankshaft layout dimensions

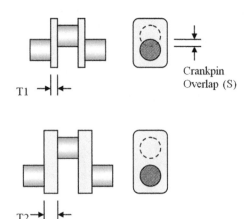

Figs. 16.14 and 16.15. TDC-Firing puts the crankpin in compression, and TDC-Overlap puts the crankpin in tension. Stress in this area is further magnified by torsional vibration, to be discussed later. Often metal improvement processes, such as shot peening or fillet rolling, are used to introduce residual compressive stress at the fillet radii to improve fatigue life.

Pin overlap, crankshaft web thickness, and journal fillet radius are the key parameters that are varied in crankshaft design. These variables are depicted in the layout in Fig. 16.16. A crankshaft having greater pin overlap (large journal diameters and short stroke) does not require as great a web thickness. A crankshaft that has little or no pin overlap will require a thicker web to gain stiffness.

One way to increase pin overlap, is to increase journal bearing diameter for a given engine stroke, as shown in Eq. 16.1. This will increase crankshaft stiffness, but increasing pin diameter increases fluid film friction at the bearings, reduces net engine power, and increases fuel consumption. It also increases crankshaft weight as the increased mass of the crankpin will need to be offset by increasing mass in the counterweights.

$$S_{Overlap} = \frac{(D_{crankpin} + D_{Main} - Stroke)}{2} \tag{16.1}$$

Where:

$S_{overlap}$ = Pin overlap
$D_{crankpin}$ = Diameter of the crankpin journal
D_{main} = Diameter of the main bearing journal

Fillet radii have enormous effects on the maximum stress in the crankshaft. Once the basic proportions of the crankshaft have been determined, the choice of these radii is critical.

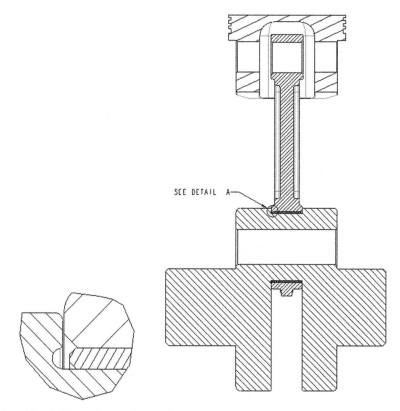

Fig. 16.17 Crank fillet radius, undercut fillet

The larger the fillet that can be used on the crankpin or main journal, the less the stress concentration notch factor will be. However, the larger the journal fillet, the less surface area is available for the connecting rod or main bearing as shown in Fig. 16.17. Several things can be done to minimize the stress at these fillet radii, but at additional cost. The geometry can be improved by using an undercut fillet as shown in the figure to increase the effective radius. The material properties can be improved by introducing residual compressive stresses by fillet rolling or shot peening the fillet radius.

When designing a new crankshaft, it is useful to have a starting point. This can come from benchmarking of successful crankshafts in similar applications, or from initial sizing values. These initial sizing values for the crankshaft were developed in the 1930's and 1940's in an era before computer aided design, and were developed by loading crankshafts and measuring displacement with extensometers or strain gauges. In modern practice, initial values will be assumed and the design will progress to three-dimensional structural analysis. Initial sizing guidelines are listed in Table 16.1. These values are given as a function of cylinder bore diameter.

Table 16.1 Initial sizing values

Feature	Initial sizing value
Cylinder bore diameter	D
Cylinder spacing	$1.20 \times D$
Crankpin diameter	$> 0.6 \times D$
Crankpin journal width	$0.35 \times D$, width/dia. > 0.3
Main journal diameter	$0.75 \times D$, $>$ pin dia.
Main journal width	$0.40 \times D$, width/dia. > 0.3
Web (cheek) thickness	$0.25 \times D$
Crankpin fillet	$0.04 \times D$, $> 0.05 \times$ journal dia.
Main fillet	$0.04 \times D$

The majority of external loads are applied to the crankshaft perpendicular to its rotational centerline, and the reaction forces are thus transmitted through the rod and main bearings. However, in addition to these loads the crankshaft is exposed to some thrust loading—loading applied along the axis of crankshaft rotation. Thrust loads occur as the clutch in a manual transmission application is engaged or disengaged. With an automatic transmission, the load transfer through the torque converter includes a thrust component. If the camshaft is driven with a gear train using helical gears a further thrust load is transmitted to the crankshaft. As the engine fires, the crankshaft throw deflects, and the main bearings spread axially. Finally, dimensional stack-up between the crankshaft and the connecting rods and cylinder bore centerlines results in a small thrust load. For all of these reasons the crankshaft must include a thrust bearing surface. This is typically provided in conjunction with one of the main bearings. Because the largest thrust loads are generated at the rear of the crankshaft the thrust bearing is often placed at or near the rear main bearing; for packaging reasons the second-to-rear main bearing is often used since the rear main bearing must also incorporate the rear oil seal. If the crankshaft stiffness is sufficient the middle main bearing may be chosen. This is done because the machined thrust surface provides the fore and aft datum for crankshaft machining. Placing this datum in the middle of the crankshaft allows the fore and aft tolerances to be split equally between the front and rear portions of the crankshaft, making machining process control easier.

16.4   Crankshaft Natural Frequencies and Torsional Vibration

Torsional vibration results whenever an unsteady or cyclical load is applied to a spinning shaft. A number of components in engines, including camshafts, water and oil pump drives, various accessory drives, and the crankshaft meet the criteria for potential torsional vibration problems. Of these the crankshaft receives by far the most attention since the combination of the length of the shaft and the magnitude of the forces result in the most severe conditions. Torsional vibration increases the stresses in the crankshaft webs, and

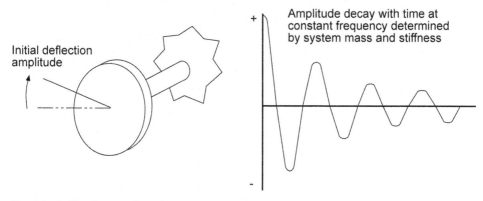

Fig. 16.18 Simple case of Torsional vibration with single degree of freedom

the resulting oscillation of the crankshaft nose loads and unloads the camshaft and accessory drives. The latter significantly increases drive wear and engine noise. In heavy-duty
engines camshaft torsional vibration is receiving increasing attention, and examples of
vibration problems in accessory drives can be readily found. The focus in this section will
be on the crankshaft, but many of the principles discussed here can be applied to other
components.

In Fig. 16.18 and Eq. 16.2 the fundamental concepts of torsional vibration are summarized. Looking at this simple case a disk of some appreciable mass is rigidly mounted at
the end of a shaft, the other end of which is mounted to prevent it from spinning. A torque
is then applied to the disk to rotate it slightly from its initial position, elastically deform-
ing (twisting) the shaft. The torque is then suddenly released, and the shaft unwinds. The
shaft seeks to return to its original non-deformed state, but the mass of the disk results
in an overshoot and the shaft is twisted in the opposite direction. The disk will oscillate
back and forth at the natural frequency of the system, with decaying amplitude in each
successive oscillation. The natural frequency is determined by the mass of the disk and the
stiffness of the shaft.

$$k_r = \frac{G \cdot J}{L} = \frac{\pi d^4 G}{32L} \tag{16.2}$$

Where:

k_r = Torsional spring constant of the shaft
d = Diameter of shaft
G = Shear Modulus
L = Length of shaft

If the shaft consists of several sections of different diameters, as in a stepped shaft or crankshaft, the equivalent torsional spring constant can be calculated in same way as for springs in series as shown in Eq. 16.3:

$$\frac{1}{k_{eq}} = \frac{1}{k_{r1}} + \frac{1}{k_{r2}} + ...$$ (16.3)

The natural frequency can be modeled using Eq. 16.4:

$$\omega_n = \sqrt{\frac{k_{eq}}{I}}$$ (16.4)

Where:

I = Inertia of Disc

The case just described can now be extended in two important ways. First, one can imagine that the shaft and disk are spinning at some constant speed and that the torque is applied to and released from the spinning disk. In this case the resulting oscillation of the disk will be superimposed on the mean speed of the shaft and disk. The next extension is to consider a case where the spinning shaft has not one but several disks, and at various points in time sudden torque impulses are applied in succession to various of the disks. The problem has now become appreciably more complex as different portions of the shaft twist and untwist relative to other portions. Where the first two cases each had one *degree of freedom* and a single *natural frequency* this third case has an additional degree of freedom and natural frequency for each additional disk. Assuming the mass of each disk, and the shaft stiffness between each disk, are known, and assuming that the torque impulses applied to each disk can be characterized as a function of time the torsional vibration can still be calculated. The resulting matrix of equations is difficult to solve by hand, but can readily be addressed with computer calculations.

The third case just described is exactly that of the crankshaft in a multi-cylinder engine. In order to address crankshaft torsional vibration it is necessary to characterize both the crankshaft system and the system excitation—in this case the torque applied to the system at each cylinder.

Turning first to the crankshaft system it is helpful to characterize this system as a series of disks connected by stiff springs as shown in Fig. 16.19. Each disk represents the rotating mass associated with a portion of the crankshaft system. The first disk represents the crankshaft nose, vibration dampener, and accessory drive pulleys. Disks two through seven represent the cylinders of this in-line six-cylinder engine. The eighth disk represents the flywheel or torque converter. It should be noted that the transmission and the remainder of the drivetrain are not represented. This is an accurate approach in automotive engines because transmission of torsional vibration is minimized through the fluid coupling in the torque converter and the clutch pack typically has a torsional compensator built in. If the

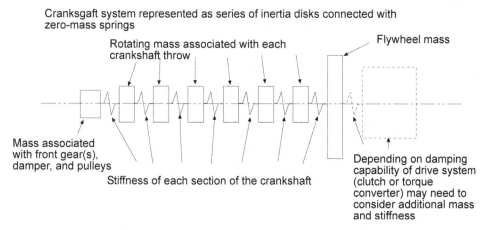

Fig. 16.19 Representation of six-cylinder crankshaft system for torsional vibration analysis

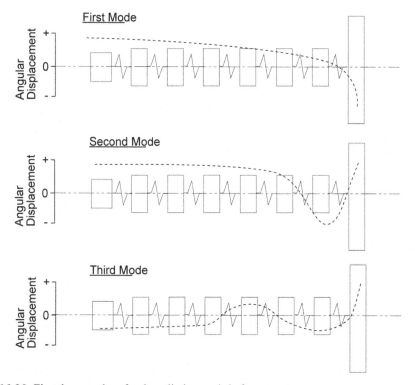

Fig. 16.20 First three modes of a six-cylinder crankshaft system

engine were rigidly mounted to the component being driven (an electric generator, for example) this additional mass would need to be included. The example eight mass system shown here has eight *degrees of freedom* and eight *natural frequencies*. The first three vibration *modes* are depicted in Fig. 16.20. For each vibration mode the crankshaft will have that number of *nodes*—locations along the crankshaft at which angular deflection

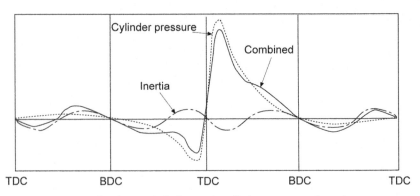

Fig. 16.21 Forces transmitted to crankshaft at each cylinder versus crank angle

relative to the mean crankshaft speed is zero. Because crankshafts are relatively stiff only the first mode or two are generally of interest. As crankshaft stiffness increases the natural frequency of each mode increases; the natural frequencies of the higher modes of a crankshaft system are typically above the frequencies at which significant forces will be seen. It should be noted that for each vibration mode the **anti-node**, or location of maximum angular displacement, is at the crankshaft nose. This results directly from the flywheel location at the opposite end of the crankshaft.

It is now important to look at the exciting forces—the net torque impulse applied at each crankshaft throw, or at each of the disks two through seven in the model shown in Fig. 16.19. The net torque impulse, as a function of crank angle is the net result of the pressure and reciprocating forces, as depicted in Fig. 16.21. Because this combination of forces cannot be directly represented mathematically it is helpful to represent it as a Fourier Series. The torque signal is expressed as the sum of a constant value and an infinite series of harmonics at various amplitudes and frequencies. Because the torque impulse is repeated every second revolution in a four-stroke engine the fundamental frequency is a half-order frequency, repeating every second revolution. The remaining harmonics then represent each half order in an infinite series. The Campbell Diagram shown in Fig. 16.22 shows the frequency of each harmonic, from the half-order through the sixth order, as a function of engine speed.

Torsional vibration requires the application of the torque impulse, now represented as a series of harmonics, to the crankshaft system described earlier. Returning to the Campbell Diagram a given crankshaft will have various natural frequencies—one for each vibration mode described earlier. Each natural frequency can be overlaid as a horizontal line on the Campbell Diagram. It follows that there will be a critical, or resonant speed for every whole and half-order harmonic. Some of these will occur well outside of the operating speed range of the engine.

The torque impulses are being applied at various points along the length of the crankshaft, and in a particular sequence based on the engine's firing order. As a result some harmonic orders assist one another and increase the vibration amplitude. These are referred to as the **major orders**. Others partially cancel one another and are termed **minor orders**. The

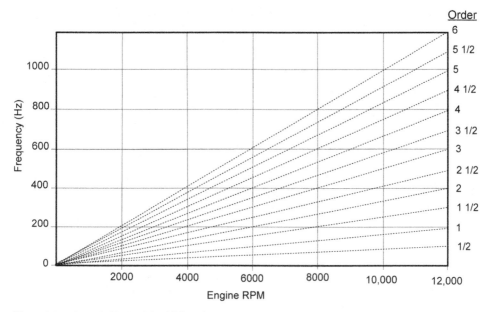

Fig. 16.22 Campbell plot identifying frequency versus engine speed for each vibration order—shown for every half order through sixth order

major orders tend to be those that are direct multiples of the number of torque pulses per revolution. However, other orders may be important depending on the particular crankshaft layout and firing order. For example, on an in-line six-cylinder engine the third and sixth orders might be expected to be important. The ninth order may be important although the magnitude of vibration is typically much lower. The 4 ½ order will also be found important on most in-line six-cylinder engines. A complete explanation is beyond the scope of this book, but lies in the choice of firing order resulting in relative high amplitude forces acting over the front versus the back halves of the crankshaft at the 4½ order.

As a summary of the concepts just discussed, the third, 4½, sixth, and ninth order frequency versus engine speed of the Campbell Diagram are re-plotted in Fig. 16.23. Overlaid on this diagram are the first- and second-mode natural frequencies of a particular crankshaft system. It can be seen that for this engine resonant speeds for the first vibration mode will occur at approximately 2700 rpm (6th order), 3600 rpm (4½ order), and 5700 rpm (3rd order). Second mode resonance will occur at 7300 rpm (6th order) and 9600 rpm (4½ order). Note that unless the engine will be run at very high speed the third crankshaft vibration mode will not be seen.

Some general trends can be observed in crankshaft geometry. Higher stiffness crankshafts have higher natural frequencies which is important for both torsional stresses and lateral (axial) vibrations. Low moment of inertia crankshafts also help raise the natural frequencies of the system. Conversely, adding counterweights adds to the moment of inertia of a crankshaft, lowering the frequency. On multi-throw crankshafts, such as V-6 or V-8 engines, shaping the counterweights is critical to keep the inertia low and the counterweight effectiveness high.

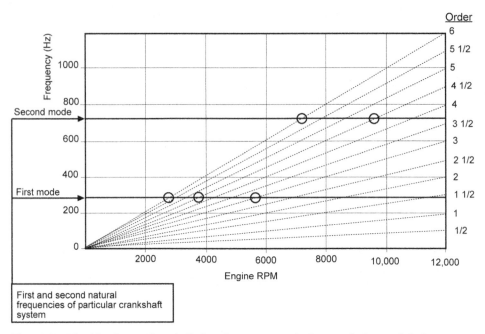

Fig. 16.23 Identification on Campbell plot of resonant speeds for a particular crankshaft system

For the reasons discussed at the beginning of this section damping is applied to reduce the torsional vibration amplitude in most automotive engines. Various damping techniques can be used but the most common are the untuned viscous shear damper and the tuned harmonic damper.

A cross-section of a typical viscous shear damper is depicted in Fig. 16.24 along with its effect on crankshaft deflection. The viscous shear damper consists of a ring of mass encased in a cavity filled with a silicon-based fluid. The casing is rigidly bolted to the crankshaft nose, and the mass is free to float in the viscous fluid. As the crankshaft nose experiences a vibration impulse the mass reacts in the opposite direction, creating a shear force in the fluid that damps the vibration impulse. The vibration energy is dissipated as heat energy from the viscous shear. This type of damper is referred to as untuned because it will damp the vibration pulses regardless of frequency, as can be seen in the plot of Fig. 16.24. The unit must be sized large enough to dissipate the vibration energy without overheating, and adequate air flow around the damper must be ensured. This type of damper is relatively expensive so is generally seen only on larger, higher cost engines. Because the running clearance is quite small it is susceptible to damage.

More commonly used in automobiles is the tuned harmonic damper shown in Fig. 16.25. In this design the damper hub is again rigidly mounted to the crankshaft nose, and a seismic mass is then mounted around the hub through a rubber isolator. The combination of

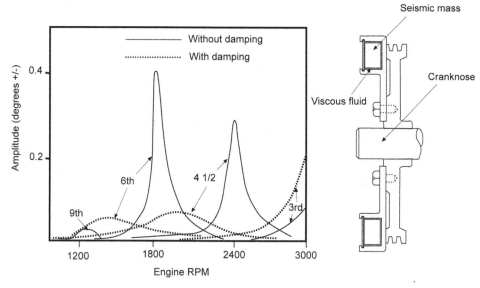

Fig. 16.24 Untuned viscous shear damper operation

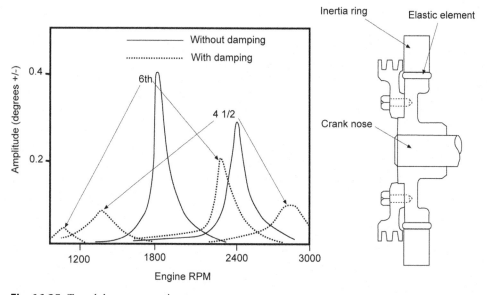

Fig. 16.25 Tuned damper operation

the mass and the stiffness of the rubber isolator is tuned to damp out vibration at a particular frequency, as shown in the figure. It should be noted that each order now has two resonant frequencies because the coupled mass acts as a second torsional system linked through the rubber isolator.

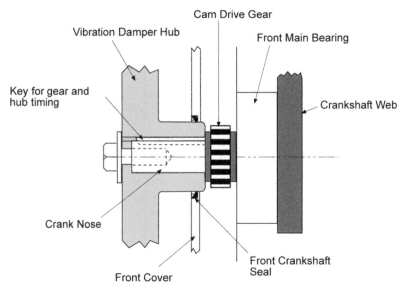

Fig. 16.26 Crankshaft nose design and assembly details

16.5   Crankshaft Nose Development (Straight, Taper, Spline Fit)

A cross-section showing an example of the crankshaft nose and the front details of the engine are shown in Fig. 16.26. A cam drive gear is fit onto the front of the crankshaft and located using a keyway or spline. From this gear the camshaft will be driven using a gear, chain, or belt drive, as discussed further in Chap. 17. This or an adjacent gear may also be used to drive the oil pump and other internal engine drives (balancer shafts or a fuel pump are examples). The crankshaft nose then protrudes through the front cover, and the vibration damper and accessory drive pulleys are mounted to the nose. The front oil seal is mounted to the front cover, and either directly contacts the crankshaft surface inboard of the damper, or contacts the damper hub as shown in the figure.

There are three drive options for accessories at the front of the crankshaft: straight fit, taper fit, or spline. The torque capacity of each these joints needs to be calculated to ensure a successful design. The straight fit and taper fit configurations will need a keyway, not to handle drive torque, but to ensure timing of the accessories to the crankshaft. The torque applied to the clamping fastener(s) provides normal force and friction in the joint, which provides the torque capacity.

The objective of the taper fit joint is again to provide sufficient fastener clamp load (F_{clamp}) to withstand torsional loading on the crankshaft. However, the taper design is benefited by the interference fit generated between shaft and hub. These equations are developed from those used for a thick wall pressure vessel, and the variables are defined in Fig. 16.27. This method assumes that the hub does not bottom against a shoulder, and

Fig. 16.27 Taper fit variables

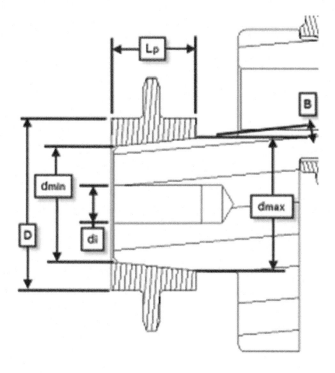

is only supported by the taper. It also assumes that the clamping fastener does not bottom on the end surface of the shaft, to ensure that all fastener clamp force is going into the hub.

Using the assumed starting value for interference fit the contact pressure is calculated as shown in Eq. 16.5.

$$P = \frac{i}{\dfrac{d}{E_o}\left(\dfrac{1+ce^2}{1-ce^2}+v_o\right)+\dfrac{d}{E_s}\left(\dfrac{1+ci^2}{1-ci^2}-v_s\right)} \tag{16.5}$$

$$d = \frac{d_{min}+d_{max}}{2} \qquad ce = \frac{d}{D} \qquad ci = \frac{d_i}{d}$$

Where:

P = Contact Pressure
i   = Assumed interference fit
d = Mean Taper diameter
E_o = Young's Modulus of outer material (hub)
E_i = Young's Modulus of inner material (shaft)
v_o = Poisson's Ratio for outer material
v_i = Poisson's Ratio for inner material

c_e = Major diameter ratio
c_i = Minor diameter ratio
D = Ouside hub diameter
d_i = Inner hole diameter

Maximum introduction and extraction forces are calculated using Eqs. 16.6 and 16.7 respectively. Maximum torque retention and required fastener torque are then calculated as indicated.

$$F_i = \pi \cdot d \cdot L_p \cdot P \cdot (f + \tan(\beta)) \qquad (16.6)$$

Where:

Lp = Length of taper fit area
 f = Coefficient of friction between components
 β = Half of taper included angle

$$F_e = \pi \cdot d \cdot L_p \cdot P \cdot (f + \tan(\beta)) \qquad (16.7)$$

$$T_{max} = F_i \cdot \frac{d}{2} \qquad (16.8)$$

$$T_f = k \cdot F_i \cdot d_f \qquad (16.9)$$

Where:

k   = Nut factor, typically between 0.18-0.22 for steel
d_f = Fastener major diameter

A splined joint can transmit more torque for its size than other types of joints as shown in Figs. 16.28 and 16.29, and can be used at the nose of the crankshaft to drive accessory loads, or more commonly used at the power take off end of the crankshaft to drive the vehicle. The objective of a spline is to drive using the teeth on the joint; it does not rely primarily on clampload. Several failure modes must be designed for: base shaft breakage, hub bursting, teeth of spline shearing off at the pitch line, teeth of the internal spline breaking at root due to bending stress, and wear on the drive surface of the flank of the spline. Splines can be either "fixed", where there is no relative or rocking motion between the internal and external teeth as in a clamped joint, or "flexible" where there is relative rocking or axial motion.

Fig. 16.28 Splined shaft

Fig. 16.29 Splined hub (*gear*)

The crankshaft nose is subject to high-cycle fatigue loading. The combination of the dampener and pulley mass and accessory drive belts create a resultant force of constant magnitude and direction. As the crankshaft spins each location on its nose experiences a complete bending load cycle once every revolution of the shaft. The outside surface of the nose experiences alternating tensile and compressive loads, and the threaded mounting hole(s) for the dampener assembly results in local stress concentrations, and the threads should be counterbored to mitigate. Rig testing can be readily devised to duplicate this load cycle and provide durability validation.

16.6   Crankshaft Flange Development

The same rationale just discussed for crankshaft nose loading and durability applies to the flywheel mounting flange as well. The flywheel or torque converter flex plate are typically mounted to the crankshaft flange using a multiple bolt pattern. The flywheel connection relies on friction to generate the shear torque capacity in the joint, but a frictionless condition must also be designed for. The steps in calculating the frictional torque capacity of the bolted joint are as follows:

1. Approximate nominal clampload per fastener
2. Calculate the shear capacity produced per fastener
3. Determine the total frictional torque capacity in the joint

In order to increase the frictional torque capacity of the joint, it may be difficult to change the fastener pitch circle radius or fastener diameter once a design is in production. If it is desired to increase the fastener pitch circle diameter, the main journal diameter may need to increase. If the main journal diameter increases, it will increase the bearing diameter, seal diameter, cylinder block support web, and oil pan rail width, depending on design of the crankshaft.

It is a frequent occurrence that the marketing organization will request an increase in engine torque or power shortly after a new design is released to production. This is one key area of engine design to package protect for future increases in output.

If a design is complete and in production, and more torque capacity is required of the joint, various treatments can be added to increase the capacity of the joint at an additional cost. The coefficient of friction at the joint surface can be increase by changing the machined surface finish, by adding abrasive coatings, or by adding additional locating dowels. Alternately, the fastener grade may be increased to allow greater clampload.

Two examples of the rear crankshaft oil seal details are shown in Fig. 16.30. The case on the left incorporates a split seal in the rear main bearing. That on the right uses a continuous seal riding on the flywheel flange.

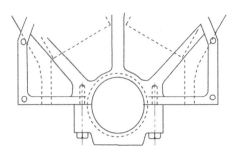

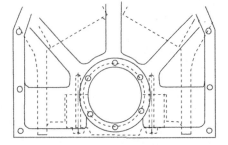

Split rear seal incorporated in rear main bearing journal

Continuous rear seal in housing bolted to rear of block

Fig. 16.30 Rear crankshaft seal designs: split seal with short skirt block, and continuous seal with extended skirt block

16.7 Crankshaft Drillings

Crankshafts are drilled for several reasons: material removal for balancing, material removal for weight reduction, and pressurized oil feed. Crankshaft drilling for balance is usually done in the counterweight webs as shown in Fig. 16.31. This is done when dimensions of the forging or casting of the base crankshaft cannot be held close enough to achieve the desired balance, or when a lighter piston is introduced that changes the required balance. This area is usually not significantly stressed, and so drilling does not present a significant stress concentration. These drillings can either be radial, as shown, to remove mass or can occasionally be axial in the same location to allow addition of more dense material such as tungsten alloy, depleted uranium, or lead. The addition of dense balance weights is more expensive, but can make the overall crankshaft lighter though a more efficient use of material, or can be used to reduce crankshaft inertia for racing applications.

Crankshaft drilling for weight reduction is typically done in the crankpin. This drilling of the crankpin reduces the amount of counterweight material required and can make the entire crankshaft lighter. It can also change the local stiffness of the crankpin, to better distribute stress in the crankshaft and reduce it in the critical journal fillet region. The crankpin is typically a highly loaded area, so attention must be paid to detailed design of the cross drilling in this area, and clearance in adjacent counterweights is typically required to allow access for this drilling. Frequently, this cross drilling is not coaxial to the crankpin.

If the crankshaft has plain bearings, it will be necessary to provide pressure lubrication via internal drilling of the crankshaft. These long drillings typically connect the main journal to the crankpin to provide pressurized lubrication and these internal drilled passages come very close the highly loaded crankpin and main journal fillets as illustrated in Figs. 1.6 and 1.8. If the drilled passage comes close to the surface in a highly loaded area, either at a journal fillet or in the crank web, a fatigue crack may initiate. Additionally, care must be taken at the location where the drilled hole breaks the surface of the journal as this may also create a fatigue crack initiation point. The drilled oil hole is typically chamfered

Fig. 16.31 Crankshaft drilling for balance, oil passages

where it contacts the journal surface to diffuse stress, and also to spread pressurized oil to a larger area under the bearing. One technique to improve fatigue resistance is to use a hardened ball, and peen the ID of the chamfer where it meets the cross-drilled hole. This will introduce residual compressive stresses in this area and reduce a crack initiation site. Finally, the location where the oil hole breaks the surface of the crankpin is important for the bearings. It should be located somewhere between 90 and 30° before TDC on the compression stroke to provide the best feed of oil to the bearing. This is the point when the load on the bearing is least. The critical pressure region of 0–45° after TDC on the power stroke must be avoided. High rod loading increases oil pressure above the system pressure and blocks the feed to the bearings.

16.8   Connecting Rod Development

16.8.1   Connecting Rod Column Forces

The connecting rod is among the highest stressed components in the engine, and load is applied several times on every revolution of the engine. The connecting rod must be designed for high cycle fatigue to withstand the high number of engine cycles, and also for stiffness in supporting fluid film bearings. The connecting rod transfers the gas and inertia loads from the piston to the crankshaft, and experiences high rates of loading and direction reversal, high temperatures, and varying degrees of lubrication. Axial inertia and connecting rod whip forces (tangential to crank radius) are a function of connecting rod weight, so the higher the weight, the higher the forces. Since inertia can be a dominant load, and the piston, piston pin, and connecting rod all contribute, a more durable connecting rod design might require less material (be lighter) instead of the typical approach of adding more material to increase life. The higher the piston and rod weight, the more crankshaft counterweight is required, which leads to a heavier engine. As a result, rod weight is of primary importance. Value added operations such as shot peening to enable a lighter connecting rod are often justified to reduce forces and weight in the rest of the powertrain. Piston normal forces due to connecting rod angle and connecting rod whip forces are translated to the engine, and require stronger support in the chassis. However, when a connecting rod fails, it typically destroys the entire engine as would a crankshaft failure. The desire for minimum weight must be balanced with rod durability, and the desire for the lowest cost manufacturing methods. Connecting rods must be cost effective, manufacturable, and serviceable.

The engine type has a significant effect on the peak loads seen by the connecting rod. At one extreme is the low engine speed and high BMEP diesel, and at the other end is the high engine speed and lower BMEP high performance gasoline engine. Gas loading and inertial loading at TDC oppose each other, reducing total loading on the connecting rod on the power stroke. At lower engine speeds, the gas forces dominate and put the connecting rod into compression. At higher engine speeds, inertial loads may dominate and stress the rod in both tension and compression.

Connecting rod loads vary as a function of crankshaft angle, and also as a function of engine cycle. At TDC-exhaust (overlap) on a 4-stroke engine, the connecting rod experiences the highest tension load case because there is very little gas pressure force to resist it. As opposed to TDC-power stroke, when there will be a combination of tensile forces from inertia, and compressive forces from combustion. At BDC-exhaust, the inertia forces put the rod into compression.

The maximum tensile forces on the connecting rod occur at TDC-Exhaust as illustrated in Eq. 16.10. Gas pressure forces can be neglected since they are near zero during valve overlap.

$$Frecip, total = -(mpiston + mconrod, recip) \cdot r \cdot \omega^2 \cdot (\cos\theta + \lambda\cos 2\theta) \qquad \left(\lambda = \frac{r}{l}\right)$$

$$(16.10)$$

The maximum compressive forces on the connecting rod occur either at TDC-Power stroke for a low speed high BMEP engine, or at BDC-Exhaust for a high speed low BMEP engine as shown in Eq. 16.11.

$$Frecip, total = -(mpiston + mconrod, recip) \cdot r \cdot \omega^2 \cdot (\cos\theta + \lambda\cos 2\theta) - F_{Gas} \quad (16.11)$$

Notice that these equations are a function of the connecting rod ratio (λ), with typical values between 0.2 and 0.35. As the connecting rod grows compared to the engine stroke, it has the tendency to reduce the peak force on the rod, and also the peak piston thrust force. However, a longer connecting rod has more weight and increases the engine deck height. These conflicting requirements must be balanced.

The main beam of the connecting rod is subject to inertial bending forces (rod whip) as it swings through TDC. For an initial analysis, it is assumed that the connecting rod is a simply supported beam, subject to a linearly varying distributed load as shown in Fig. 16.32. Once the connecting rod geometry is complete, finite element analysis can be performed and the load can be spread to all of the elements of the beam in proportion to their distance from the piston pin end.

The total bending force due to the triangular distribution is calculated in Eq. 16.12, and the bending moment at any distance 'x' from the piston pin centerline is calculated in Eq. 16.13.

$$F_{beam} = \frac{2}{3} \cdot m_{rod,upper} \cdot r \cdot \omega^2 \qquad (16.12)$$

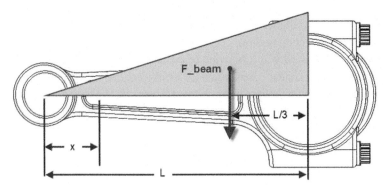

Fig. 16.32 Simply supported beam with uniform increasing load

Where:

$m_{rod, upper}$ = Mass of connecting rod between two pins

$$M = F_{beam} \cdot \frac{x}{3} \cdot \left(1 - \frac{x^2}{L^2}\right)$$ (16.13)

Since gas pressure loading dominates at low engine speed, the rod must be designed for compression. There is also a high compressive load at BDC due to inertia at high engine speed. In addition to designing for compressive stress, column buckling must also be considered and in high BMEP Diesel engines this is a very important consideration. The rod can buckle in one of two directions, parallel to the plane of connecting rod motion, and perpendicular to it.

The end constraint on the connecting rod significantly influences its buckling resistance. If the rod is pinned at either end the least resistance to buckling is present, and if the rod is fixed in 6° of freedom at either end the most resistance to buckling is present. Depending on the exact end constraint, the column is said to have an equivalent length (L_e) to a column pinned at both ends. Column buckling is also sensitive to the geometry of the cross-section, and typically requires three-dimensional analysis.

16.8.2 Connecting Rod Crankpin Bore Cylindricity

The crankpin end of the connecting rod is primarily designed for stiffness, as the connecting rod needs to support the crankpin journal bearing and maintain roundness. As the engine progresses through its cycle, tensile inertia forces attempt to stretch the rod and deform the crankpin end of the connecting rod. Stretching of the connecting rod along the cylinder bore causes the connecting rod to pinch in along the cap split due to the Poisson effect, which may lead to the oil film breaking down and the bearing contacting the crankshaft. If rolling element bearings are used, pinching of the rolling elements may occur leading to skidding or spalling.

Since downward compressive forces load the upper half of the bearing, and not the bottom half, only tensile forces are considered. Now, both the upper half and lower half of the rod mass are counted for inertial calculations. At TDC exhaust, the entire rod/piston system is creating a force away from the crank equal to the reciprocating force and the rotating force from the lower end of the rod as shown in Eq. 16.14. This force is reacted by the lower end of the connecting rod. In high speed engines, this can be the limiting load on the connecting rod because of ovalization of the crankpin bore, bolt limits, or cap strength.

$$F_{TDC-Exh} = -(m_{rot} + m_{recip}) \cdot r \cdot \omega^2 (1 + \lambda) \qquad (16.14)$$

Where:

m_{rot} = Rotating mass of connecting rod
mrecip = Reciprocating mass of connecting rod and piston assembly

In order to achieve a true deflected shape, the bearing press fit and support from the fluid film or rolling element bearing must be incorporated in this analysis. It is important to work with the bearing manufacturer to establish limits of cylindricity.

The outermost surface of the split line has the largest stress. This is because as the rod cap is loaded in tension, the bolt flange will rotate about this outmost line of contact. As this line of contact is moved further away from the crankpin bore, the lever arm increases on the flange, and the unit load decreases. For this reason, most connecting rods have additional material added outboard of this flange to decrease unit loading and increase flange stiffness. Ideally the connecting rod bolt would be as close to the bore as possible, to reduce this flange rotation due to cantilevering.

16.8.3 Connecting Rod-to-Cap Alignment

Connecting rod-to-cap alignment is critical to a successful rod design, and ensuring crankpin journal bearing life. Angular split lines present more of a challenge because of the direction of the applied forces in relation to the split line, but are sometimes necessary. Some alignment methods are:

- *Increased* fastener shank diameter at split line, to provide alignment
- Ring dowels, coaxial with the fastener in a counter-bore
- Separate dowel pins next to the connecting rod bolt.
- Specially machined serration at the rod-to-cap split surface.
- Stepped surface, specially machined rod-to-cap split surface.
- Cracked, manufactured by cracking the end of the rod off by force at the intended parting plane.

Each method has pros and cons, which is why there are so many methods in use. The increased shank fastener is inexpensive to manufacture and is compact in design, however if the alignment deviates from ideal there may be interference between the bolt and rod at the split line. This may add bending stress to the bolt, or upset material in the rod into the bearing bore and degrade circularity. Ring dowels are very similar in performance, with the added disadvantage of moving the rod bolt further away from the bearing bore. For packaging reasons, this may increase the overall rod size. Separate dowel pins are inexpensive, but may also increase the size of the connecting rod and add weight.

A serrated or stepped surface between the connecting rod and cap is more expensive to manufacture due to the increased sensitivity to manufacturing tolerances, but this type increases the shear resistance in the rod-to-cap joint. Stepped surface rod caps operate in a similar manner, using a lip to limit cap motion. Angle split rods increase the amount of shear force on the joint, and often use one of these types of alignment.

The cracked design is now widely used in high volume production, in both forged and powdered metal rods. This method is cheap, repeatable, and insensitive to tolerances. Alignment and shear resistance are excellent. There is very low distortion of the split surface, since both surfaces are symmetrical. However this method dictates the parent material of the connecting rod, and may limit the choice of manufacturer since high production volumes are typically needed to justify the expense of special equipment to form and crack the rod.

16.8.4 Connecting Rod Bushing Press Fit and Journal Bearing Crush

The small end, or pin end of the connecting rod is almost invariably a continuous bearing surface—no separation plane is needed as the piston pin can be inserted through the connecting rod from either side. Two designs are commonly seen for retaining the piston pin. One is the fixed pin, where the pin is pressed into the connecting rod bore, and rotates freely in the piston pin bore. The other is the floating pin, where the piston pin is free to rotate in both the piston and connecting rod bores, and is held in place with snap rings fitted in grooves on both outer edges of the piston pin bore. The fixed pin reduces cost and weight but has lower load carrying capability and requires special assembly provisions. The degree of required press fit generally requires the connecting rod to be heated. Because the connecting rod does not provide an acceptable bearing surface the floating pin requires the addition of a bushing to the connecting rod. This and the required snap rings and machined grooves significantly increases cost relative to the fixed pin. Traditionally the floating pin was seen only in high performance and heavy-duty engines, but due to the demand for increased loads it is now being adopted in many engines.

Interference fits are used to retain the piston pin bushing, as well as the crankpin journal bearing. Both are based on the same fundamentals of thick walled pressure vessels, but their application and assembly method are different. This requires different approaches to design. The thick wall pressure vessel equations are valid as long as the wall thickness-to-

Table 16.2 Press conditions to evaluate

		Temperature		
		Maximum temperature	Operating temperature	Minimum temperature
Tolerance (fit)	Maximum interference	X		Y
	Nominal interference		X-Y	
	Minimum interference	Y		X

radius ratio is greater than 0.1. Bearing press fits must be designed for operating loads, but also extremes of temperature. The goal is to have sufficient press fit to retain the bearing in its bore under all operating conditions as presented in Table 16.2, while not exceeding the stress limits for the given materials.

Depending on the material used for the housing and the material used for the bushing, differential thermal expansion may help or hurt the design. For a low thermal expansion coefficient housing (steel) and a high thermal expansion coefficient bushing (aluminum), the "X's" represent the extreme conditions to evaluate. For a high thermal expansion coefficient housing (aluminum) and a low thermal expansion coefficient bushing (steel), the "Y's" represent the extreme conditions to evaluate.

The final hoop stress on the crankpin end of the rod is developed in much the same way as the press fit bushing example above, except the bearing is split into two different pieces for installation in the upper rod and cap. The diameter of the journal bearing is slightly larger than the bore in the connecting rod (or main bearing), and is compressed during installation. Since the bearing is neither uniform in thickness, or constant in radius, a direct measurement of the diameter is misleading. A different method of measurement is required to describe the press fit. This is called crush height and is measured by the overstand test, the amount of which affects the press fit. The overstand test consists of placing the bearing shell in a half circle gauge, flush at one end and extending above the other. The amount the shell protrudes from the gauge is the amount of circumferential crush the bearing will be under when assembled.

The crush height is measured under a gauge force. This is effectively measuring the circumference or length of the journal bearing shell, rather than diameter. A bearing with sufficient crush height ensures that the bearing is preloaded enough to have uniform contact with the housing. This ensures good heat transfer from the bearing to the housing, and prevents fretting. Typical values of the crush height of automotive bearings are 0.05–0.10 mm.

16.8.5 Connecting Rod Computational Stress Analysis

Typically, the entire connecting rod is not analyzed. A half model with symmetry, or even a quarter model, is used. The mesh is typically refined around the bolted joint area, and around critical fillets and transitions. The connecting rod is analyzed in the following steps.

The assembly load cases are determined by applying the bolt preload to the connecting rod cap. Deformations are incorporated into the initial geometry of the model to simulate machining of the crankpin bore when clamped together. Bushing and bearing assembly load cases are applied by simulating the interference due to press fit of the piston pin bushing and crankpin journal bearings.

The dynamic load cases produced by the engine are analyzed by applying maximum tensile load at TDC-Exhaust stroke and maximum compressive load at BDC-Power stroke. The piston pin and crankpin are modeled as separate cylinders and used as boundary conditions since they contribute to the overall system stiffness. Finally, the loading due to rod whip is analyzed.

The final step of the analysis is to evaluate the connecting rod for critical column buckling in the plane of connecting rod motion, and perpendicular to it. This enables a more refined solution to the analytical method, since detailed geometry is represented.

16.9 Flywheel Design Considerations

The purpose of the engine flywheel is to smooth the cyclical speed variation of the engine within a given engine revolution. It absorbs energy during the power stroke, and distributes it during the exhaust, intake, and compression strokes. The transmission of instantaneous torque spikes developed by the engine will be reduced to the driveline.

The flywheel or torque converter mass is important in controlling idle speed fluctuation within limits acceptable to the driver. As speed increases fluctuations become much less apparent and the need for this mass decreases. An acceptable idle speed fluctuation limit is identified and flywheel mass is chosen to be sufficient to reduce fluctuation within that limit. Additional mass beyond the minimum required will penalize the engine's response to transient requirements.

This storage and release of energy has other uses in the engine and the vehicle, when the engine average speed changes. Large flywheel inertia will enable a low engine idle speed. A low engine idle speed, by reducing the total number of revolutions the engine makes at idle, will reduce fuel consumption and improve emissions during an engine operating regime that does little useful work. The lower the speed variation, the easier it is to calibrate the fuel injection at low engine speeds as the piston approach to TDC is more consistent.

Like in heavy machinery, the flywheel will allow sudden loads to be placed on the engine without stalling it. This is helpful when starting a vehicle from rest, or during sudden load changes. The inertia of the flywheel can be balanced against the inertia of the vehicle. A large engine flywheel will also make the engine less responsive during transient drive modes, as the engine will not accelerate quickly with the application of throttle and will not decelerate quickly during coasting/sail or removal of the throttle. This may have a beneficial effect in a work truck or for cruising on the highway at steady speeds, but may not be desirable in a sports car where quick engine response is desired.

In many cases the flywheel and crankshaft are balanced as a unit, and a non-symmetric bolt circle or locating dowel is used to ensure that if the flywheel is removed it is again mounted in the same position. The clutch disc and pressure plate are then mounted to the flywheel with a series of cap screws around the perimeter of the pressure plate. A pilot bearing or bushing mounted in the rear of the crankshaft at the shaft centerline supports the nose of the transmission input shaft. In transverse installations this shaft may not feed directly into the transmission, but drives a chain that then transfers load to the transmission.

In the case of an automatic transmission a flex plate, bolted to the same bolt circle on the crankshaft flange, replaces the flywheel. The torque converter is then bolted to the flex plate with three or four bolts near its perimeter.

Equation 16.15 for kinetic energy stored in a flywheel is shown below, and is equivalent to the amount of energy released if the flywheel speed is changed from its current rotational speed to a full stop:

$$E = \frac{I\omega^2}{2} \quad or \quad \frac{Wv^2}{2g} \tag{16.15}$$

Where:

E   = Energy stored in the flywheel
I   = Polar moment of inertia
ω = Rotational velocity
W = Weight of flywheel rim
v   = Linear velocity at mean radius

A flywheel is designed to reduce the speed fluctuation to a desired amount. This is measured in percent change in speed called the Coefficient of Fluctuation (C_f) as represented in Eq. 16.16, or in absolute speed change. If a lower speed fluctuation is desired, a larger flywheel will be required. However, the larger the flywheel, the more difficult to package and the heavier the engine will be. Since the flywheel is usually directly coupled to the crankshaft, it is not possible to spin the flywheel faster than engine speed, removing this as a design option. To gain smoothness, the flywheel diameter or thickness will need to be increased. A thin disc of large diameter has the most efficient use of material, and a hub with most material concentrated near the rim would be even better. Typically dense materials such as steel or cast iron are used for flywheels.

$$C_f = \left| \frac{\dot{u}_2 - \dot{u}_1}{\dot{u}_{mean}} \right| \tag{16.16}$$

If the velocity of a flywheel changes, the energy it absorbs or discharges will be proportional to the difference of initial and final speeds per Eq. 16.17:

$$E = \frac{I(\omega_2^2 - \omega_1^2)}{2} \quad or \quad \frac{W(v_2^2 - v_1^2)}{2g} \tag{16.17}$$

The traditional method for calculating the required amount of energy in the flywheel is to determine the peak energy needed by the work operation and subtract the average energy provided by the engine or motor providing energy input. The difference is the flywheel energy needed to maintain speed as represented in Eq. 16.18. This method is relevant to the vehicle side, as a load is applied to the vehicle, the required flywheel inertia to maintain engine speed can be calculated.

$$E_{work} - E_{motor} = E_{Flywheel} \tag{16.18}$$

Once the desired flywheel energy storage is determined, and the desired speed fluctuation is set, the equation can be rearranged to determine inertia required utilizing Eq. 16.19:

$$I = \frac{2E}{(\omega_2^2 - \omega_1^2)} \tag{16.19}$$

16.10   Crankshaft and Connecting Rod Construction

Crankshaft steels are typically medium carbon (0.3–0.4%) and heat treated to increase the tensile strength. Forged crankshafts have inherently higher material strength, especially in critical regions through close control of grain flow. Cast iron alloys vary over a wide range in tensile strength depending on material and heat treat; their values can be found in many references. Casting a crankshaft can enable other desirable geometry for little cost, such as hollow crankpins and detailed shaping of the counterweight webs. The development of the metallurgy and casting methods has allowed the use of cast crankshafts in more and more applications. Where packaging dimensions are not strictly constrained, cast crankshafts can be a reasonable alternative to forged steel shafts.

As discussed when covering bearing design, an important variable in determining the minimum oil film thickness for hydrodynamic operation of plain bearings is that of shaft surface finish. As the finish is improved unit loading can be increased, and lower viscosity lubricants can be used to improve fuel efficiency. In order to achieve the required main and rod bearing surface finishes the machining is done in three steps. Lathe cutting is followed by grinding, and finally by lapping or polishing.

In high output applications steps may be taken to increase crankshaft strength in the critical fillet transitions. These value-added operations are ion nitriding, shot peening, induction hardening, and fillet rolling or roll hardening. These operations locally increase the strength of the crankshaft by modifying the material properties or introducing compressive residual stresses. These additional operations cost additional money, and careful design can often avoid their requirement. However, the designer must compare the added manufacturing costs to the larger implications to the engine and vehicle downstream. In order to increase the power output of the engine, these processes can be added without having to redesign the engine. Enabling smaller journal diameters can reduce engine friction. These operations can be beneficial when considering the system as a whole.

Historically connecting rods have been cast, forged, or machined from billet. Recently, connecting rods have been manufactured from forged powdered metal for high volume automotive applications. This process can be lower cost, enabling more consistent parts and a beneficial rod cap alignment surface. Heavy-duty engines still primarily use forgings.

The materials used are cast iron, steel, aluminum, and titanium. Cast iron is typically used in cost sensitive applications, with forged steel being used in high load applications. Aluminum connecting rods are being used in light-duty engine applications, and recent improvements in metallurgy have increased the suitability of this material. Recall from earlier discussion that the connecting rods own weight acts against itself in high speed applications, so lighter materials enable higher engine speeds. Titanium is typically reserved for racing applications due to its high cost.

Connecting rods proceed from casting or forging, to rough machining of the upper rod and rod cap separately—or as a single piece if the cracked rod cap is used. The two components are then bolted together for the finish machining of the assembly and precision honing of the bores. It is important to perform this precision machining in the assembled state, as bolt clamp loads will distort the big end bearing shape, and cylindricity is key to fluid film bearing life. Occasionally, the piston pin bushing and crankpin journal bearing are assembled to the rod, and an additional boring and honing operation is performed if a higher class of tolerance is to be maintained.

Frequently, value added operations are applied to the connecting rod to increase component life, or reduce component weight. Trimming of a casting or forged connecting rod blank will often leave score marks in the finished part. These score marks can act as stress risers in the key loaded areas of the connecting rod. They may be ground or polished smooth to reduce the stress concentrations. Additionally, shot peening of the connecting rod will introduce compressive residual stresses in the component. While these operations add cost, the designer must balance this against the potential added benefit to the engine and vehicle.

16.11   Analysis and Test

The typical analysis path is to evaluate each component individually, and then once a level of maturity is reached, combine the individual components into a system model. The crankshaft analysis typically starts with analysis of a simply supported, single throw. Once that is complete, a full crankshaft model is analyzed. Often a course mesh model of the crankshaft is analyzed to determine system natural frequencies for torsional calculations. Then the crankshaft, connecting rod, and cylinder block are combined in a system analysis. This is typically more complex, as the fluid film must be modeled as gap elements. Some system stiffness is contributed by the crankshaft, and some by the cylinder block.

There are two opposing design philosophies for crankshaft-cylinder block design, one of cylinder block guided stiffness, and another of crankshaft guided stiffness. Each of the different methods has one component significantly stiffer than the other. Regardless of the method chosen, the goal is to have sympathetic deflection in the same direction between the housing and the shaft, to prevent edge loading of the bearings.

If a crankshaft or connecting rod fails during engine testing, the results usually include extensive progressive damage. If either component fails many other engine components are damaged beyond repair. For this reason, crankshafts and connecting rods are typically subjected to component tests to prove their reliability, prior to exposing an expensive prototype engine to destruction.

Various servo-hydraulic rigs are used for component level testing. The crankshaft is usually evaluated for torsional strength and bending resistance at the crankpin. For torsional testing, it is usually supported at the main bearings, fixed at the power take off, and a firing pressure load applied at the crankpin. The bending load at the crankpin can either be simulated by fixing the crank near TDC and loading via a driven connecting rod, or can be subjected to a moment by bending the crankshaft from one main bearing to another as represented in Fig. 16.34. A single throw section is rigidly mounted in a heavy fixture that is then suspended as shown. By making the fixture mass large it will resonate at a very high natural frequency. Once this frequency is identified high fillet stress can be achieved with a very low driving force. Strain gauges are added in the fillet region to drive the testing apparatus, or to assess the impact and sensitivity of different fillet radii under a given load. As shown in Fig. 16.35 the location of maximum alternating stress along the fillet is identified. By varying the magnitude of loading at the resonant frequency an 'S-N' diagram can be generated specific to the crankshaft section. The cycle accumulation rate is so high that the number of cycles the crankshaft would accumulate over its entire life in an engine can be accumulated in a matter of hours. The 'S-N' diagram results can be compared with the actual stresses seen by the crankshaft in the engine to determine the actual fatigue life. The rig test method described here is so easily done that it can be used for production quality checking as well. Production samples can be tested and overlaid on the 'S-N' diagram to quickly identify shifts that might occur due to casting or forging problems, material alloy changes, or grinding, hardening, shot peening, or cold rolling process control problems.

Fig. 16.33 Connecting rod column bending

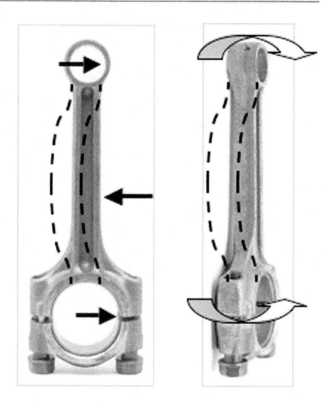

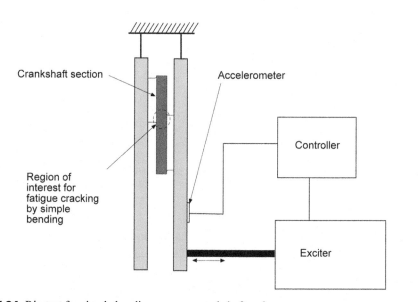

Fig. 16.34 Rig test for simple bending across a crankshaft web

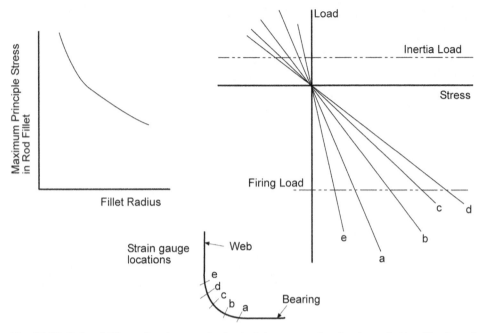

Fig. 16.35 Role of fillet radius in control of principle stress at bearing journal; identification of maximum stress location

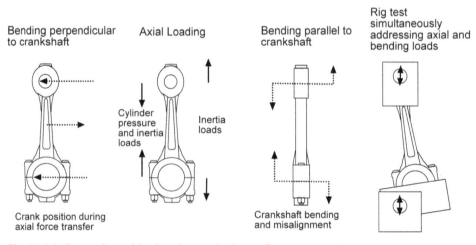

Fig. 16.36 Connecting rod loads and example rig test fixture

In connecting rod rig testing a load is applied at the piston pin bushing, axial to the main beam of the rod. The crankpin end of the rod is fixed by a shaft of the same diameter of the crankpin to limit bore distortion. These tests are sometimes accelerated by inputting a force at the natural frequency of the component. An enhancement of this test is to mount the crankpin end of the rod off center from the piston pin end, and this can simulate axial as well as bending load due to rod whip as presented in Fig. 16.36.

Once individual component testing is passed, system level tests can begin. These can again be servo-hydraulic, where a hydraulic cylinder applies load to simulate combustion to a short block assembly of crankshaft, connecting rods, and cylinder block. The short block can also be rotated on a motoring rig, where the engine is spun by an electric motor without combustion taking place. This also allows the verification of the lubrication and breathing systems, and the measurement of engine friction. A disadvantage of a motoring rig test is the lack of combustion forces. This will under-represent the force due to combustion, but will amplify the loading due to inertia since gas forces will not counteract inertia during TDC power stroke.

These system level tests allow easier access to components for strain measurement. The use of strain gages and telemetry or slip rings are also methods to obtain information. Finally, a specialty device known as a "grasshopper linkage" can be used to take direct measurement of strain on a moving connecting rod. This is a two link assembly with one end attached to the connecting rod, and one end attached to a fixed point in the engine. This removes the need to have strain gauge wires rotate around the crankshaft, and once assembled and moving resembles the rear leg of a grasshopper.

Once these components have completed testing, they are cleared for use on the engine dynamometer. Typically, low speeds are evaluated first, and engine load and speed are gradually increased as confidence grows.

16.12   Recommendations for Further Reading

The following paper provides a design and development flowchart specifically addressing crankshafts. The level of detail provided will be especially appreciated by a first-time crankshaft designer (See Dubensky 2002).

Although this is an older paper it provides a good summary of the dimensional details and design variables that determine crankshaft fatigue life (See Shaw and Richter 1979).

The following three papers summarize crankshaft development for new automobile engines. The first is for a V6 engine and the second an in-line four. The third paper, again for a V-6 engine, emphasizes NVH considerations (See Paek 1999; Fujimoto et al. 2001; Druschitz et al. 1999).

The following papers discuss crankshaft durability. The first discusses resonant bending fatigue testing, and the second discusses crankshaft forging and grain flow (See Yu et al. 2004; Shamasundar 2004).

Shot peening is an important process for applying compressive residual stress to crankshaft fillets, thus improving their fatigue life. This paper presents a detailed look at the process, the resulting surface characteristics, and the effect on fatigue life (See Wandell 1997).

This presentation, available through the Gas Machinery Institute provides an extremely detailed look at Torsional vibration in engines, and the effects of various vibration dampers (see Feese and Hill 2002).

The following paper presents a recent study of connecting rod fatigue in powdered metal and forged designs. The next paper discusses fracture-split connecting rod design. The text then listed devotes a chapter to connecting rod fatigue, and rig testing for rod development (See Afzal and Fatemi 2004; Park et al. 2003; Wright 1995).

References

Afzal, A., Fatemi, A.: A comparative study of fatigue behavior and life predictions of forged steel and PM connecting rods. SAE 2004-01-1529 (2004)

Druschitz, A.P., Warrick, R.J., Grimley, P.R., Towalski, C.R., Killion, D.L., Marlow, R.: Influence of Crankshaft material and design on the NVH characteristics of a modern, aluminum block, V-6 engine. SAE 1999-01-1225 (1999)

Dubensky, R.G.: Crankshaft concept design flowchart for product optimization. SAE 2002-01-0770 (2002)

Feese, T., Hill, C.: Guidelines for preventing torsional vibration problems in reciprocating machinery. Presented at Gas Machinery Conference, Nashville, Tennessee, 2002

Fujimoto, T., Yamamoto, M., Okamura, K., Hida, Y.: Development of a Crankshaft Configuration Design. SAE 2001-01–1008 (2001)

Paek, S.-Y.: The development of the new type Crankshaft in the V6 engine. SAE 990051 (1999)

Park, H., Ko, Y.S., Jung, S.C., Song, B.T., Jun, Y.H., Lee, B.C., Lim, J.D.: Development of fracture split steel connecting rods. SAE 2003-01-1309 (2003)

Shamasundar, S.: Prediction of defects and analysis of grain flow in Crankshaft forging by process modeling. SAE 2004-01-1499 (2004)

Shaw, T.M., Richter, I.B.: Crankshaft design using a generalized finite element model. SAE 790279 (1979)

Wandell, James L.: Shot peening of engine components. ASME Paper No. 97-ICE-45, (1997)

Wright, Donald H.: Testing Automotive Materials and Components. SAE Press, Warrendale (1995)

Yu, V., Chien, W.Y., Choi, K.S., Pan, J., Close, D.: Testing and modeling of frequency drops in resonant bending fatigue tests of notched Crankshaft sections. SAE 2004-01-1501 (2004)

Camshafts and the Valve Train

17

17.1 Valve Train Overview

The poppet valve, as previously detailed in Fig. 9.5, is now used universally in four-stroke vehicular engines—both to draw fresh charge into the cylinder, and to exhaust the spent products. The valves face an especially harsh environment. Because they are exposed directly to the combustion chamber, and provide very restrictive heat transfer paths they operate at especially high temperatures. The demand for rapid opening and closing results in high impact loads, and a requirement for high hardness valves and seats. The combination of high hardness and high temperature requirements drives the selection of special steel alloys, typically with high nickel content for both the valve head and the valve seat. Most automobile valves are made as a single piece, while the valves in heavy-duty engines generally have the nickel alloy head inertia welded to a mild steel stem. Hollow stem two-piece valves are generating interest in automobile applications, for savings of both weight and cost. A valve spring and retainer assembly as shown in Fig. 9.5 completes the installation. The retainer is typically stamped from mild steel, and holds the spring in a partially compressed position with two hardened steel keepers fitted near the top of the valve stem.

While the poppet valve assembly just described shows little design variation throughout the engine industry there are many design options for the remainder of the system. As of this writing a camshaft, driven at one-half crankshaft speed is used to actuate the valves on virtually all production engines. The system through which the camshaft is driven, and the train between camshaft and poppet valve, are among the most rapidly changing mechanisms in engine design today. The interest is driven by a desire to optimize valve lift and timing over the entire operating range of the engine. Trends in valve train system design are discussed in Sect. 17.6.

Looking first at today's production systems there are a variety of linkage arrangements between the camshaft and valves. Several examples are depicted in Fig. 17.1. The goals

© Springer Vienna 2016
K. Hoag, B. Dondlinger, *Vehicular Engine Design,* Powertrain,
DOI 10.1007/978-3-7091-1859-7_17

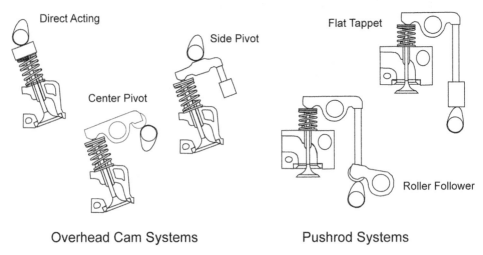

Overhead Cam Systems   Pushrod Systems

Fig. 17.1 Example valve train arrangements

of any such system are to provide sufficient stiffness throughout the valve train so that the valve events do not move around as speed and load are changed, and to keep its natural frequencies high; to keep the mass of the system sufficiently low that the valves can be opened as rapidly as desired at the maximum required engine speed; and to minimize the cost of the system.

Typically lowest cost are the pushrod systems. Such systems result in the simplest camshaft drives (to be discussed in Sect. 17.5), and allow a single camshaft to be used in vee and horizontally opposed engine configurations. The primary disadvantage is the lower stiffness and lower natural frequencies of pushrod systems as compared to overhead cam systems. Another disadvantage is that the required additional mass compromises the valve opening rates and may limit the top speed of the engine. The various overhead cam systems each have the goals of increased stiffness and reduced mass. While the direct acting arrangement provides the simplest configuration it results in increased engine height as compared to the side- and center-pivot systems. Each of the pivot systems also allows increased leverage so that the valve lift is greater than that of the cam lobe—typically 1.2–1.5 times. The direct acting system requires the cam lobe to provide the entire lift, in turn requiring a larger camshaft diameter, higher lobe stresses, or some sacrifice in the lift profile.

Contact between the valve train components include ball and socket joints, sliding contact, and in some cases rolling contact. The sliding contact between the rocker lever and valve stem sees mixed film lubrication, as does that at the rocker lever pivot. A flat tappet in sliding contact with the camshaft experiences hydrodynamic lubrication. The use of a roller follower at the cam lobe interface allows the load at this interface to be approximately twice that with a flat tappet, and measurably reduces friction, but with some cost increase. A hydrodynamic oil film must be maintained between the roller and the cam lobe.

Another design consideration in the valve train is that of dimensional stack-up. It must be ensured that the valves can completely close under all operating conditions, while too much clearance between components will increase impact loads and wear. As engine

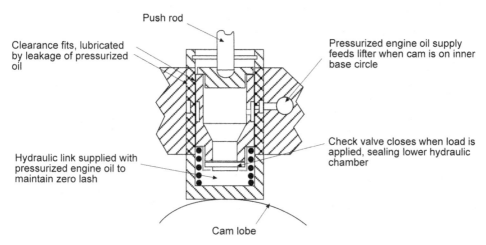

Push rod

Clearance fits, lubricated
by leakage of pressurized
oil

Pressurized engine oil supply
feeds lifter when cam is on inner
base circle

Hydraulic link supplied with
pressurized engine oil to
maintain zero lash

Check valve closes when load is
applied, sealing lower hydraulic
chamber

Cam lobe

Fig. 17.2 Hydraulic lifter construction and operation

temperatures change the varying rates of thermal expansion among the valve train components will change the clearance. Wear between the valve head and seat decreases the
clearance over time, and wear at each of the other contact points increases clearance. In
heavy-duty diesel engines a valve lash adjustment mechanism is included somewhere in
the valve train—most often at the pushrod end of the rocker lever. In automobile engines
the use of hydraulic lash adjusters that automatically take up the clearance with a column
of engine oil are nearly universally used. The operation of the hydraulic lifter is explained
in Fig. 17.2. The valve train will be further discussed in Sect. 17.4 when the topic of development for durability is taken up.

17.2   Dynamic System Evaluation and Cam Lobe Development

Camshaft lobe design and dynamic analysis of the valve train will need to be addressed
together as the dynamic capability of the valve train provides an important constraint on
the lobe profiles.

The cam lift profile is depicted in Fig. 17.3 along with plots of follower velocity and
acceleration. The profile is divided into an opening and closing, or 'lift' and 'fall,' region,
each of which can be further divided as shown (ramp, flank, and nose portions). The initial
ramp period is designed to ensure that any system lash is taken up prior to the high acceleration of the flank portion. At the transition from the ramp to the flank portion a high acceleration is applied to begin opening the valve as rapidly as possible. This is the concave
portion of the profile, and the maximum acceleration rate is limited by hertz stress normal
to the cam lobe at the cam and follower interface. It may also be limited by lobe grinding
restrictions against concavity of any portion of the lobe. Geometry between the cam lobe
and follower is a limiting factor on the flank portion of the cam profile. This is shown in
Fig. 17.4 as the eccentricity with a flat tappet or pressure angle with a roller follower.

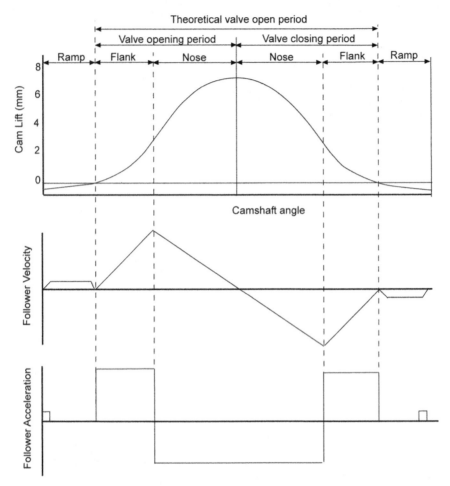

Fig. 17.3 Cam lobe construction, showing lobe regions, velocity, and acceleration

The transition from the flank to the nose portion of the cam lobe is geometrically the inflection point between the concave and convex portions of the lift curve. It is the point at which maximum velocity is reached and the system begins decelerating. Another constraint comes into play at this point as a rapid deceleration may result in "no follow." This is the limiting point at which valve train inertia results in separation occurring somewhere in the system. The inertia force associated with valve train component mass has become sufficient to overcome the valve spring force and creates the separation. It is this constraint that provides the impetus for reduced valve train mass. For a given valve train the engine speed at which no follow occurs can be increased with increased spring force, but this in turn increases the cam lobe stress. A maximum deceleration rate is determined to avoid no follow at the desired "redline" or maximum engine speed. This may in turn result in the need to reduce the preceding acceleration rate.

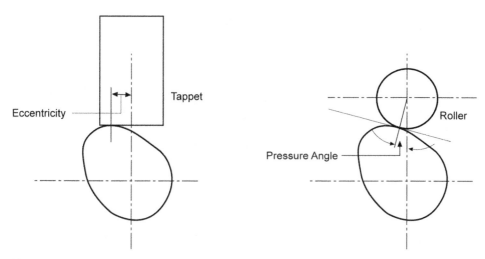

Fig. 17.4 Camshaft flank acceleration limiting parameters with flat tappets and roller followers

The closing, or fall portion of the cam lobe reverses the sequence of the lift portion. An objective during closing is to maintain high lift as long as possible with an extended nose portion, and then rapidly close the valve during the flank portion. Here again the rate of closing is limited by lobe geometry. The valve is brought nearly to closure, and an extended constant-velocity closing ramp is then used to minimize the impact force of valve seating.

Various approaches can be taken to determine the cam lift profile. Examples include geometric and curve matching methods, and mathematical methods such as multi-sine and polynomial techniques. An increasingly common method is based on acceleration curve shaping. A series of acceleration rates are pieced together based on the constraints discussed in the preceding paragraphs. The resulting acceleration profile shown as a series of constant accelerations in Fig. 17.3 is integrated once to obtain the velocity and again to obtain the lift. Further lobe profile decisions required for durability will be discussed in Sect. 17.3.

As was mentioned at the beginning of this section cam lobe design must be closely tied to dynamic analysis of the valve train. The system can be accurately modeled if the geometry is provided, and the mass and stiffness of each component is measured or calculated. Example dynamic representations for an overhead cam and pushrod system are shown in Fig. 17.5. One aspect of the dynamic system analysis was previously described in the discussion of the role of component inertia in no-follow. Another important aspect is that of resonance or natural frequency of the system.

It can be surmised from Fig. 17.5 that any given valve train system possesses several degrees of freedom, and will thus have several natural frequencies. In many cases only the first mode needs to be considered as the higher modes have little energy. In extremely stiff systems higher orders must be considered, and more complex models are required. Ap-

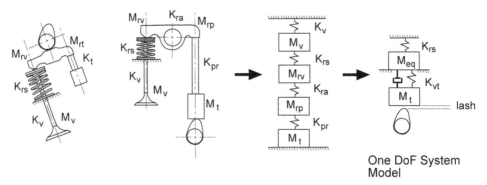

Fig. 17.5 Valve train dynamic system model construction

proximating the actual system with a one degree-of-freedom model is shown in Fig. 17.5. The natural frequency is thus estimated as

$$f_n = \frac{1}{2\pi}\sqrt{\frac{K}{m_{eq}}} Hz$$

The typical range of natural frequencies for various types of valve trains is as follows:
Pushrod 500–900 Hz
Direct Acting OHC 1000–2000 Hz
Indirect Acting OHC 800–1500 Hz

An example of predicted performance for the pushrod system modeled in Fig. 17.5 is shown at one engine speed in Fig. 17.6. The calculations have been provided by R. Bruce Dennert, CamCom Inc. The geometric valve lift based on the cam profile and rocker ratio is shown along with the actual valve lift trace. Valve velocity, cam torque, and the force acting on the valve are also plotted. The impact of system vibration at its natural frequency is clearly seen in the torque, velocity, and force traces. It should be noted that the natural frequency can be directly identified by assessing the length of each cycle in cam degrees. Because the natural frequency is constant for a given system, the number of cycles completed during one valve event drops as engine speed is increased. It should also be noted that an experimental measurement of strain versus time allows the actual natural frequency to be determined and compared to the simplified model calculation.

A speed sweep from 4000 to 6000 rpm is shown for the same valve train system in Fig. 17.7. Once again the constant natural frequency can be seen to stretch the time required for each vibration cycle in degrees cam angle. As the speed increases the force can be seen to momentarily become negative. This is evidence of the no follow condition discussed earlier. The speed at which the force sees excursions to zero is the maximum speed at which the valve train should be operated.

Another important consideration is the effect of system vibration on valve closing velocity. It can be seen in Fig. 17.7 that the instantaneous valve velocity is also impacted by

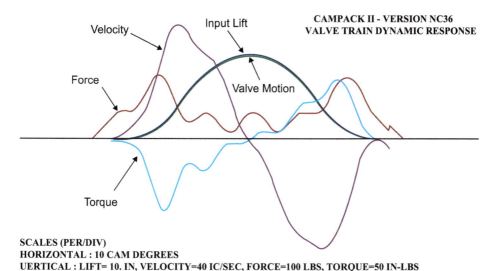

ENGINE FILE - test.ENG
VALVE LIFT FILE - test.VAL
VALVE WEIGHT - 0.500
IAPPET WEIGHT - 0.400
ENGINE RPM - 5000

SPRING PRELOAD - 125
VALVE LASH - 0.000
GAS FORCE - 0
LEAKDOWN - 0.003
STATIC FRICTION -   0

RETURN SPRING RATE - 300
V-TRAIN STIFFNESS - 200000
NATURAL FREQUENCY - 600
% CRITICAL DAMPING - 2.5
SELDING FRICTION - 0

CAMPACK II - VERSION NC36
VALVE TRAIN DYNAMIC RESPONSE

Velocity

Input Lift

Force

Valve Motion

Torque

SCALES (PER/DIV)
HORIZONTAL : 10 CAM DEGREES
UERTICAL : LIFT= 10. IN, VELOCITY=40 IC/SEC, FORCE=100 LBS, TORQUE=50 IN-LBS

Fig. 17.6 Dynamic system model calculations showing results at one speed. Figure provided by R. Bruce Dennert, CamCom Inc.

vibration, and that this impact changes with engine speed. It follows that at some speeds the velocity resulting from vibration will counter the closing velocity, thus slowing it down. At other speeds the effect of vibration will be to significantly increase the closing velocity. This can be seen in Fig. 17.8 where the valve closing region of the case shown in Fig. 17.7 is expanded. This must be carefully considered over the range of typical engine operating speeds to ensure that the system's natural frequency does not adversely impact valve closing velocity at speeds that will be regularly seen during operation.

The single degree-of-freedom analysis presented here is generally quite sufficient for initial valve train analysis. In cases where higher vibration modes or spring surge are important the same calculation techniques can be expanded to include the higher orders. Several commercial software tools for such analysis are readily available.

17.3 Camshaft Durability

The camshafts used in automobile engines are cast or forged from steel alloys or cast from iron; those in heavy-duty engines are most often machined from steel alloy bar stock. Of increasing interest in light-duty engines, for reduced cost and weight, are composite

ENGINE FILE - test.ENG **SPRING PRELOAD - 125** **RETURN SPRING RATE - 300**
VALVE LIFE FILE - test.VAL **VALVE LASH - 0.000** **V-TRAIN STIFFNESS - 200000**
VALVE WEIGHT - 0.500 **GAS FORCE - 0** **NATURAL FREQUENCY - 600**
TAPPET WEIGHT - 0.400 **LEAKDOWN - 0.003** **% CRITICAL DAMPING - 2.5**
STARTING RPM - 4000 **STATIC FRICTION - 0** **SELDING FRICTION - 0**
MAX RPM, INC - 6000, 200

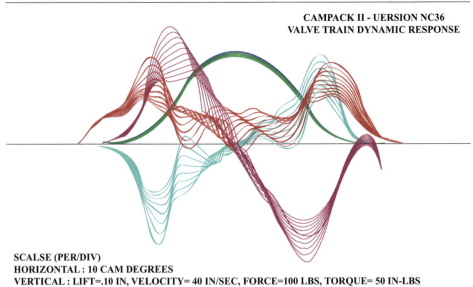

CAMPACK II - UERSION NC36
VALVE TRAIN DYNAMIC RESPONSE

SCALSE (PER/DIV)
HORIZONTAL : 10 CAM DEGREES
VERTICAL : LIFT=.10 IN, VELOCITY= 40 IN/SEC, FORCE=100 LBS, TORQUE= 50 IN-LBS

Fig. 17.7 Dynamic system model calculations showing speed sweep. Figure provided by R. Bruce Dennert, CamCom Inc.

camshafts. Powdered metal or forged lobes are fitted in place on a hollow shaft through which a mandrel is then pressed to expand the shaft against the lobes.

Initial camshaft machining involves lathe cutting of the bearing journals and lobes, followed by case hardening. The bearing journals are then ground and lapped, and the final lift profiles are ground.

The interface between the cam lobe and either a roller follower or flat tappet is intended to be under fully hydrodynamic lubrication. If the cam and follower were both perfectly stiff they would experience line contact and an infinite oil film pressure. The actual contact patch results from elasticity of both the cam and follower, which must be considered in order to size the components to handle the peak loads.

Cam lobe stress is summarized in Fig. 17.9. Maximum tensile stress occurs at the lobe surface. The differing tensile stress magnitudes in different planes, along with the differing rates of drop-off below the surface, result in maximum shear stress beneath the surface. This stress distribution in combination with the hardened surface results in micro-spalling when the lobe is overloaded. Tensile cracking at the surface combines with ductile cracking under the high shear loading beneath the surface, resulting in material removal. It is a progressive failure mechanism since the remaining material is then further overloaded, increasing the spalling.

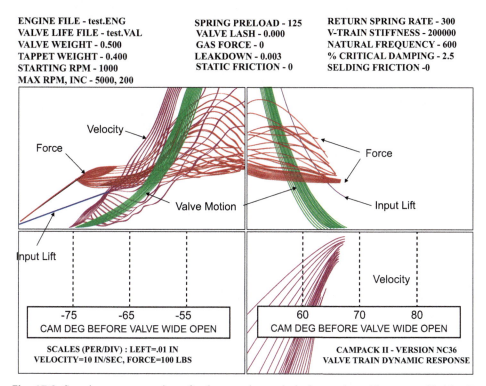

ENGINE FILE - test.ENG
VALVE LIFE FILE - test.VAL
VALVE WEIGHT - 0.500
TAPPET WEIGHT - 0.400
STARTING RPM - 1000
MAX RPM, INC - 5000, 200

SPRING PRELOAD - 125
VALVE LASH - 0.000
GAS FORCE - 0
LEAKDOWN - 0.003
STATIC FRICTION - 0

RETURN SPRING RATE - 300
V-TRAIN STIFFNESS - 200000
NATURAL FREQUENCY - 600
% CRITICAL DAMPING - 2.5
SELDING FRICTION -0

Fig. 17.8 Speed sweep expansion of valve opening and closing regions. Figure provided by R. Bruce Dennert, CamCom Inc.

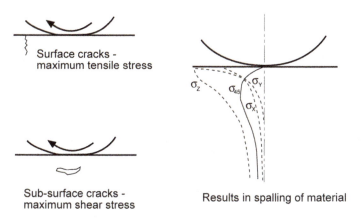

Fig. 17.9 Cam lobe stress, and mechanism of spalling failure

The lobe to follower interface is depicted in Fig. 17.10a for a flat tappet and Fig. 17.10b for a roller follower. In the case of the flat tappet the tappet is offset slightly from the lobe centerline to promote tappet rotation; the lobe is cut on a slight angle as shown to reduce

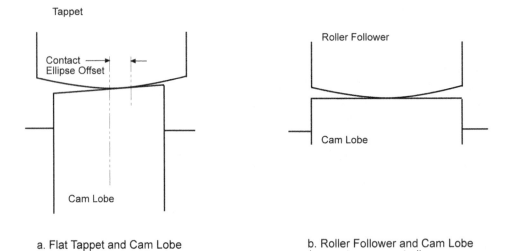

Fig. 17.10 Expanded look at lobe and follower geometry with flat tappet and roller follower

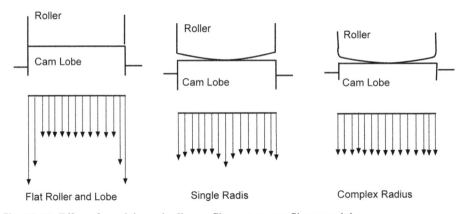

Fig. 17.11 Effect of cam lobe and roller profile on stress profile across lobe

edge stresses. Figure 17.11 shows the stress profile across a cam lobe in contact with a roller follower. If both the roller and follower are machined at a constant diameter across the contact patch high edge stresses are seen at the discontinuity resulting from the lobe edge. By machining a radius on the roller and in some cases the cam lobe the stress profile can be improved—edge stresses are reduced, but stress at the lobe center is increased. As machining technology has improved complex radii can now be specified—the varying radius can be selected to maintain a nearly constant stress across the interface.

Roller followers are becoming increasingly widely used as they result in significantly higher load carrying capability as compared with flat tappets. Approximate allowable hertz stresses in the cam lobe with various follower designs are summarized below:

Flat sliding tappet 100–125 ksi

Crowned sliding tappet 140–150 ksi
Roller follower—constant radius 225–240 ksi
Crowned roller follower 250–265 ksi
Complex radius roller follower 280–300 ksi
Micro-finished lobes and rollers 300–325 ksi

The approximate stresses quoted here are intended to provide a relative comparison between the designs. Actual limits will be dependent on material and process choices, and the required durability for the given engine.

Another important concern with roller followers is ensuring that they actually roll and do not slide on the cam lobe. Sliding in an interface intended to have rolling contact results in significant overloading and high friction heating. The roller material is typically softer than that of the cam, and locally melts, transferring to the cam lobe in circumferential streaks. This is termed *galling*, and quickly results in failure.

17.4   Valve Train Development

Durability validation of the camshaft itself was discussed in the previous sections. Various aspects of design and durability for the remainder of the train are discussed in the paragraphs of this section.

Valve Beat-in   As has been discussed in previous sections of this chapter the valve and seat interface is subjected to impact loads that must be minimized to ensure acceptable wear levels. The wear rate is dependent on the valve and seat materials and operating temperature, the actual seating velocity (taking into account both the cam profile and valve train resonance), dimensional alignment and stem-to-guide clearance, valve and seat angle, and the presence of abrasives in the gas flows. Materials and seating velocity were previously discussed. Dimensional alignment between the valve stem and valve seat centerlines, and perpendicularity between the valve stem and valve head are of crucial importance. The valve guide bore and valve seat bore should be machined simultaneously from a single axis during cylinder head machining. Once the seat and guides have been pressed in place final guide machining should be completed, and then the seats are cut or ground using the guide to position the tool. Valve guide wear leads progressively to excessive valve beat-in, so another important aspect in minimizing valve beat-in is minimizing guide wear.

It can generally be assumed that unless special steps are taken the valves rotate very little if at all. Measurements show that after initial assembly the valves will rotate less than a single revolution, to a geometrically preferred position based on small dimensional irregularities. Valve rotators can be installed, and are often effective in reducing valve beat-in. Most valve rotators are ratcheting devises incorporated in the spring retainers. Each time the valves are actuated the ratchet indexes the valve slightly. Because of their added expense rotators are seldom used unless other solutions to excessive beat-in cannot be found.

Another important design variable is that of valve and seat angle. Increasing the angle promotes effective sealing, and as a rule improves the valve flow characteristics at all but the lowest lifts. However, as the valve angle is increased it increases the sliding contact between the valve and seat, thus increasing beat-in.

Valve beat-in is not predictable, and because of the importance of temperature and gas composition it cannot be accurately quantified through rig testing. Careful measurement of the valve and seat profiles before and after engine testing provides the best measure of valve beat-in. Because of the importance of dimensional stack-up several measurements on different cylinders, and different engines, should be made. High speed, high load testing results in the most rapid wear rates. Valve beat-in wear cannot be eliminated, so must be quantified to ensure that the wear rates are low enough to meet engine life expectations.

Valve Burning This valve failure mode results from deposits on the valve that reach temperatures exceeding the melting temperature of the valve. Deposits, on the valves and throughout the combustion chamber, are functions of the chemical constituents and their temperature and time history. Deposit build-up is difficult to predict, and the nature of the deposits is dependent on temperature (and thus load), as well as the fuel chemistry and the presence of lubricant and lubricant additives in the combustion chamber. Deposits tend to be formed at lower engine operating temperature and low loads. The deposit heating resulting in valve burning occurs when the deposits are exposed to extended high load operation. Deposits resulting from the oil and oil additives are especially problematic; the first step in minimizing the possibility of valve burning is good oil control—especially past the valve stems, but also from the crankcase.

Valve Sticking Another problem closely associated with oil control and deposit build-up is that of valve sticking. In this case oil fragments or combustion products become trapped between the valve stem and guide. The build-up, in combination with tight running clearances, results in sticking. The valve spring force is insufficient to overcome the sticking, and the valves remain open until the piston approaches TDC and aids in the closing process. This quickly results in progressive damage.

Component Fatigue The valve train components are subjected to cyclic loading, repeated with each operating cycle of the engine. Durability validation thus includes fatigue analysis based on the high-cycle loads. Because the loads are well understood both structural analysis and rig testing are very well suited to this work. It should be noted that it is important to fixture the part in a rig test such that it is loaded in the geometric position representative of maximum load along the valve lift profile. An S-N curve is generated from rig testing specific to the particular part of interest. In this way one can be assured that the roles of design features and manufacturing processes specific to the part are included in the fatigue property database. By doing this the resulting database can be used to check production samples and quickly identify material or process control problems.

Lubrication and Wear While the cam bearings and cam lobe and follower experience fully hydrodynamic lubrication the remaining joints in the valve train (rocker pivots, rocker-to-stem, stem-to-guide) can be expected to see mixed film lubrication and metal-to-metal contact. Each of these joints experience stop-and-start motion, so even if pressurized oil is supplied a hydrodynamic film cannot be maintained. Because metal-to-metal contact cannot be avoided wear will occur not only under start-up but during operation. The wear rates must be quantified to ensure that they are low enough to be acceptable over the desired engine life. Careful measurements of the surface profiles before and after testing are required. High speed, overloaded tests, conducted with "dirty" oil, allow the wear rates to be accelerated. A historic database of acceptable wear rates in production engines is required to relate the test results to acceptable levels in actual operation.

Over-speed Capability As discussed previously the valve train is one of the limiting parameters in the maximum speed capability of an engine. It is important that the maximum anticipated speed is determined, and that the system then be validated over the resulting engine speed range. With non-governed engines such as are typical in spark-ignition automobile engines the "redline" speed often serves as this limit. But one may ask whether momentary excursions above redline may occur—due for example to a missed shift in a high performance application. While the crankcase may be capable of withstanding such a momentary over-speed, the valve train may be much more vulnerable to rapid damage; no-follow may result in a valve contacting a piston, or valve keepers coming off, and a "dropped valve." Such simple failures due to very short high speed excursions result in catastrophic progressive damage. This possibility confirms the need to validate the valve train for short excursions to speeds above redline.

Diesel engines have high-speed governors that reduce the fueling at speeds above rated power, thus governing the engine to a maximum speed even at zero load. However, here too it is important to recognize that in automotive applications the governed speed can be exceeded. This may result from the operator choosing too low a gear during a downshift, or due to the vehicle over-speeding the engine on a long downhill grade. Again, the result is a need to ensure that the valve train is capable of withstanding over-speed conditions above the governed speed.

Dynamic computations were presented in Sect. 17.2, and are quite effective in assessing over-speed capability. Another important tool is the use of a variable-speed electric motor to drive the complete valve train over the entire speed range of interest. If the motor is placed at the crankshaft drive the entire valve train and drive system can be assessed as well. Such a rig lends itself well to detailed instrumentation and high speed photography. Playing the resulting movies back at lower speeds allows various effects such as valve train resonance and spring surge to be clearly seen. It is important to recognize that the rig may not capture all of the relevant loads. For example, exhaust valve opening loads are significantly increased because of the cylinder pressure at the time of the opening event; this effect is difficult to directly include in a rig test. Comparing measured strains on the rig test to those at the same speed with an operating engine allow the differences to be better identified.

Valve Springs   The valve springs are selected to provide sufficient closing force such that a positive force is maintained throughout the valve train over the entire operating speed range of the engine (including any over-speed requirements as discussed in the preceding paragraph). Increasing the closing force increases the maximum speed capability of the valve train, but also increases the loads on the camshaft and throughout the train. Older engines often used double or even triple springs under a single retainer. Spring material capabilities and geometry modifications now allow single springs to meet most engine requirements. The geometry of the spring allows the force change as the spring is compressed to be modified. With constant geometry over the entire spring height the force increases linearly as it is compressed. Modifying the coil spacing or the spring diameter versus height allows the force profile to be modified, providing a progressive force increase with increased spring compression.

In addition to the spring force important design considerations include the fatigue life of the spring, coil bind, and spring surge. The first aspect of fatigue life is that of high cycle fatigue due simply to continued actuation over the engine life. Coil bind significantly reduces fatigue life, leading to spring failure. Coil bind is defined by the solid height of the spring at full compression. If insufficient spring height is available to accommodate the cam lift and rocker ratio, coils may be forced to bind. Another source of coil bind may result from improper design of variable geometry springs—coils in one section of the spring may bind before other coils have adequately compressed. Like that of geometric stack-up this problem is easily avoided and easily identified through static measurements. Coil clash, resulting from spring surge, is more difficult to identify. Spring surge is a resonance within the coils of the spring causing waves of compression and expansion to occur through the spring. Because the phenomenon is frequency dependent it will be engine speed dependent. At resonant speeds the spring will undergo significantly more fatigue cycles, and may experience coil clash. The spring manufacturer should be able to supply resonant frequency specifications, and high speed photography allows it to be directly identified. If an engine is to be operated over a wide speed range it may not be possible to eliminate resonance, but it is important to ensure that if it occurs it will be at speeds where the engine spends little time.

Rocker Shaft Deflection   In engines where all of the rocker levers on a given cylinder head pivot along a single shaft the deflection of this shaft becomes important. As one rocker lever is loaded the resulting shaft deflection may change the timing of another valve event along the shaft, and may cause another valve to crack open when it is intended to remain closed. In some diesel engines the fuel injectors are driven by rocker levers mounted between the valve rockers on the same shaft. Shaft loading is often significantly higher during fuel injection, and shaft deflection may also change fuel injection timing.

The amount of deflection can be measured or predicted computationally, and the shaft stiffness must be sufficient to keep the amount of deflection within required limits.

17.5   Drive System Development

In the four-stroke engine the camshaft is driven by the crankshaft at one half the crankshaft speed. In addition to driving the camshaft at its intended speed over the entire range of desired engine speeds the drive system must address several further challenges. It must be able to handle a reversing torque load, due both to crankshaft (and possibly camshaft) torsionals, and resulting from load transmission back to the camshaft during valve closure. It must be able to compensate for dimensional changes due to thermal growth, and in the case of overhead cam engines dimensional stack-up due to block and head machining and head gasket crush. In many cases it must also drive various accessories. Design considerations include durability, cost, drive system resonance, and the contribution of the drive system to engine noise.

The drive system may be entirely through gears, a cogged belt, or a roller or inverted-tooth chain. Each of these systems will be discussed in turn in the paragraphs that follow.

Gear Drives   In engines where the camshaft is located low in the cylinder block it is possible to drive the camshaft directly from the crankshaft, using only two gears. This results in the lowest cost and certainly the simplest drive system. A far more complicated gear system may be required if the torque required to drive the cam is high, and if long life is required. System cost and its noise contribution are significant challenges with gear systems. Straight-cut spur gears reduce the cost of the gear train, but because the load is carried by only a single tooth at a time, and is continually "passed" from one tooth to the next the noise level is quite high. Helical gears result in significantly lower noise, as the angle cut distributes the load over several teeth at any point in time. However, these gears are more costly to manufacture, and introduce a thrust load which must be addressed through additional bearing surfaces. More recently the high-contact-ratio spur gear has been introduced, combining the best features of both systems. By modifying the tooth geometry of the spur gears more than one tooth carries the load at any time with these straight-cut gears.

Another important design consideration is the tendency of gears to walk apart when loaded. A component of the load is transferred toward the centerline of the meshing gear. The gear mounts must be sufficiently stiff to minimize this phenomenon—an especially important consideration when the gears are carried on an aluminum housing. In cases where the reversing torque load is high anti-backlash scissors gears are sometimes used. These are split gears with internal springs to keep one portion of the gear in contact with the opposite face of each tooth as the load is transferred through the other portion. This adds cost and increases the required gear widths, but reduces noise and tooth wear.

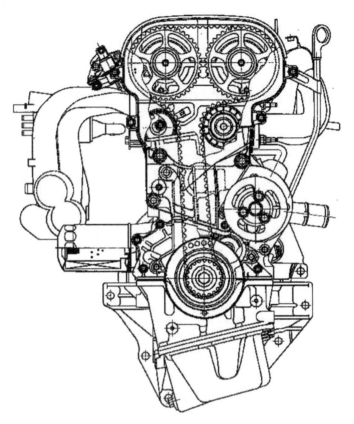

Fig. 17.12 Cogged belt drive. Provided by Ford Motor Company

Chain Drives An example of inverted tooth chain drives is shown in Fig. 1.7c. The chain drive is lower cost and typically results in lower noise than the gear drive system, but cannot be expected to provide the same load carrying capacity or durability. Reversing torque requires the use of chain tensioners and damping strips which can also be seen in Fig. 1.7c.

Cogged Belt Drive A cogged belt drive is shown in Fig. 17.12. It provides the lowest cost, lowest noise, but also lowest load carrying and lowest durability drive system. The cogged belt drive system requires more space than the chain or gear-drive systems. A jockey tensioner uses a preloaded spring to maintain belt tension and address torque reversal.

17.6   Future Trends in Valve Train Design

As stated at the beginning of this chapter the valve train is among the most rapidly evolving systems in engine design today. This evolution is driven by potential performance, fuel economy and emission improvements that might be gained through valve timing and lift profile flexibility with changes in engine or vehicle operating conditions.

The first step that might be considered is camshaft phasing. In double overhead cam engines the intake and exhaust lobes are on separate camshafts, so variable phasing of one or both camshafts relative to the crankshaft can be considered. Improvements can be realized because for sufficiently long valve overlap engine performance is relatively insensitive to changes in overlap. The entire exhaust cam lobe can be phased for optimum exhaust valve opening timing for power and efficiency. The intake cam lobe can be phased to reduce variation in volumetric efficiency with speed. Camshaft phasing is limited by the need to ensure piston to valve clearance during valve overlap. The concept is depicted in Fig. 17.13, showing the geometric limitations and performance impact. Phasing is most often controlled with a device in the cam gear; in some cases the cam follower is adjusted relative to a fixed phase camshaft. Camshaft phasing is now widely used in production automotive engines.

A more flexible but more complex system seen on several production engines is to change the lift profiles in two or three steps with changing engine operating conditions. This approach is used on the intake valve lobes to optimize volumetric efficiency, reduce part load throttling, and control valve overlap and thus exhaust gas residual for emission reduction. These step-change, variable valve lift systems use separate intake lobes on the camshaft and a mechanism for switching the valve train from one lobe to another during engine operation.

Still greater flexibility is achieved with continuously variable intake valve lift profiles, and at the same time the ability to control event phasing. The greatest further advantage

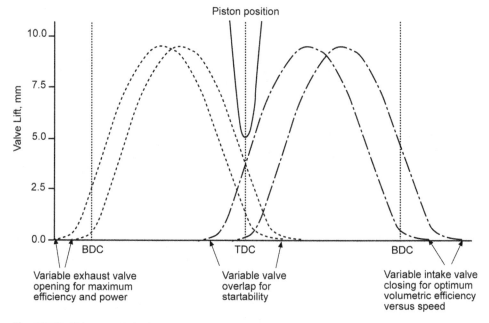

Fig. 17.13 Valve event phasing

provided by such systems is the ability to control engine load with the intake valve event. Intake valve opening timing is fixed, and load is controlled by closing the valve early—at some point during the intake stroke. During the remainder of the stroke the mixture is expanded and recompressed, resulting in significantly lower pumping work than that of a throttled engine.

Eliminating the mechanical camshaft, and actuating the intake and exhaust valves with electronic, hydraulic, or pneumatic actuators is the next possible step toward fully flexible valve event control. Resulting further benefits must be traded off versus several factors, the combination of which has precluded production of such "cam-less" engines to date. The first consideration is that of cost of the replacement system as compared with a mechanical, cam-driven system. For many years this factor alone ruled out most alternative systems, but the costs are becoming continually less prohibitive. The next consideration is that of energy consumption—will the benefits be offset by additional energy requirements to actuate the valves. It should be noted here that with the cam-driven system a not insignificant portion of the valve opening energy is returned as the valves are closed. The very low power requirements of electronic solenoids have increased the attractiveness of the proposed systems. The remaining considerations that are keeping the cam-less engine from production include control of valve event timing as engine load is changed, and the difficulty of managing impact loads. Valve event timing is a difficult challenge because as engine speed and load change the pressure against which the valves must open changes significantly, and with it the power requirements. In most cam-less systems this impacts the valve opening profile versus crank angle. The final concern is that of increased impact loads. The cam lobes profiles of a mechanical valve train are designed to nearly close the valve and then slow the velocity down considerably before the valve impacts the seat. This level of control has not yet been achieved with any of the alternative actuation systems.

17.7   Recommendations for Further Reading

Shortly after the first edition of this book was published a detailed book focused entirely on valvetrain was released. Recommendations for further reading must certainly start with this detailed and highly readable account (Wang 2007).

The further readings recommended for this chapter begin with dynamic system analysis tools. The list begins with a recent summary of advances in measurement techniques and their application. The next two papers listed describe system analysis tools and their capability. This is followed by a paper looking at a specific aspect of cam lobe design. The next two papers address specific components of the valve train—chain load prediction, and valve spring dynamics (Kerres et al. 2012; Keribar 2000; Colechin et al. 1993; Norton et al. 1999).

Specific component modeling (Takagishi et al. 2004; Schamel et al. 1993)

The following paper addresses camshaft design and durability (Druschitz and Thelen 2002).

The following papers discuss various aspects of valve and valve train durability development. The first two papers address valve durability—fatigue, and beat-in, or guttering—and the next paper addresses the specific problem of roller sliding (Roth 2003; Arnold et al. 1988).

System durability (Duffy 1993)

Noise and vibration were introduced in Chap. 10. The following paper addresses the contribution of the valve train to engine noise (Suh and Lyon 1999).

The following papers focus on analysis of the various cam and accessory drive systems. Two papers on chain drive modeling are followed by one each on gear and belt drive systems (Sakaguchi et al. 2012; Rodriguez et al. 2005a, b; Sandhu et al. 2003).

Of increasing interest in valve train development are systems that allow valve timing to be varied. It was beyond the scope of this book to discuss specific approaches to variable valve timing, so the following papers are recommended for their discussion of various approaches. Systems in production at the time of this writing are camshaft driven. Some change the phasing of the entire camshaft relative to the crankshaft; others use mechanical devices that allow an individual event to be more closely controlled. The first paper listed below provides a nice summary of the various approaches that can be taken in new automobile engines. The papers that follow begin with production systems, and then include a look at "camless" fully electronic systems and potential resulting performance improvements (Kirsten 2011).

Current Production and Cam Phasing Systems (Hannibal et al. 2004; Kramer and Phlips 2002; Sellnau and Rask 2003):

Continuously Variable Systems (Flierl and Klüting 2000; Kreuter et al. 2003):

Camless Systems and Performance Impact (Allen and Law 2002; Turner et al. 2004; Pischinger et al. 2000):

References

Allen, J., Law, D.: Production electro-hydraulic variable valve-train for a new generation of I.C. engines. SAE 2002-01-1109 (2002)

Arnold, E.B., Bara, M.A., Zang, D.M., Tunnecliffe, T.N., Oltean, J.: Development and application of a cycle for evaluating factors contributing to diesel engine valve guttering. SAE 880669 (1988)

Colechin, M., Stone, C.R., Leonard, H.J.: Analysis of roller-follower valve gear. SAE 930692 (1993)

Druschitz, A.P., Thelen, S.: Induction hardened ductile iron camshafts. SAE 2002-01-0918 (2002)

Duffy, P.E.: An experimental investigation of sliding at cam to roller tappet contacts. SAE 930691 (1993)

Flierl, R., Klüting, M.: The third generation of valvetrains—new fully variable valvetrains for throttle-free load control. SAE 2000-01-1227 (2000)

Hannibal, W., Flierl, R., Stiegler, L., Meyer, R.: Overview of current continuously variable valve lift systems for four-stroke spark-ignition engines and the criteria for their design ratings. SAE 2004-01-1263 (2004)

Keribar, R.: A valvetrain design analysis tool with multiple functionality. SAE 2000-01-0562 (2000)

Kerres, R., Schwarz, D., Bach, M., Fuoss, K., Eichenberg, A., Wüst, J.: Overview of measurement technology for valve lift and rotation on motored and fired engines. SAE 2012-01-0159 (2012)

Kirsten, K.: The variable valve train in the debate on downsizing and hybrid drives. 32nd International Vienna Motor Symposium (April 2011)

Kramer, U., Phlips, P.: Phasing strategy for an engine with twin variable cam timing. SAE 2002-01-1101 (2002)

Kreuter, P., Heuser, P., Reinicke-Murmann, J., Erz, R., Peter, U., Böcker, O.: Variable valve actuation—switchable and continuously variable valve lifts. SAE 2003-01-0026 (2003)

Norton, R.L., Eovaldi, D., Westbrook III, J., Stene, R.L.: Effect of valve-cam ramps on valve train dynamics. SAE 1999-01-0801 (1999)

Pischinger, M., Salber, W., van der Staay, F., Baumgarten, H., Kemper, H.: Benefits of the electromechanical valve train in vehicle operation. SAE 2000-01-1223 (2000)

Rodriguez, J., Keribar, R., Fialek, G.: A comprehensive drive chain model applicable to valvetrain systems. SAE 2005-01-1650 (2005a)

Rodriguez, J., Keribar, R., Fialek, G.: A geartrain model with dynamic or quasi-static formulation for variable mesh stiffness. SAE 2005-01-1649 (2005b)

Roth, G.: Fatigue analysis methodology for predicting engine valve life. SAE 2003-01-0726 (2003)

Sakaguchi, M., Yamada, S., Seki, M., Koiwa, Y., Yamauchi, T., Wakabayashi, T.: Study on reduction of timing chain friction using multi-body dynamics. SAE 2012-01-0412 (2012)

Sandhu, J.S., Wehrly, M.K., Perkins, N.C., Ma, Z.-D., Design kit for accessory drives (DKAD): Dynamic analysis of serpentine belt drives. SAE 2003-01-1661 (2003)

Schamel, A.R., Hammacher, J., Utsch, D.: Modeling and measurement techniques for valve spring dynamics in high revving internal combustion engines. SAE 930615 (1993)

Sellnau, M., Rask, E.: Two-step variable valve actuation for fuel economy, emissions, and performance. SAE 2003-01-0029 (2003)

Suh, I.-S., Lyon, R.H.: An investigation of valve train noise for the sound quality of I.C. engines. SAE 1999-01-1711 (1999)

Takagishi, H., Shimoyama, K., Asari, M.: Prediction of camshaft torque and timing chain load for turbo direct injection diesel engine. SAE 2004-01-0611 (2004)

Turner, J.W.G., Bassett, M.D., Pearson, R.J., Pitcher, G., Douglas, K.J.: New operating strategies afforded by fully variable valve trains. SAE 2004-01-1386 (2004)

Wang, Y.: Introduction to engine valvetrains. SAE International, Warrendale, PA (2007)

Index

Acceleration, 23–25, 75, 183, 222, 364
  piston, 74, 80, 85
Acoustics, 55
Additive package, lubricant, 126, 217, 223
Adhesion, 42, 43, 137
After cooler, 7, 8
Age hardening, 101
Air-to-fuel ratio, 22, 62, 66, 67
Alternating load
  crankshaft web, 35, 180
  head gasket, 37, 180
  piston, 40
Alternating stress, 35, 37–40, 356
Aluminium, 53, 97, 105, 135, 146, 149, 220,
  234, 268, 272, 351
Angular deflection, crankshaft, 335
Annealing, casting, 37, 104
Antinode, torsional vibration, 336
Application
  automobile, 9, 24, 26, 64, 149, 168, 361
  heavy truck, 8, 25, 30, 67
  marine, 8, 9, 29, 30
  off-highway, 27–30
Articulated pistons, 304
Aspect ratio, combustion
  chamber, 76
Asperity contact, 195, 210
Assembly loads, cylinder head, 38, 137
Balance, engine, 52, 125, 323
Balance shafts, 89
Bank, cylinder, 92, 93, 141, 142, 168, 250
Base stock, 216–218, 220
Bearing
  analysis, 207, 213
  caps, 48, 54, 70, 120, 133, 135, 178

materials, 207, 208
  shells, 48, 144, 193, 196, 206, 212, 351
Beat-in, valve, 49, 371, 372
Belt drive, camshaft, 254, 340
Bi-metal bearings, 208
Block casting, 102, 104, 114, 121
Block loading
  high cycle, 37, 40, 177, 178, 188
  low cycle, 37, 40, 183
Block rig test fixtures, 36
Block stiffness, 111, 113, 135, 182, 184, 201,
  317
Blowdown, exhaust, 157
Bore
  distortion, 128, 137, 277
  polishing, 46
Bore-to-stroke ratio, 73
Boring, surface finish, 113
Boundary lubrication, 44, 45, 218
Breadboard testing, 58, 59
Broaching, surface finish, 47
Bronze bearings, 209
Bulkhead loads, 232
Bulkheads, 107, 117, 120–122, 144, 145, 179
Campbell diagram, 336, 337
Camshaft
  machining, 368
  placement, 53, 143, 144, 147
Camshaft bearing bore, 144
Cap separation, 135, 197, 212
Carbide, 98, 99
Casting solidification, precipitation sites, 37
Catalyst placement, 62, 167, 223
Cavitation
  bearings, 49, 212

© Springer Vienna 2016
K. Hoag, B. Dondlinger, *Vehicular Engine Design*, Powertrain,
DOI 10.1007/978-3-7091-1859-7

Printed by Printforce, the Netherlands